Adaptations of Desert Organisms

Edited by J.L. Cloudsley-Thompson

Springer
Berlin
Heidelberg
New York
Barcelona
Budapest
Hong Kong
London
Milan
Paris
Santa Clara
Singapore
Tokyo

Volumes already published

Ecophysiology of the Camelidae and Desert Ruminants
By R.T. Wilson (1989)

Ecophysiology of Desert Arthropods and Reptiles
By J.L. Cloudsley-Thompson (1991)

Plant Nutrients in Desert Environments
By A. Day and K. Ludeke (1993)

Seed Germination in Desert Plants
By Y. Gutterman (1993)

Behavioural Adaptations of Desert Animals
By G. Costa (1995)

Invertebrates in Hot and Cold Arid Environments
By L. Sømme (1995)

Energetics of Desert Invertebrates
By H. Heatwole (1996)

Ecophysiology of Desert Birds
By G.L. Maclean (1996)

Plants of Desert Dunes
By A. Danin (1996)

Biotic Interactions in Arid Lands
By J.L. Cloudsley-Thompson (1996)

Structure-Function Relations of Warm Desert Plants
By A.C. Gibson (1996)

Physiological Ecology of North American Desert Plants
By S.D. Smith, R.K. Monson, and J.E. Anderson (1997)

Ecophysiology of Small Desert Mammals
By A.A. Degen (1997)

In preparation

Ecophysiology of Amphibians Inhabiting Xeric Environments
By M.R. Warburg (1997)

Homeostasis in Desert Reptiles
By S.D. Bradshaw (1997)

A. Allan Degen

Ecophysiology of Small Desert Mammals

With 102 Figures and 56 Tables

 Springer

Prof. Dr. A. ALLAN DEGEN
Desert Animal Adaptations and Husbandry
Jacob Blaustein Institute for Desert Research
Ben-Gurion University of the Negev
Beer Sheva 84105, Israel

Front cover illustration: *Cynomys ludovicianus* (prairie dog).
Photograph taken by G. Shenbrot, Mitzpe Ramon, Israel

ISBN 3-540-59259-8 Springer-Verlag Berlin Heidelberg New York

Library of Congress Cataloging-in-Publication Data. Degen, A. Allan, 1946– Ecophysiology of small desert mammals/A. Allan Degen. p. cm. – (Adaptations of desert organisms) Includes bibliographical references (p.) and index. ISBN 3-540-59259-8 1. Mammals – Ecophysiology. 2. Desert animals – Ecophysiology. 3. Mammals – Adaptation. 4. Desert animals – Adaptation. I. Title. II. Series. QL739.8.D44 1997 599'. 052652 – dc20 96-32227

Cover design: Design & Production GmbH, Heidelberg

Typesetting: Thomson Press (India) Ltd., Madras

SPIN: 10121159   31/3137/SPS – 5 4 3 2 1 0 – Printed on acid-free paper

This book is dedicated to two wonderful and remarkable women whose ability to look at things from a different viewpoint from mine has added an important dimension to my life:

my mother, the late Sara Rozenblatt Degen (1920–1995)

who could never quite understand why I chose to live in a desert town in Israel but readily supported it, and who would have enjoyed seeing this book, and

my wife, Lily

who enjoys living with me and our sons, Yaniv and Doron, in Omer in the northern Negev desert and who, hopefully, will find pleasure in reading this book.

Preface

Deserts have always intrigued and fascinated, yet challenged and frustrated man. Man has always felt a need to conquer the desert. This has led to a very active campaign of the "war against desertification", and phrases such as "settling the desert" and "making the desert bloom" have become common.

The word desert appears 248 times in the Bible, often referring to its desiccating and harsh nature: "To you, O Lord, I call. For fire has consumed the pastures in the wilderness (desert), and scorching heat has devoured all the trees of the countryside. The very animals of the field cry out to you; for the watercourses are dried up and fire has consumed the pastures in the wilderness (desert)" (Joel 1: 19–20; translated from the Hebrew). Yet, although deserts conjure an image of a barren, desolate and lifeless wilderness, they do possess a special beauty and tranquility that have intoxicated man throughout the ages. Deserts can suddenly be transformed into a lush colourful carpet of wild flowers of unparalleled beauty after rain. Dry riverbeds (*wadis*) become violent rushing waters, creating spectacular floods, and dried waterholes fill up quickly and teem with life. These contrasts not only occur in the landscape, but are evident in the different lifestyles "adapted" to the desert. Deserts are inhabited by nomads who barely eke out an existence but live in harmony with the environment, and also by "nouveaux riche" who maintain a very high standard of living but in an artificial environment. Camels and donkeys are still used for transportation, but dune buggies and four-wheel drive vehicles have become common.

I was introduced to the research of animals inhabiting the desert by the late Michael (Mike) Morag in 1972. The first studies we carried out concerned the water and energy requirements of small ruminants in the Negev desert. Tragically, Mike was killed in a road accident in 1973. About this time, I was fortunate to meet Amiram Shkolnik and Roger Benjamin, two exceptional scientists and teachers. I have learned much from them over the years – about desert animals from Amiram and desert agriculture from Roger – and have thoroughly enjoyed their friendship. I have also benefitted from many colleagues who shared my interest in the adaptations of desert animals. In particular, Jan Hoorweg spent numerous hours discussing their energy and nutritional requirements, and Berry Pinshow collaborated in many studies on small animals in the Negev desert. Other colleagues include (in alphabetical order) Zvika Abramsky, Philip Alkon, Klaus Becker, Alistair Dawson, Gary Duke, David Galili, Boris Krasnov, Irina Khokhlova, Kenneth Nagy, Konstanin Rogovin, Georgy Shenbrot, Marina Spinu, Ovidiu Spinu, David Thomas and Bruce Young. My graduate students contributed to much of the data presented in the book, and they include (also in alphabetical order) Avner Anava, Nurit Carmi-Winkler, Arava Hazan, Michael Kam, Aaron Leeper, Meira Levgoren, Avi Rosenstrauch, Iris Sneider, Simi Weil, Yaffa Zeevi. Michael (Miki) Kam joined the unit which I head after he received his Ph.D. Many of the ideas expressed in this book were formulated as a result of our frequent discussions. Excellent technical assistance and data collection were provided by (in chronological order) Gideon Raziel, Victoria Meltz, Hana Arnon, Yael Bertleson, Debbie Jurgrau and Abdullah Abou-Rachbah.

I thank John Cloudsley-Thompson for inviting me to write this book, and the superb manner in which he edited it. John has an uncanny ability to clarify ambiguous statements and can often express what I am attempting to express better than I can. His encouragement has made the writing of this book a pleasant experience. Special thanks are extended to the Kreitman Fund for its financial support over the years. I am extremely grateful to Irene and Hyman Kreitman for their generosity which supported much of the research described in this book.

Ben-Gurion University of the Negev   A. ALLAN DEGEN
September 1996

Contents

CHAPTER 1

Introduction

> "It comes as rather a surprise to learn that most hot deserts of the world have a rich and diverse fauna of small mammals. These small mammals (adult weight up to 3 kg) are rarely seen by the casual observer as they are mainly nocturnal and widely dispersed. Some unfavourable parts of the Sahara are completely devoid of mammals while other favourable habitats support flourishing populations of several species. The most successful orders of small mammals are the rodents, carnivores and bats and, to a lesser extent, the hares and the hyraxes. All of them are adapted to the harshness and unpredictability of the desert environment."
>
> (Happold 1984)

Ecophysiology of Small Desert Mammals, at first glance, seemed like a straightforward title for a book. After all, this topic has been researched, studied and discussed by a large number of scientists worldwide, and consequently, one would think that there is a general consensus as to what is being examined and how it should be done. However, with the exception of the world "of", each word proved to be problematical and has been interpreted in a variety of ways.

1.1
Ecophysiology

Ecophysiology is a relatively new discipline, having gained prominence in the last 40 years. It is also known by a variety of other terms, including "physiological ecology", "ecologically relevant physiology", "comparative physiology" and "evolutionary physiology". Its past achievements, present status and possible future directions have been widely and vigorously discussed (Jorgenson 1983; Calow 1987; Feder et al. 1987; Tracy 1988; Feder and Block 1991; Sibley 1991; Block and Vannier 1994).

In a workshop entitled *New Directions in Ecological Physiology*, ecophysiology was termed as the study of "physiological diversity in relation to the environments in which organisms live or have lived" and "embraces aspects of behavior, morphology, biochemistry, and evolutionary biology, among other fields" (Feder et al. 1987). This discipline combines the physiological, morphological and behavioural adaptations and the natural history of an organism in order to examine how that organism is in equilibrium with its environment: that is, how the adaptations of an organism allow it to survive and reproduce in its natural environment. Block and Vannier (1994) defined ecophysiology as "the science concerned with interactions between organisms and their environments" and "involves both the descriptive study of the responses of organisms (individuals or groups) to ambient conditions and the casual analysis of the corresponding ecologically dependent physiological mechanisms, at every level of organization. These physiological mechanisms are examined not so much in their own right but with respect to their adaptive significance. For this reason, the ecophysiological approach must take into account polymorphism in individual responses, which are largely responsible for the adaptive capacity of any given population. In this respect, ecophysiological study yields information which is fundamental for an understanding of the mechanisms underlying adaptive strategies".

Thus far, ecophysiology has focussed primarily on the physiological adaptations of organisms to extreme environments and has examined the limits of their physiological responses (Feder 1987). Traditionally, experimental study species were chosen on the basis of possessing some unique or exceptional physiological, behavioural or morphological adaptation (Feder et al. 1987). In addition, species were chosen based on their distribution, either in extreme environments or across environmental gradients. This resulted in the accumulation of a physiological database on a large number of individual species, and in particular, physiological responses that allow animals to survive in hostile environments.

In the 1940s with the emergence of the New Synthesis, emphasis in ecophysiology shifted from descriptions of unique individual species well adapted to their habitats to the description of general patterns that allow animals to adapt to their environments. Bartholomew (1987) discussed four types of interspecific comparisons that can be used in these studies: (1) convergence and divergence in physiological responses among animals; (2) adjustments that allow closely related species to inhabit dissimilar environments; (3) related species inhabiting the same extreme environment; and (4) distantly related organisms adapting to the same extreme environment. Adaptive strategies revealed by these comparisons can provide generalizations about the interrelationship among organisms and their biotic and abiotic environment that are not only species-specific but can be used to transcend taxa (Tracy 1988). Furthermore, Tracy suggested that the capacities and characteristics of an organism, as traditionally examined in ecophysiology, can provide generalizations about the organism's constraints or opportunities according to its environment. For example, body size is related to the capacity of feed intake of an organism and can put constraints on or offer opportunities regarding dietary habits.

The shift to generalizations in physiological responses was accelerated with the emergence of ecological studies at the level of ecosystems and mathematical modelling in systems analysis. Empirical data on ecological fluxes required in ecosystem analysis, in particular, fluxes of matter and energy, were stressed as integral measurements in the physiological discipline of bioenergetics (Jorgensen 1983).

1.2
Small

Terrestrial mammals range in body mass from less than 2 g for the Etruscan (*Suncus etruscus*) and pygmy (*Sorex tscherskii*) shrews to more than 5 tons for the African elephant (*Loxodonta africana*). However, distribution according to body mass is uneven, with the great majority of mammals weighing between a few grams and several kilograms. In addition, there are a large number of mammals above 20 kg, but a paucity of species between 5 and 20 kg.

The definition of a small mammals is rather arbitrary. In their article on the energetics of small mammals, Grodzinski and Wunder (1975) restricted body mass to the range between 3 and 300 g, and Happold (1984) in his article on small mammals of the Sahara used a body mass up to 3 kg. Heusner (1991) used 20 kg in dividing mammals into small and large sizes. The International Biological Programme (IBP) Small Mammals Working Group decided that mammals weighing up to 5 kg are to be classified as small (Boulière 1975). This definition will be adopted in the present text. Mammals thus classified include most Rodentia and Chiroptera, which together contain more genera and species than all the other orders combined, and also include most Insectivora and Lagomorpha (Table 1.1). Using this guideline, Artiodactyla such as the 1.6 kg

Table 1.1. Number of genera and species of mammalian orders, their main food habits and body sizes. Taxonomy and number of genera and species after Wilson and Reeder (1993) and food habits and body size category based on Boulière (1975)

Order	Number of genera	Number of species	Food habits[1]	Size category[2]
Monotermata	3	3	A	S
Didelphimorphia	15	63	A	S
Pauciturberculata	3	5	A	S
Microbiotheria	1	1	A	S
Dasyuomorphia	17	63	A	S
Peramelemorphia	8	21	V	S
Notoryctemorphia	1	2	A	S
Diprotodontia	38	117	V	S
Xenarthra	13	29	A/V	L
Insectivora	66	428	A	S
Scandentia	5	19	A/V	S
Dermopera	1	2	V	S
Chiroptera	117	925	A	S
Primates	60	223	V	S/L
Carnivora	129	271	A	S/L
Cetacea	41	78	A	L
Sirenia	3	5	V	L
Proboscidea	2	2	V	L
Perissodactyla	6	18	V	L
Hyracoidea	3	6	V	S
Tubulidentata	1	1	A	L
Artiodactyla	81	220	V	L
Pholidota	1	7	A	S
Rodentia	443	2021	V	S
Lagomorpha	13	80	V	S
Macroscelididea	4	15	A	S

[1] A = animal; V = vegetable
[2] S = small; L = large

lesser mouse deer (*Tragulus javanicus*), 4 kg dik-diks (*Madoqua phillipsi* and *M. guentheri*), duiker (*Cephalus dorsalis*), and suni (*Neotragus pygmaeus*) are included as small mammals, but Rodentia such as the 9 kg agouti (*Agout paca*) and the 15 kg Indian crested porcupine (*Hystrix indica*) are excluded.

1.3
Desert

The term desert has been discussed widely and has been defined in a variety of ways. As a consequence, different authors define some areas similarly while these areas are, in fact, very different. Some of this confusion was well summed up by McGinnies et al. (1970) who stated, "As glibly as scientists and laymen use the term, there is, nevertheless, no common agreement as to what constitutes a desert. This fact becomes important when one searches the literature, and one author limits publications on the desert to areas with less than 2 inches of rainfall annually, while others include areas with 10, 15 inches, or even more. In addition, it is recognized that rainfall alone is not an adequate criterion for a desert; however, adding other parameters is not a complete solution, because opinion diverges widely about the elements to be included and the relative weight of each in the aridity formula."

Part of the problem in defining deserts lies in the difficulty in delineating areas of arid and semi-arid zones as no easily recognizable natural division points exist under desert environmental conditions. "The range of conditions is a continuum from those extremely dry, under which no vegetation can grow, to the environment where climate and soils no longer limit, but rather compete among themselves. Any attempt to mark absolute dividing points along the complex continuum of conditions is in a sense arbitrary, because there are no natural boundary points along the line" (McGinnies et al. 1970). A more detailed discussion on deserts and their definitions are presented in the next chapter.

1.4
Mammal

The definition of mammal is widely accepted, but the taxonomic relationships among mammals are disputed and differ among reference sources. This can be a source of difficulties as accurate taxonomic relationships can be important in the use of comparisons to explain the adaptations of animals and the evolution of their responses (Pagel and Harvey 1988; Harvey and Purvis 1991; Harvey and Nee 1993).

The taxonomy of mammals has been based mainly on morphological and karyological data (Nevo 1989). However, recent studies have challenged some well established phylogenetic relationships. Such was the case when Geffen et al. (1992) examined ten fox-like canid species from five genera. They analyzed mitrochondrial DNA and concluded that Urocyon and Otocyon and phalogenetically distinct, but that they suggested that *Alopex lagopus* (arctic fox) and *Vulpes Vulpes* (kit fox) are closely related, although they are of different genera. In addition, *Fennecus zerda* (fennec fox) is more closely related to *Vulpes cana* (Blandford's fox) than is its co-generic, *Vulpes vulpes*.

The genus *Acomys* has been examined intensively and is another good example in which phylogenetic relationships have been questioned. Following immunological studies, it was suggested that the taxonomic position of *Acomys* within the family Muridae should be reconsidered (Fraguedakis-Tsolis et al. 1993). Furthermore, electrophoretic analyses demonstrated that *Acomys subspinosus* and *Acomys spinosissimus* were allozymatically distinct and different from other taxa of *Acomys* (Janecek et al. 1991), unlike in previous classifications (e.g. Dippenaar and Rautenbach 1986).

In this text, the taxonomy of Wilson and Reeder (1993) described in their book, *Mammal species of the world – a taxonomic and geographic reference*, will be used primarily. The many changes occurring in taxonomic classification are evident when noting the numerous changes made in the number and classification of orders and in the number of genera and species within orders between the first (1982) and second (1993) editions of their book.

1.5
Small Desert Mammals

Small mammals in the world's arid and semi-arid areas are represented by 9 orders, 41 families and 164 genera (Mares 1993). However, classifying some mammals as desert or non-desert has not been equivocal. This has arisen as a consequence of two main points: (1) the difficulty regarding the definition of desert; and (2) many small mammals that inhabit deserts also inhabit non-deserts, or only inhabit the margins of deserts under certain conditions. Shenbrot et al. (1994b), in their study of desert rodents, concluded that in North America only 39% (27 of 70) of the species (Table 1.2) or 38% (6 of 16) of the genera that occur in deserts can be characterized as being strictly or mainly desertic (Table 1.3); these respective percentages are higher in Asian deserts where they reach 67% (65 of 97 species) and 64% (23 of 36 genera). In Australia, Morton (1982) reported that 44% of desert-dwelling rodent species are confined only to the arid zone.

Table 1.2. Number of species in North American and Asian deserts that are strictly desert species, or that also occur in other life zones. (Shenbrot et al. 1994b)

Continent	No. of species		
	Strictly desert	Partly desert	Total
North America	27	43	70
Asia	65	32	97

Table 1.3. Number of genera in North American and Asian deserts with all species strictly desert, most species strictly or mainly desert, or most species non-desert in life zone distribution. (Shenbrot et al. 1994b)

Continent	No. of genera			
	Strictly desert	Mainly desert	Mainly non-desert	Total
North America	2	4	10	16
Asia	18	5	13	36

In the present text, mammals inhibiting arid and semi-arid areas will be considered as desert species. Decision as to whether a mammal is desert or non-desert will be based, where possible, on the description of the mammal's habitat in the publication in which it is reported.

CHAPTER 2

Deserts

"... in folklore the arid areas that cover a seventh of the globe's land surface are a
forbidding wasteland – sun-seared and wind-scoured, waterless and endless, empty
of shelter and, except for venomous creatures lurking under the rocks, largely devoid of life. The legendary image of the desert is an utterly hostile one."

(Leopold 1961)

Much has been written about deserts, and although all authors agree that deserts
are areas which are arid and have little precipitation, there is widespread disagreement in defining the terms which describe the level of aridity. Terms such
as "desert", "true desert", "extreme desert", "total desert," "absolute desert",
"semi-desert", "hyper-arid", "extreme arid", "arid", "semi-arid" and "steppe" are
frequently used but are defined differently. This has led to considerable confusion and has made comparisons among deserts quite difficult. In fact, according
to McGinnies et al. (1970), this confusion in the definition of desert may "hin-
der rather than further communication" in research and discussions of deserts.

2.1
Definition of Deserts

Emberger (1942) defined a true or extreme desert as an area in which there are
periods of more than 1 year without precipitation and as an area that is characterized by not having a regular seasonal rhythm of rainfall. This definition
has been, to a large extent, adopted by many (for example, Meigs 1953). Evenari
(1981) also agreed with this definition and stated that the most outstanding fea-
ture common to all deserts is the unpredictability of rainfall due to its extreme
quantitative, temporal and spatial irregularities. Quantitative uncertainty can
be expressed by the ratio of maximum to minimum annual rainfall. In the Negev
Desert, this ratio ranges between 5 and 8, but can surpass 100 in extreme deserts.
Taking these characteristics into account, Evenari et al. (1971) suggested the
following definition for a desert:

"a desert is a region where the moisture index is below − 40 (Thornthwaite's moisture index, see
below) and that is also characterized as BW (desert climate) by Koppen's climate formula. Its
rainfall is scanty and highly irregular, with a quotient of variation of 5 or more. Under these conditions soil will be immature, and, since vegetation integrates these factors, it will be sparse and
of the contracted type. Desert pavement and desert varnish are common phenomena, and defla-
tion is most pronounced."

Because of the significance of rainfall, deserts frequently have been classified solely according to total annual precipitation. One of the first attempts was
made by Koppen who defined deserts as areas receiving less than 10 inches

(254 mm) annual rainfall and generally with high air temperatures (Leopold 1961). Furthermore, Koppen defined steppes as areas with an annual rainfall between 10 inches (254 mm) and 20 inches (508 mm), and with large daily and annual temperature ranges. These two rainfall areas were designated as arid and semi-arid, respectively, and each comprised approximately 14% of the world's land area.

Extreme deserts have been defined as areas receiving an annual rainfall of less than 70 mm, true deserts as areas receiving between 70 and 120 mm and semi-deserts as areas receiving between 120 and 400 mm (Shmida 1985). In addition, areas with up to 100 mm annual rainfall have been described as hyper-arid, areas with 100 to 400 mm as arid and areas with 400 to 600 mm as semi-arid (Wilson 1989). Sud et al. (1993) limited deserts to areas with less than 250 mm of annual rainfall. These areas were considered as arid and semi-arid regions and comprise more than 30% of the land surface. In Australia, Morton (1982) delimited the arid zone by the 250 mm isohyet in the south and southeast and by the 550 mm isohyet in the north and northeast. Classifying deserts on the basis of annual rainfall does have some advantages as this is a relatively simple measure and often the only data available from a number of desert sites.

However, although annual rainfall is generally a good indication of aridity (Noy-Meir 1973), aridity is also dependent upon a number of other factors. Thornthwaite (1948) developed a moisture index or index of aridity based on the difference between precipitation and potential evapotranspiration of natural vegetation, in which zero meant that the two variables are equal. An index below -40 was considered as arid, between -20 to -40 as semi-arid, and between 0 and -20 as subhumid.

Meigs (1953) incorporated Thornthwaite's index, categorized the arid and semi-arid areas of the world and, in his description, included season of rainfall, maximum air temperature in the hottest month and minimum temperature in the coldest month. Symbols used by Meigs included degrees of aridity like E, extremely arid, A, arid and S, semiarid; and seasonal rainfall as a, no marked seasonal precipitation; b, summer precipitation; and c, winter precipitation. Two digits followed these codes, the first represented the mean temperature of the coldest month and the second, the mean temperature of the warmest one: $0 = <0\,°C$; $2 = 10$–$20\,°C$; $3 = 20$–$30\,°C$; and $4 = >30\,°C$. Meigs' system received widespread acceptance, and in many cases, the world's deserts are usually located in his arid and extremely arid homoclimates (Dick-Peddie 1991). Furthermore, the three levels of aridity depicted by Meigs (1953) roughly coincide with the three levels described by Shmida (1985).

Nonetheless, because of the importance of desertification and the confusion created by the number of terms used, UNESCO undertook a program to define arid and semi-arid zones, basically to refine Meigs' classification further (UNESCO 1977). New climatic maps were published with the hope that the definitions and terminology would be universally adopted. Level of aridity was based on the ratio between the mean total annual precipitation (P) and the mean annual potential evapotranspiration (ETP). The latter was calculated using Penman's equation, which includes solar radiation, atmospheric humidity and wind. Four main degrees of aridity were classified according to P:ETP ratios as follows: (1) < 0.03 as hyper-arid zone in which interannual rainfall variability reaches 100%; (2) 0.03–0.20 as arid zone in which annual rainfall is 80–350 mm and interannual rainfall variability is 50–100%; (3) 0.20–0.50 as semi-arid zone in which annual rainfall is 200–500 mm in winter rainfall areas and 300–800 mm in summer rainfall areas, with an interannual rainfall variability between 25 and 50%; and (4) 0.50–0.75 as sub-humid, in which interannual rainfall variability is less than 25%.

2.2
Formation of Deserts

Most deserts are located around the Tropic of Cancer and Tropic of Capricorn, 15°–35° north and south of the equator. Hot deserts are usually located between latitudes 20° and 35° and cold deserts between 65° and 90° (Fig. 2.1).

Deserts are formed as the result of four main causes (Table 2.1):

1. *Subtropical Anticyclones (Hadley Circulation).* These deserts occur mainly along the 30° latitude and are caused by descending dry, warm air due to high pressures which result in dry, stable air masses. This air had previously gathered moisture, ascended and then underwent adiabatic heating. Following cooling, the air condenses, precipitation occurs, and dry, high altitude air results. Subtropical anticyclones are primarily responsible for the formation of the Saharo-Arabian and Thar deserts of North Africa and Asia and also the Mohave and Sonoran deserts of North America in the northern hemisphere. In the southern hemisphere, they are primarily responsible for the Kalahari Desert in Africa, Atacama Desert and parts of the Patagonia Desert in South America and some of the Australian deserts.

2. *Orographic Effects (Rain Shadow Effects).* These deserts are the result of the impediment of air mass movements by high mountains. Air masses are cooled adiabatically as they rise up mountain sides, and most of the water is precipitated at high altitudes on the windward slopes. The air masses are thus relatively dry on the leeward side, and this air warms as it descends. The Monte and Patagonia deserts of South America and most of the North American deserts are mainly formed by orographic effects.

3. *Inter-Continental.* These deserts are formed due to their position in a continent. Winds far from the oceans have usually released most of their moisture before reaching the interior of the continent. This causes hot deserts in central Australia and in North America and cold deserts in central Asia.

4. *Cool Coastal.* These deserts are a result of onshore winds from the western rims of continents. These winds are cooled by cold ocean currents and as a result drop only little water. Interestingly, advectional fogs occur as a result of the temperature inversions. Foggy deserts occur in the Namib and Atacama deserts of the southern hemisphere and in Baja, California, and the western Sahara in the northern hemisphere.

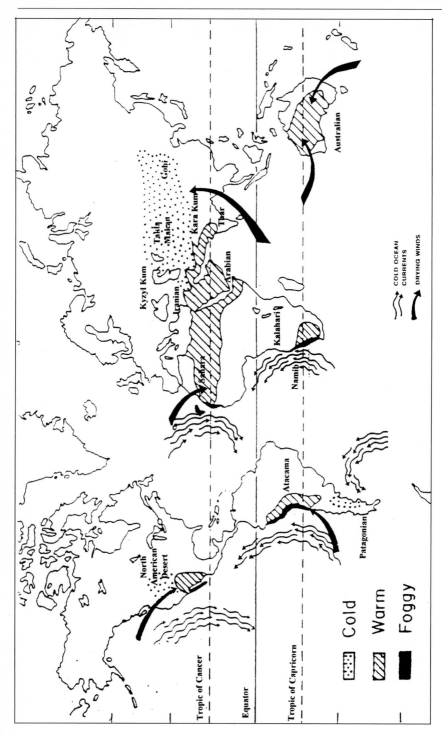

Fig. 2.1. Deserts of the world depicting whether they are either cold, warm or foggy and showing the cold ocean currents and drying winds

Table 2.1. Classification and description of deserts of the world

Continent and desert	Latitudinal range(°)	Cause of formation[a]	Temperature	Aridity[b]	Vegetation
North America					
Great Basin	35–42 N	RS	Hot	E, A	*Artemesia-Artiplex* steppes and pygmy open woodland
Sonora	25–33 N	SB, RS	Hot	S, A	Thorny succulent savanna
Chihuahua	23–32 N	SB, RS	Hot	S, A	Open shrublands and thickets
Baja California	23–32 N	CC	Hot-cool	S, A	Thorny succulent savanna and dwarf shrubs
South America					
Patagonia	37–47 S	SB, RS	Cold	S	Perennial grassland
Peru-Chile	7 N–35 S	SB	Cold	A, S, E	Succulent desert dwarf shrubs
Monte	23–30 S	RS		SD	Thorny succulent savanna
Asia					
Gobi	42–47 N	IC	Cold	S, A	Chenopod-Tamarix desert shrubland
Takla Makan	36–43 N	RS, IC	Cold	A, S, E	Chenopod desert dwarf shrubs
Turkestan	36–47 N	IC, RS	Cold	S, A	*Artemesia-Stipa* steppes and pygmy open woodland
Iran	27–36 N	SB, RS	Hot	S, A	Chenopod-*Artemesia* steppes and pygmy open woodland
Thar	24–31 N	SB, RS	Hot	S	Thorny-rattanoid savanna
Syrian	31–37 N	SB, RS	Hot	A, S	*Artemesia*-chenopod steppes and desert dwarf shrubs
Arabian	15–31 N	SB	Hot	E, A	Chenopod-*Zygophyllum* desert dwarf shrubs; *Artemesia* and Poaceae on sands
Africa					
Sahara	22–32 N	SB, IC	Hot	E, A	Chenopod-*Zygophyllum* desert dwarf shrubs; *Artemesia* and Poaceae on sands
Sahel	13–22 N	SB, RS	Hot	S	Thorny-rattanoid savanna
Somali	3–13 N	SB	Hot	S,A,E	Thorny savanna and desert dwarf shrubs
Namib	18–30 S	CC	Hot-cool	S,A,E	Succulent desert dwarf shrubs
Kalahari	21–28 S	SB	Hot	S	Thorny-succulent savanna
Australia	21–33 S	RS, IC	Hot	S, A	Chenopod shrublands and sclerophyll evergreen low woodland

[a]RS, Rain shadows; SB, Subtropical anticyclones; CC, cool coastal; IC, interior continental.
[b]E, extreme; A, arid; S, semi-arid.

2.3
World Deserts

Thirteen major desert areas have been recognized in the world including: Kalahari-Namib, Sahara, Somali-Chalbi, Arabian, Iranian Thar, Turkestan, Takla-Makan, Gobi, Australian, Monte-Patagonian, Atacama-Peruvian and North American (McGinnes et al. 1970). Of the world deserts, 88% occurs in the Old World (Africa, Asia and Europe) and 12% in the New World (North America, South America and Australia; Table 2.2). Furthermore, of the extreme deserts of the world, 5.62 million km^2 or 97% of the total occurs in the Old World and only 0.20 million km^2 or 3% in the New World.

The Sahara is the most arid and largest of the deserts, stretching across the 5120 km width of North Africa and covering 9.07 million km^2. Most of the interior averages less than 2.54 cm (1 inch) of rainfall annually. Some 49% of Australia is covered by desert, but none is considered extreme. The driest areas in Australian deserts average about 12.7 cm (5 inches) of rainfall annually. Similarly, most North American deserts are not extreme, except for Death Valley and Yuma Desert, both formed as a result of rain shadows. The Atacama-Peruvian is the smallest of the world deserts, 0.36 million km^2, but has the least mean annual precipitation, averaging less than 1.27 cm (0.5 in.) per year.

The hottest deserts are those lowest in altitude and closest to the equator. The low-lying southern part of the Sahara is the world's hottest desert. The highest recorded air temperature in the shade was 58 °C in Azizia, Libya, in the Sahara Desert. Temperatures of 50 °C are common in many deserts, and ground temperatures often exceed air temperatures by 15 to 30 °C. In addition to high air temperatures, hot deserts are characterized by high solar radiation, low humidity and large differences in air temperature between day and night. Meteorological data describing some of these characteristics in the Negev Desert are presented in Table 2.3. Of the six sites included in Table 2.3, only Elat

Table 2.2. Area (in million km^2) and percentage of semi-deserts, true deserts and extreme deserts in different world desert regions. (modified after Meigs 1953; Shmida 1985)

	Extreme deserts		True deserts		Semi-deserts		Extreme true deserts		Total		Area of continent
	Area	(%)	Area	(%)	Area	(%)	Area	(%)	Area	(%)	
Africa	4.56	78	7.30	34	6.08	24	11.86	43	17.94	35	38
Asia	1.06	18	7.91	36	7.52	30	8.96	33	16.48	32	28
Europe	0	0	0.17	1	0.84	3	0.17	1	1.02	2	1
North America	0.03	0.5	1.28	6	4.50	18	1.31	5	5.81	11	10
South America	0.17	3	1.21	6	3.50	14	1.39	5	4.39	8	8
Australia	0	0	3.87	18	2.52	11	3.86	14	6.38	12	49
Old World	5.62	97	15.38	86	13.60	62	20.72	88	33.42	70	
New World	0.20	3	2.49	14	8.00	38	2.70	12	10.20	30	

Note the small percentages of extreme desert areas, except in Africa and Asia.

Table 2.3. Temperatures, relative humidity, evaporation, and rainfall data, and index of dryness for six places in the Negev, 1960 and 1961. (Evenari et al. 1971)

Quantity	Nir Yitzkhak	Beersheba	Mitzpeh Ramon	Elal	Avdat	Shivta
Elevation (m)	80	250	890	25	610	350
Temperature (°C)						
Mean annual	20.2	19.7	17.8	25.9	18.7	19.5
Maximum	44.9	43.8	38.5	47.4	46.4	39.8
Minimum	2.0	1.4	0.5	5.1	0.2	2.2
Mean of hottest month	26.1	26.3	24.7	38.8	25.7	26.9
Mean of coldest month	12.7	11.7	9.3	15.9	10.8	11.9
Mean relative humidity (%)	63.5	55.5	48.5	34.5	51.7	55.2
Total annual evaporation[a] (mm)	2154.8	2897.2	3278.6	4867.3	3031.3	3137.4
Total annual rainfall (mm)	169.9	162.2	65.2	20.2	68.8	89.5
Index of dryness[b]	12.6	17.8	50.4	243.0	43.9	34.8

[a]Measured with a Piche evaporimeter.
[b]Ratio of annual evaporation to annual rainfall.

(which is located in the southern Arava, an extreme desert) is considered climatically a hot desert; Mitzpeh Ramon and Avdat sites are designated as temperate (Evenari et al. 1971). Global radiation averages about 200 kcal cm^{-2} year^{-1} in the Negev desert where there are 3000–3600 h of sunshine annually (Evenari 1981). Low humidity allows heat to dissipate, and air temperatures may drop by 50 °C or more in the evening (Leopold 1961). Studies described in this book will concentrate mainly on small mammals in hot deserts.

In contrast, cold deserts are found at higher altitudes and at greater distances from the equator. Of these, the high altitude Gobi Desert of Mongolia is the coldest. Cold deserts are characterized by severe climate and extreme seasonal air temperature fluctuations. In the arid steppes of Central Mongolia, permafrost occurs in many areas. Average monthly air temperature is -24 °C in January, the coldest month, and averages $+16$ °C in July, the warmest month. The yearly amplitude spans -45 to $+36$ °C. Annual precipitation is close to 250 mm, with 90% falling in the summer when the relative humidity is very low and evapotranspiration extremely rapid (Weiner and Gorecki 1981).

Although deserts have little precipitation, they may have well defined winters and summers. In the Mediterranean-type climate, as found in North Africa and southern California, all rain occurs in the winter. Winters are mild, and summers are hot and dry. In the continental-type climate, as in the deserts of Central Mexico, rain occurs as thundershowers in the summer. The southern fringe of the Sahara has monsoon-type summer rainfall. These deserts have strong seasonal contrasts and highly variable weather.

2.4
Desert Soils

Desert soils are immature and have either no profiles or weakly developed ones. The soils are heavily impregnated with sodium and potassium salts released from the earth's rocky crust. Because the potential evaporation greatly exceeds precipitation, minerals in deserts may move upward with moisture that travels to the surface via capillary action. As a result, some deserts become heavily saturated with minerals and contain little nitrogen and organic matter. Consequently, many deserts are devoid of plants, as is the case in large regions of the Atacama. However, depending on the mineral content, some deserts may be quite fertile, as in California's Imperial valley.

CHAPTER 3

Body Size and Allometry

"Body size is the single most important characteristic of an organism; it influences the physical environment faced; the likely food and predators encountered, and the organism's responses to these circumstances. The other fundamental characters of organisms, including their anatomy, cost of reproduction, and means of locomotion, vary with body size. The relationships existing between these characters and body size are quantitative, which leads to the concept of 'scaling' i.e. the quantitative change in the parameters of an organic character with a change in body size. Such size dependent relationships are called allometric."

(McNab 1988)

3.1
Allometric Scaling

A large number of biological functions and structures (dependent variables) have been related to body mass (an independent variable). Some are related linearly (e.g. lung volume, Gehr et al. 1981; gut volume, Demment and van Soest 1985); however, many are non-linear. Least squares regression analysis on the logarithmic transformation (base 10) of both the dependent variable (Y) and body mass (m_b) yields the linear relationship:

$$\log y = \log a + b \log m_b,$$

where a is the Y-intercept and b is the slope.

This log-log transformation is convenient in that it usually fits the data well and allows statistical calculations of regression coefficients and confidence intervals. The relationship is generally expressed as the power function:

$$Y = am_b{}^b,$$

where the exponent b is called the scaling factor.

In linear relationships, the exponent $b = 1$, whereas in cases in which body mass does not affect the independent variable, $b = 0$. In relationships in which a change in the independent variable is proportionately less than in body mass, b lies between 0 and 1, and when the change is greater than the change in body mass, $b > 1$.

Allometric scaling was first introduced by Huxley (1924, 1932) when he proposed a power function to describe relative growth. Since then, a large number of empirical allometric relationships have been generated, and much has been written interpreting the biological implications of the observed slope, the possible biological reasons for the residual variation, and for individuals' or species' points lying furthest from the regression line (Peters 1983; Calder 1984; Schmidt-Nielsen 1984).

The empirically derived equations have been used extensively to: (1) predict characteristics in unmeasured animals; and (2) compare observed measurements with those predicted for an animal of that body size. The former is convenient in that the body mass of an animal is generally relatively easy to measure. The latter is often used to explain animal adaptations.

Smith (1984) describes two approaches in scaling biological characteristics to body size. In the first approach, called "broad allometry", a characteristic is measured over a wise range of body masses. Attempts can then be made to account for the residual variation which may be due to phylogenetic factors (taxonomic affiliation) as well as to ecological behavioural factors (food habits, climate, activity level; McNab 1992b). In the second approach, called "narrow allometry", size effects are minimalized by studying a number of animals of approximately the same size. Differences among the animals can then be attributed to adaptations. Narrow allometry can be very useful in comparing animals, but has been relatively neglected. Its main advantage is that the regression equation refers to the range of body sizes of the animals in question, which make the results more relevant. Narrow allometry will be stressed among small mammals in this text.

Allometric scaling, although used extensively, has a number of drawbacks and should be employed with caution (Smith 1980, 1984; McNab 1988). Logarithmic transformations reduce the statistical problems related to outliers. In addition, numbers on a log scale that are relatively close together can appear greatly different when expressed on a linear scale, and thus it is often difficult to envision the true scatter of points. Furthermore, of 60 cases examined by Smith (1984) in which allometry was used to describe intraspecific and interspecific relationships, in only 12, all interspecific, was the correlation coefficient significantly higher for the log-log transformed data than for the untransformed data.

3.1.1
Correlation Coefficient and Standard Error of Estimate in Allometric Relationships

Often, the only statistics provided for the regression equation describing allometric relationships is either the correlation coefficient (r) or the coefficient of determination (r^2). However, a high r value does not ensure that the regression equation provides a good prediction, only that the regression line fits the data well. The r^2 describes the proportion of the total variance of Y that can be contributed to the variance in body mass. For example, an $r^2 = 0.90$ signifies that 90% of the variance in Y can be explained by the variance in body mass. However, both r and r^2 are affected by the range of values for the dependent and independent variables and by the slope of the regression line; that is, they are dependent on the units used. The values of r and r^2 increase with a greater range of the X and Y variables and with a steeper slope of the line.

To evaluate the predictive ability of a regression equation, the standard error of the estimate, $S_{y \cdot x}$, should be presented. $S_{y \cdot x}$ is independent of units and of slope and is basically equivalent to the standard deviation of the regression of Y on X (Smith 1984; McNab 1988). For a log-log curve, $S_{y \cdot x}$ is the log of the residuals of the mean standard deviation of the residuals in Y from the fitted curve at a given mass. The range within which 68% of the measurements will fall when using the regression equation can be calculated by the antilog of $(2 \pm S_{y \cdot x})$, as suggested by Brody (1945). For example, if $S_{y \cdot x} = 0.078$, then 68% of the measurements will fall between 0.84 and 1.20 times the predicted value (calculated from antilog 1.922 and 2.078).

3.1.2
Interspecific and Intraspecific Allometric Relationships

Interspecific relationships with large ranges in body mass usually yield higher correlation coefficients than intraspecific relationships that have much smaller ranges. However, that does not mean that the predictability is improved; in fact, often it is not. Smith (1984) illustrated this point well when comparing different studies. For example, in an interspecific allometric study of basal metabolism in snakes, the ratio of maximum to minimum body mass was 1995:1, r was 0.974, and the mean prediction error was a high 35.9%. In an intraspecific allometric study of the basal metabolic rate in rats, the ratio of maximum to minimum body mass was only 1.83:1, r was 0.754, but the mean prediction error was a low 5.4%.

Furthermore, an interspecific regression line describing an allometric relationship between a parameter and body mass often does not describe intraspecific relationships (or a group of animals with a small range of body mass) of the same parameters. Heusner (1982) presents several examples in which this can occur (Fig. 3.1). In Fig. 3.1A, the intraspecific lines have different slopes and intercepts; in Fig. 3.1B, the intraspecific slopes are the same, but the intercepts are different, and the group means do not fall on the interspecific line; while in Fig. 3.1C, the lines have the same slopes but different intercepts, and the intraspecific means fall on the interspecific line. In these three cases, the overall regression lines do not represent the intraspecific relationships. Only in Fig. 3.1D, where the intraspecific relationships have the same slope and intercepts, does the overall regression reflect the allometric relationship between the parameter and body mass.

3.2
Body Size and Environment

Body size is usually the central measurement employed when making comparisons among animals and is used extensively in developing hypotheses related to physiological and behavioural responses and the distribution of animals. Accordingly, Peters (1983), in his book, listed more than 1000 allometric relations describing how different animal characteristics are related to body mass.

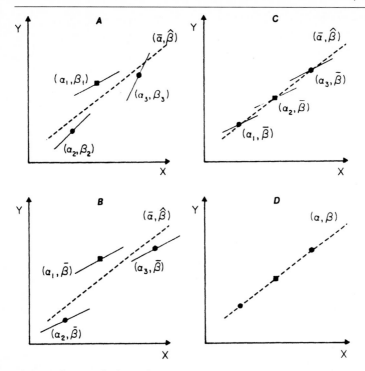

Fig. 3.1. Theoretical relationships between the logarithm of body mass (X) and the logarithm of basal metabolism (Y) in different animal species. **A–D** (Heusner 1982)

3.2.1
Bergmann's Rule and Body Size of Animals

According to Bergmann's rule, animals decrease in body mass with an increase in aridity. The explanation offered for this trend relates to the principles of heat exchange (Steudel et al. 1994). The lower surface to body mass of larger animals leads to greater thermal efficiency and allows better conservation of heat, an advantage in colder climates. In contrast, the higher surface to body mass in smaller animals offers a greater relative area for heat dissipation, an advantage in warmer climates.

Indeed, this trend of body size has been reported in a number of mammals. Spiny mice (*Acomys cahirinus* and *A. russatus*) generally decrease in body mass with a decrease in rainfall and latitude and an increase in air temperature (Fig. 3.2; Nevo 1989). Subterranean mole rats (*Spalax ehrenbergi*) also decrease in body mass with increased aridity (Nevo et al. 1988).

However, the validity of Bergmann's rule has been widely questioned (Geist 1987, 1990) because a large number of mammals do not conform to it. For example, body mass was measured in three granivorous gerbilline rodents in Israel along a north-south gradient which was characterized by increasing aridity and

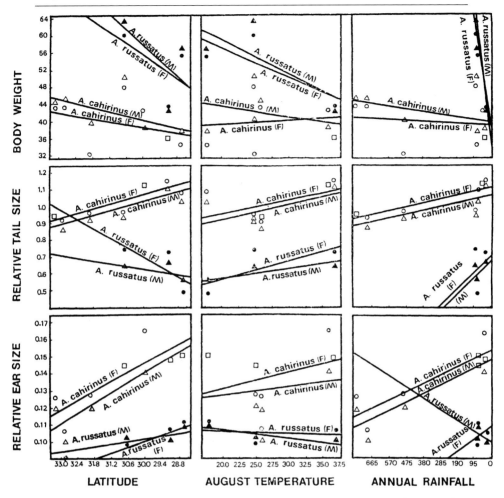

Fig. 3.2. Regressions of body mass and of relative ear and tail size of *Acomys cahirinus* and *A. russatus* on latitude, August temperature and annual rainfall. Δ population mean of *A. cahirinus* males, o population mean of *A. cahirinus* females, ▲ population mean of *A. russatus* males, ● population mean of *A. russatus* females, □ both sexes of *A. cahirinus*, M males, F females. (Nevo 1989)

increasing air temperature (Brand and Abramsky 1987). *Gerbillus pyramidum* showed a decrease in body mass along this north-south gradient which agreed with Bergmann's rule; *Gerbillus gerbillus* showed an increase in body mass along this gradient which was in disagreement with the rule, and *Gerbillus allenbyi* showed no difference in body mass along the gradient.

Moreover, it has been suggested that some mammals which conform to Bergmann's rule, such as a number of carnivores (McNab 1971; Mendelssohn 1982) and granivores (McNab 1971), do so as a consequence of food availability, and not for thermal exchange. It was argued that the decrease in body size with increasing aridity is simply a reflection of seed availability for granivores (McNab 1971) and prey size for carnivores (Rosenzweig 1968; McNab 1971).

3.2.2
Allen's Rule and Thermoregulatory Effective Surfaces

According to Allen's rule, extremities or appendages increase in size with an increase in aridity and air temperature. Most notable are the large ears in desert rabbits and foxes compared with those of their non-desert counterparts. Because of the large surface to mass ratios, appendages can be important sites in heat exchange with the environment (Conley and Porter 1985). These appendages generally have sparse pelage, increased blood flow and temperatures that facilitate heat dissipation. A convincing example of surface area as a function of air temperature was exhibited in littermate pigs maintained at either 5 to 35 °C from 4 to 11 weeks of age and fed sufficient amounts to maintain similar body mass. Pigs reared in the warm had larger ears, longer limb bones, less hair and fewer blood vessels in the skin than pigs reared in the cold (Mount 1979). In a study on two species of spiny mice inhabiting areas of different aridity, Nevo (1989) found that, in general, relative tail size and ear size increased with a decrease in latitude and annual rainfall and an increase in air temperature (Fig. 3.2).

The difference in thermoregulatory effective surfaces among closely related species which inhabit areas differing in aridity has been demonstrated among foxes. In foxes, the thermoregulatory surfaces include the area of the face, nose, dorsal head, pinnae, lower legs and paws (Fig. 3.3), areas that are characterized

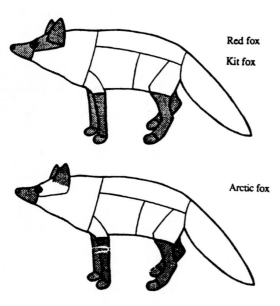

Red fox

Kit fox

Arctic fox

Fig. 3.3. The thermoregulatory effective surface areas (*shaded*) in the red fox (*Vulpes vulpes*), arctic fox (*Alopex lagopus*) and kit fox (*Vulpes macrotis*). (Klir and Heath 1992)

by having relatively little cover (e.g. short fur) in all seasons. These surfaces are basically the same in all species of foxes but differ significantly in relative size, being larger in species inhabiting more arid area. Of the total surface area, the thermoregulatory effective surface is 21.6% in the arctic fox (*Alopex lagopus*), a species that inhabits extremely cold environments; 32.9% in the red fox (*Vulpes vulpes*), a species that occurs in all temperature regions of the world; and 38.2% in the kit fox (*V. macrotis*), a species that inhabits hot desert areas (Klir and Heath 1992; Table 3.1). Thus, the desert fox has a relative area that is about twice that of the arctic fox, while the temperate fox falls in between the two.

3.2.3
Dehnel Phenomenon and Seasonal Body Mass Changes

Seasonal changes in body mass have been observed in many rodent species (Iverson and Turner 1974; Viro 1987; Korn 1989; Ellison et al. 1993). Possible reasons for these changes include: (1) an adaptive strategy that reduces energy requirements during unfavourable seasons (Stebbins 1978); (2) a redistribution in population age structure and/or adherence to the "youth conservation" principle (in *Clethrionomys glareolus*; Olenev 1979); and (3) responses to fluctuations in food resource (Brand and Abramsky 1987). The adaptation hypothesis received support from a study on *Microtus montanus* and *Phodopus sungorus sungorus* in which body mass was found to be regulated by photoperiod (Petterborg 1978; Heldmaier and Steinlechner 1981) and another on individuals of *Cricetus cricetus* born in the autumn: these were unable to increase their body mass in the winter even with an excess of food under stable vivarium conditions (Canguilhem and Marx 1973).

The reduction in body mass in the winter of holoarctic mammals, known as the Dehnel phenomenon, is an effective means of reducing total feed requirements during harsh non-breeding periods since smaller mammals require the intake of less absolute energy than do larger animals. Mass-specific energy requirements, however, increase. This decrease in body mass in the winter compared with the summer as a strategy for reducing energy expenditure has been described in shrews (Dehnel 1949; Mezhzherin 1964; Mezhzherin and Melnikova 1966; Hyvarinen 1984; Genoud 1985, 1988; Merritt 1995), where even a reduction in skeleton and braincase was also noted (Pucek 1970; McNab 1991).

Table 3.1. Relative proportion of the thermoregulatory-effective and thermoregulatory-ineffective body surface areas in the red fox (*Vulpes vulpes*), arctic fox (*Alopex lagopus*) and kit fox (*Vulpes macrotis*). (Klir and Health 1992)

	Percentage of total surface areas		
	Red fox	Arctic Fox	Kit fox
Thermoregulatory effective areas	32.85	21.61	38.16
Thermoregulatory ineffective areas	67.15	78.39	61.84

This reduction in body mass is most pronounced in shrews inhabiting colder climates where the smallest species of *Sorex* are generally found (Mezhzherin 1964). The reduction in energy needs is concomitant with a reduction in the availability of soil invertebrates, the main dietary items of shrews (Aitchison 1987).

The most extreme example is found in the Djungarian hamster, *Phodopus sungorus*, which occupies arid steppes and semi-deserts of central eastern Asia. In the Siberian steppe, this species faces extreme air temperatures ranging from above 30 °C in summer to below − 40 °C in winter and can survive long-term exposure to − 70 °C (Heldmaier et al. 1982). To reduce energy requirements in winter, this hamster reduces its body mass some 45%, from about 45 to 25 g (Fig. 3.4). This reduction in body mass is triggered mainly as a response to photoperiod and, to a lesser extent, low air temperatures (Heldmaier and Steinlechner 1981; Heldmaier et al. 1982; Masuda and Oishi 1988; Ruf et al. 1993). In spite of the loss in body mass, brown adipose tissue (BAT), which is important in non-shivering thermogenesis, increased during the short photophase (Heldmaier and Hoffman 1974).

Body mass change and energy intake in response to photoperiod have been reported for meadow voles (*Microtus pennsylvanicus*; Dark et al. 1983). Over a 10-week period, meadow voles maintained on a short photoperiod (10L:14D) consumed 15% less feed and weighed 20% less than voles maintained on a long photoperiod (14L:10D). Reduction in body mass was due to a decrease in total body water and in lean body mass, as BAT remained unchanged. However, the activity of white adipose tissue lipoprotein lipase reduced. The loss in body mass during short photophase is in agreement with other studies on winter-acclimated voles (Wunder et al. 1977).

The above studies on voles were carried out on species inhabiting mesic areas characterized by cool climates. Do voles inhabiting more xeric regions respond similarly? Effect of photoperiod was studied in the palearctic Levant vole (*Microtus guentheri*), an arvicolid that inhabits an area bordering semi-arid and arid regions characterized by long, dry and warm summers (Banin et al. 1994), and in the fat jird (*Meriones crassus*), a gerbilline rodent that inhabits extremely arid areas of the palaearctic deserts (Haim and Levi 1990). Responses in energy intake and body mass change in the Levant vole were similar to those of mesic voles. Energy intake was lower in voles adapted to long scotophase (LS voles; 8L:16D) than in those adapted to long photophase (LP voles; 16D: 8L), but there was no difference between the two groups in the digestibility of the diet. The LS voles lost body mass, whereas the LP voles gained body mass (Table 3.2).

Different responses to photoperiod, however, were displayed by the fat jird. Body mass appeared not to be affected by photoperiod in this species. Furthermore, in contrast to the studies on the voles, fat jirds exposed to long scotophase (8L:16D) had an energy intake that was 2.24 times that of gerbils exposed to a long photoperiod (16L:8D). From these results, it appears that *Mi. guentheri* reduces its body mass in winter as a strategy to reduce absolute

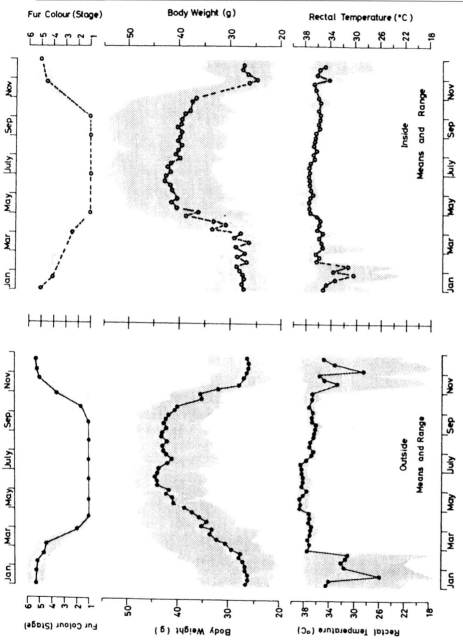

Fig. 3.4. Seasonal changes in fur coloration, body mass and rectal temperature of *Phodopus sungorus* living in natural photoperiod inside (○) or outside (●). Values are given as means ± range of data for 10 hamsters living under either condition. Records were started in October 1977 and continued until September 1978. (Heldmaier and Steinlechner 1981)

Table 3.2. Gross energy intake (GEI), digestible energy intake (DEI) and daily change of body (m_b) in Levant voles (*Microtus guentheri*) acclimated to two different photoperiod regimes at a constant ambient temperature ($T_a = 25$ °C). (Banin et al. 1994)

		m_b (g)	GEI (cal g^{-1} day^{-1})	DEI (cal g^{-1} day^{-1})	Digestibility (%)	m_b/day (%)
LS voles	8L:16D	60.0 ± 9.1	451.7 ± 88.6	406.7 ± 80.2	88.4 ± 3.4	− 2.94
LP voles	16L:8D	47.5 ± 8.1	691.7 ± 132.3*	618.3 ± 118.1*	89.5 ± 2.1	4.11

All values are means +SD of $n = 6$, *$P < 0.05$.

energy requirements, but when sufficient feed is available, *Me. crassus* does not lose body mass in winter.

3.2.4
Body Mass Changes in Free-Living Small Mammals

Seasonal changes in body mass have been described in free-living rodents of subtropical areas. In eight rodent and one shrew species of the Transvaal that breed seasonally, body mass declined during the harsh, dry, non-breeding season by 13.8–39.9% (Korn 1989). Ellison et al. (1993) concluded that the annual cycle in body mass, at least in the pouched mouse (*Saccostomus campestris*) is a consequence of reduced body mass in young animals "in winter as an adaptation to limit their energy requirements when food availability declines."

Does this pattern also hold true for rodents in extreme deserts? Seasonal body mass changes were monitored in three free-living gerbil species inhabiting sandy habitats, *Gerbillus pyramidum*, *G. allenbyi* and *G. gerbillus*. In all species, the maximum body mass was recorded in the spring when food resources were relatively abundant, and minimum body mass in the winter when food resources were relatively scarce (Brand and Abramsky 1987). In a further study on three species of gerbils, *M. crassus*, *G. dasyurus*, and *G. henleyi*, and a spiny mouse, *Acomys cahirinus*, in the Negev desert, similar results were reported, in general. In the three gerbil species, there was a significant reduction in body mass in the winter, ranging from 8.9% to 20.2%, but no change was found in *A. cahirinus* (Khokhlova et al. 1994). It would appear, then, that the pattern of reduced body mass during the harsh, non-breeding season is applicable to desert rodents as well. However, more studies are required to examine whether these changes are an adaptational strategy to reduce energy requirements or simply a result of reduced food availability.

3.2.5
Fasting Endurance and Body Size

Millar and Hickling (1990) proposed that food availability and fasting endurance are the main determinants of body size. They based their theory on intraspecific allometric scalings of energy expenditure and the capacity to store body fat. The common intraspecific exponent relating body fat reserves to body mass in seven species was 1.45, while that of food requirements to body mass was lower, 1.09. Smaller animals, therefore, use their energy reserves at a proportionately faster rate than do larger animals. As a result, larger animals are able to endure food shortages for longer periods (see also Peters 1983; Linstedt and Boyce 1985). Millar and Hickling (1990) argued that selection favours larger animals in regions where long periods of starvation, which can be endured by large animals but not by small animals, are encountered. Losses in body energy reserves by the large animals can be recouped later. In contrast, smaller body size is favoured if food is always available but in short supply.

Four scenarios are presented which may favour different body sizes (Fig. 3.5): (1) Fig. 3.5a does not favour directional size selection since the reduced food availability in period 2 is still sufficient for energy maintenance in small animals, while large animals have sufficient energy to overcome short negative balances and recoup losses during the next period; (2) Fig. 3.5b favours larger

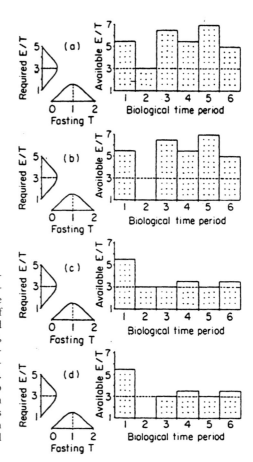

Fig. 3.5a–d. Hypothetical interactions between energy requirements (*vertical normal curve*) and fasting endurance (*horizontal normal curves*) for animals of different size, in relation to available food resources (*E*, energy; *T*, time). In all cases, an average-sized animal has requirements = 3 and fasting time = 1. Requirements and availability are scaled similarly. Fasting time and biological time are also scaled similarly. **a** Favours no change in size; **b** favours large body sizes; **c** favours small body sizes; **d** is untenable for both small and large body sizes. (Millar and Hickling 1990)

animals because smaller animals do not have sufficient energy reserves to overcome period 2, while larger animals do (unpredictable food supply); (3) Fig. 3.5c favours smaller animals since the food availability cannot support large animals (predictable smaller supply); and (4) Fig. 3.5d presents an untenable situation because the food availability cannot support a large animal, and the small animal cannot overcome the food shortage in period 2.

Based on Bergmann's rule, scenario 3 should fit the situation for desert mammals in that small desert mammals are favoured. However, this scenario assumes that a small, but predictable, supply of food is available, and this does not characterize most deserts. Indeed, most deserts are generally characterized by an unpredictable food supply. This can perhaps, in part, explain why so many desert mammals do not conform to Bergmann's law and may suggest that factors other than energetics are also involved in the evolution of body size (Speakman 1992, 1993b).

CHAPTER 4

Heat Transfer and Body Temperature

"One of the primary means by which the environment influences animals is through the exchange of energy. If the animal takes in more energy than it gives out it will get warmer, overheat and perish. If the animal loses more energy than it gains it will cool and not survive. An animal may warm or cool for a limited period of time, but on average, over an extended period of time, an animal must be in energy balance with its environment."

(Porter and Gates 1969)

Mammals and birds are homeotherms, that is, they maintain their body core temperatures within a relatively narrow range (usually between 36 and 40 °C) over a wide range of environmental temperatures and activities. This is achieved by keeping heat gains and losses in balance.

The heat inputs of an animal derive from heat from the environment, that is, from solar and infrared radiation, and from heat generated from metabolic processes. Heat is lost by sensible heat loss through infrared radiation and by insensible heat loss through evaporation, both respiratory and cutaneous. In addition, heat is both gained and lost via convection and conduction. The heat gains and losses are in a continual dynamic state, but in addition, heat can be stored or evacuated, resulting in a rise or fall of the core body temperature. However, these occurrences can only be temporary and persist until thermoregulatory responses restore thermal equilibrium. Animals which adhere more strictly to the maintenance of their body temperature are referred to as 'thermostable', whereas those animals that allow their body temperature to fluctuate to some degree are referred to as 'thermolabile'.

4.1
Heat Exchange Between Animal and Environment

To describe the heat exchange of an animal with its environment, the driving variables or boundary conditions must be measured or calculated (Fig. 4.1). These include heat fluxes from the sun, sky, ground and air. In Fig. 4.1, the velocity profile, or boundary layer, contains smaller layers. These consist of changing wind speeds at different heights, and their thickness varies with the wind speed and with the dimensions of the object being investigated (Porter 1989).

Porter et al. (1994) described the development of heat exchange models between an animal and its environment as progressing from a linear slab model to present day "non-linear internal heat generation model coupled to a porous fur model" (Fig. 4.2). The early mathematical models (slab models) of heat

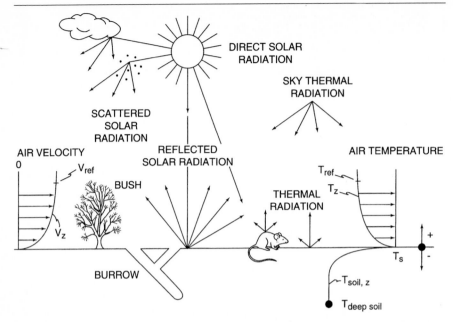

Fig. 4.1. Heat exchange variables between an animal and its environment. The boundary layer thickness, which depends on the length of the surface over which air flows, is represented by the different wind speeds at different heights. Large objects such as the earth's surface have thick boundary layers. Objects as small rodents have very thin boundary layers because of their small size. (Porter 1989)

exchange assumed that heat was generated and released at the centre of the animal and then conducted linearly through its body "as though it were a slab of tissue". In these models, there is no temperature gradient in the flesh. The fur temperature is lower and calculated heat loss from the animal is smaller than in the other models. In the next stage of development (cylinder model, solid fur), non-linear heat production from the core of the animal and the influence of the animal's shape were included. Cylindrical geometry is used, and it is assumed that heat is generated within the flesh and conducted radially in the shape of a wedge or cone extending from the centre of a cylinder, sphere or ellipsoid. This model results in a "concave downward-temperature-profile from the centre of the animal to the edge of the heat-generating tissue. Heat conduction through the fat layer and the solid fur layer occurs in non-linear fashion." The conductivity of the fur is modelled as a solid whose conductive value is constant and is not dependent on other variables. Temperature gradients exist across the radius of the body, and calculated heat flux is highest in this model. The current model (cylinder model, porous fur) takes into account the porous nature of fur. In it, heat is transferred by conduction "along hair fibres and through the air columns between the fibres, as well as by infrared radiation diffusely emitted, absorbed and re-emitted between the hair fibres in the fur in

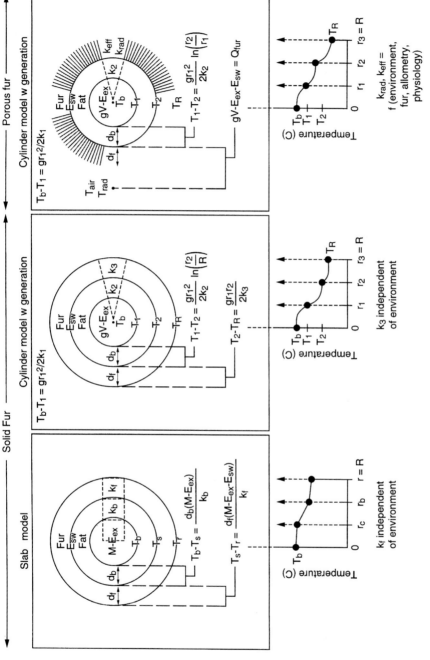

Fig. 4.2. Evolution of endotherm models from linear solid-fur models to non-linear porous-fur models. Different assumptions about geometry affect computed heat fluxes. 'Thermal conductivities' for radiation and conduction through fur, modelled as a porous medium, are not constants, but depend on many variables simultaneously. Comparisons of heat flux calculations for the same sheep show that the porous model has intermediate estimates of heat flow relative to the slab (lowest) and cylinder (highest) computed flux for the same animal and environmental parameters. *Left most panel* is from Porter and Gates (1969). (Porter et al. 1994)

the outgoing and incoming direction." The effective conduction and radiation, K_{eff} and K_{rad}, depend upon a number of variables such as environmental conditions, the animal's body size and its pelage characteristics. Computed heat fluxes are intermediate in this model (for further details, see Porter 1989; Porter et al. 1994; Steudel et al. 1994).

An example is presented in Fig. 4.3 of the parameters needed for the cylinder model with fat and porous insulation in order to determine the metabolic rate of an animal that maintains a constant core body temperature. There is only one rate of metabolic heat production that can satisfy all the conditions if the following are known: the radial dimensions of the animal from its centre to the edge of the generating tissue (in this case the skin), the properties of its fur, its core temperature and the environmental conditions (solar and infrared radiation, air temperature, wind speed and humidity). Furthermore, the posture of an animal can influence its metabolic rate (Porter 1989; Porter et al. 1994).

4.1.1
Radiation

Radiation is the flow of heat through space, including a vacuum. Heat proceeds in the form of electromagnetic waves from all objects above absolute zero (-273 °C). The flow of energy depends upon the temperatures of the surfaces radiating towards each other, their relative positions and their emissivity.

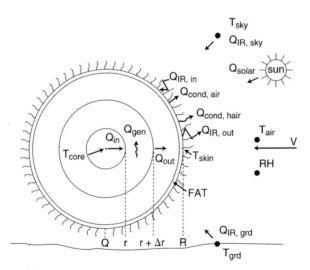

Fig. 4.3. System diagram for describing the energy balance of an endotherm with fur. The system specifies what must be known to solve the energy and mass balance equations. Variables required are core temperature, dimension of the animal, fur properties (if fur is present) and environmental conditions. Surface and silhouette are also needed. The energy balance equation is solved for the rate of heat generation as a function of core temperature function. (Porter 1989)

The spectrum of energy levels ranges from ultraviolet (UV) and visible light to invisible infrared. The wavelength for the maximum emission of radiation from a surface α_{max} (µm) is inversely proportional to the absolute temperature (T, K) of the emitting surface. This is expressed by Wien's displacement law as:

$$\alpha_{max} = 2897/T.$$

All living organisms radiate energy in the long wavelength infrared spectrum, whereas the sun's wavelength is much shorter, mainly falling between infrared and the visible spectrum. The rate of heat transferred by thermal radiation from an animal is proportional to the fourth power of the animal's absolute surface temperature. It is described by the Stefan-Boltzmann Law which takes the form:

$$Q_r = \varepsilon\sigma T_s^{\,4},$$

where Q_r is the rate of energy radiated in W m^{-2}; ε is the emissivity of the surface, ranging between 0 and 1 (with most animals close to 1); σ is the Stefan-Boltzman proportionality constant, 5.67×10^{-8} W m^{-2}; and T_s is the absolute temperature (K) of the radiating surface. In order to calculate the rate of total radiation from the animal, Q_r is multiplied by the effective radiating area of the animal (Bakken 1976).

Emissivity is a relative value indicating to what extent the surface of an animal emits radiant energy. Radiant energy striking a surface can be absorbed, reflected or transmitted. For an opaque object, the sum of absorbed and reflected radiation equals 1, and for a partially transparent object, the sum of absorbed, reflected and transmitted energy also equals 1. A perfectly black, opaque object which absorbs all incident radiation and reflects none would have a maximum value of 1. In contrast, a perfectly opaque reflector which reflects all incident radiation and absorbs none would have a value of 0.

Emissivity varies with the wavelength of the radiation surface. For long-wave radiation (greater than 3 µm) most animal surfaces behave like a black body, and the emissivity is close to unity, independent of colour or covering; whereas for short-wave radition, emissivity is directly dependent on colour.

The net radiant heat exchange for an animal is the difference between the radiant energy leaving the body and that absorbed from the environment (Fig. 4.4). This can be expressed by the equation:

$$Q_{r\ net} = \sigma A\ (\varepsilon_1 T_1^{\,4} - \varepsilon_2 T_2^{\,4}),$$

where σ is Stefan-Boltzmann's constant; ε_1 and ε_2 are emissivities of the surfaces of the animal and its surroundings; T_1 and T_2 are absolute temperatures of the surface of the animal and its surroundings; and A is effective radiating area. In Fig. 4.4, σ is 1.175 kcal day^{-1}, the body surface temperature (T_s) is 33 °C, T_R is the surrounding wall temperature, and it is assumed that the body surface and surrounding walls behave as black bodies.

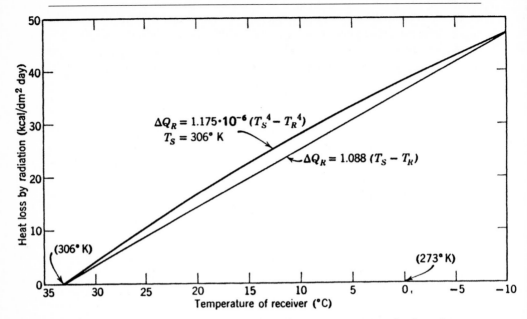

Fig. 4.4. Heat loss by radiation from a black surface of 33°C according to Law of Stefan-Boltzmann. (Kleiber 1975)

4.1.2
Convection

Convection is the transfer of heat in a fluid medium, usually air or water, and depends upon the redistribution of the molecules within the fluid. Heat transfer by convection depends upon the surface temperature of the animal, its shape, surface characteristics and size, as well as on the temperature and velocity of the air that impinges on the surface. It occurs at the interface boundary layer surrounding the warm body of a mammal. Wind velocity changes from zero at the ground surface (due to friction which limits molecular movement) and increases with height. This wind profile is dependent upon the texture of the surface and the speed and turbulence of the fluid.

Free convection occurs when the movement of molecules is due primarily to density or buoyancy differences from temperatures in the fluid. Forced convection occurs when an external force, such as wind or the movement of an animal through its medium acts upon an object. At high wind speeds, forced convection is predominant, whereas at low wind speeds, natural convection predominates.

Convective heat transfer can be expressed as follows (Robbins 1983):

$$Q_c = h_c A(T_s - T_a),$$

where Q_c is quantity of convective heat transferred from an object of area A; T_s and T_a are the surface and fluid temperatures; h_c is a mathematical description of the rate at which heat moves across the temperature gradient. The direction of heat flow is dependent upon T_s and T_a; and h_c is dependent upon the size, shape and surface qualities of the animal; fluid velocity, viscosity and turbulence in forced convection; and the temperature gradient in free convection.

4.1.3
Conduction

Conduction is the transfer of heat between the surface of the body and the surface with which it is in contact. This could be either solid, liquid or gas. The transfer of heat is always from a region of higher temperature to one of lower temperature and consists of a direct transfer of the kinetic energy of molecular motion. It can be expressed as:

$$Q_k = h_k A(T_1 - T_2)/L,$$

where Q_k is the quantity of heat transferred by conduction per unit of time and area; $(T_1 - T_2)$ is the temperature gradient driving conductive heat transfer; L is the length of the conducting path or the depth of insulation; and h_k is the conductivity coefficient describing the rate or efficiency with which the heat differential described by the temperature gradient can be conducted within or between the contacting surfaces.

Conduction can be negligible in a standing animal due to the virtual absence of contact between the animal and the ground, but may be significant if an animal is lying on the ground, and there is a considerable temperature gradient between the two. Heat flow through still air within the pelage of a mammal is referred to as conduction. The transfer of heat which takes place in the warming of ingested food and water within the body of a homeothermic animal provides another example of conduction (Degen and Young 1984).

4.2
Operative and Standard Operative Environmental Temperatures

Air temperature is often used as a thermal measure, but it fails to account for the effects of sun and wind on heat exchange between the animal and its environment. The operative and standard operative environmental temperatures (T_e and T_{es}, respectively) have been developed to include the thermal environment that the animal experiences and have proven to be useful in heat transfer analyses. These measures are "thermal indices that allow single-number representations of the complex thermal environment" (Bakken 1992).

T_e is the effective temperature of the animal when in thermodynamic equilibrium with its environment and includes conduction, convection and radiation, but does not include the physiological processes of metabolism and evaporation. It can be calculated as detailed by Bakken (1992):

$$T_e = T_a + \Delta T_r,$$

where: $\Delta T_r = (Q_a - A_e \sigma \varepsilon T_a^4)/(H+R)$ and T_e is the operative temperature (°C); T_a is the air temperature (°C); Q_a is total (solar plus thermal) absorbed radiation (W animal^{-1}); A_e is effective thermal radiation area of the animal (m^2); $\sigma = 5.67 \times 10^{-8}$ (W m^{-2} k^{-4}); ε is thermal emittance ($0 < \varepsilon < 1$); H is convective conductance (W animal^{-1} °C^{-1}); and R, radiation conductance, is $4 A_e \sigma \varepsilon T_a^3$ (W animal^{-1} °C^{-1}). The net sensible heat transfer from an animal can then be expressed as:

$$M - E - C dT_b/dt = K_e(T_b - T_e),$$

where M is metabolic heat production (W animal^{-1}); E is evaporative cooling (W animal^{-1}); C is heat capacitance of the animal (J °C^{-1} animal^{-1}); K_e is overall thermal conductance (W animal^{-1} °C^{-1}); and T_b is body temperature (°C).

T_e can be measured with the aid of a taxidermic mount consisting of a cast in metal of high conductivity, such as copper, covered with the pelt of the animal. The metal chosen for the core should have a heat capacity per unit volume near to that of the animal so that it has similar thermal properties to the animal but is thermally passive. This enables accurate measurements to be made of surface temperature distribution and of an isothermal body core temperature.

From the above equation, $T_b = T_e$ when $(M - E) = 0$ and T_b has reached equilibration so that $dT_b/dt = 0$.

T_{es} provides a thermal index of relative rate of heat exchange by combining wind, sun and air temperatures. It represents a temperature "of a still dark enclosure that would cause an equal rate of heat exchange" (Vispo and Bakken 1993). T_{es}, estimated by correcting the operative temperature (T_e) of an animal, describes the rate of heat flow under field conditions and provides an equivalent air temperature for a live animal under laboratory conditions, taking into account insolation, air and surface temperatures and convective heat exchange (Bakken 1976; Hainsworth 1995). It can be expressed as (Bakken 1992):

$$T_{es} = T_b - (K_e/K_{es})(T_b - T_e),$$

and the net heat flow from an animal with constant T_a can be expressed as:

$$T_{es} = T_b - (M - E)/K_{es},$$

where T_{es} is the operative temperature of the reference environment (°C); T_e is the operative temperature in outdoor environment (°C); K_{es} is overall thermal conductance of the animal in reference to the environment (W animal^{-1} °C^{-1}); and K_e is overall thermal conductance in the reference environment (W animal^{-1} °C^{-1}).

4.3
Body Temperature and Mammals

Mammals, like birds, are able to maintain relatively high body temperatures as a result of their high metabolic rates. Body temperature is dependent upon the metabolic rate of an animal and on its thermal conductance. According to the physical laws of heat exchange (that is Newton's Law of Cooling), body temperature is determined by the rate of heat production (HP; Watts), the temperature difference between the body of the animal (T_b, °C) and the air (T_a, °C) and the overall thermal conductance (C, Watts °C). This relation can be expressed as:

$$HP = C(T_b - T_a).$$

The range of air temperatures at which the metabolic rate of an animal is minimal is considered to be the animal's thermal neutral zone (Fig. 4.5). This zone is bounded by the lower critical temperature (T_{lc}) at its colder end and by the upper critical temperature (T_{uc}) at the hotter end. Animals exposed to air temperatures below the T_{lc} must increase their metabolic rate in order to maintain body temperature and/or allow their body temperature to decrease. The maximum rate of metabolic heat production is called summit metabolism. The air temperature at which an animal becomes hypothermic, leading to death, is described as the lower lethal temperature (T_{ll}). At temperatures above the T_{uc}, animals must increase evaporative water loss and/or allow their body temperature to increase. The quantity of heat removed from an animal when water changes from liquid to gas is the latent heat of vapourization of water. The latent heat of vapourization is dependent upon water temperature and equals 2427 J g^{-1} at 30 °C. Concomitantly with the increase in evaporative water loss, there is often also an increase in the metabolic rate engendered by thermoregulatory processes which require energy for dissipating heat.

Small mammals are generally thermostable, showing little fluctuation in their body temperature. This is not the case with many large mammals, in particular desert species when they are dehydrated. A passive labile body temperature enables an animal to be less dependent upon evaporative cooling. This strategy is employed by several large, desert-adapted mammals. For example, daily fluctuations of body temperatures in excess of 6–8 °C have been reported in the camel (Schmidt-Nielsen 1964) and in the oryx (Taylor 1970). The advantage to a large animal of using this mechanism is easily apparent in the considerable saving of evaporative water needed to maintain thermostability. Although smaller animals cannot afford to use this mechanism so effectively, thermolability has been reported in medium-sized species such as Sinai goats (Finch et al. 1980) and Grant's gazelles (Taylor 1970). All these mammals store heat and allow their body temperature to rise during the hot part of the day, and lose heat to the environment, thereby lowering their body temperature during the night and the cooler part of the day.

Small mammals cannot use thermolability effectively as a mechanism to save body water on account of their large surface to mass ratio and therefore their low thermal inertia. At high air temperatures, small mammals quickly overheat and perish. In general, their rise in body temperature is due to an inability to dissipate heat quickly enough to maintain body temperature.

However, differences between small desert and small non-desert animals in their regulation of body temperatures have been reported. In a study of 10 sub-

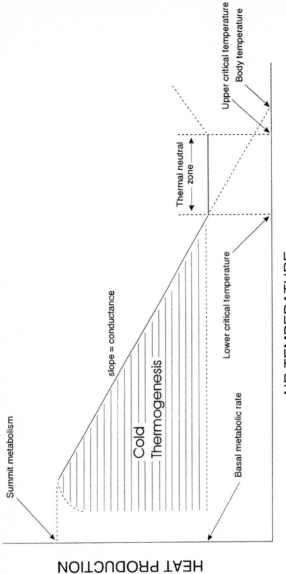

Fig. 4.5. The effect of air temperature on heat production of a homeotherm indicating the thermal neutral zone, upper and lower critical temperatures, and basal and summit metabolic rates. The slope of the line describing the points below the lower critical temperature is a measure of the conductance of an animal, and the line crosses the abscissa at an air temperature near the body temperature of the animal

species of *Peromyscus*, all were able to regulate their body temperatures at moderate air temperatures between 10 and 30 °C. Desert subspecies, however, were less efficient in regulating at low air temperatures between 1 and 5 °C but were better regulators at high air temperatures like 38 °C (McNab and Morrison 1963). Furthermore, small desert mammals usually have a greater ability to tolerate higher body temperatures than do non-desert species.

A rise in the body temperature of small desert mammals often commences while the animal is still within its thermal neutral zone, and in this way some heat is stored. This is more common among the larger of the small diurnal mammals such as *Spermophilus leucurus* and *Citellus leucurus* (Hudson 1962), but has been reported in some nocturnal rodents such as *Acomys cahirinus* (Shkolnik and Borut 1969), *Dipus aegiptus* (Kirmiz 1962), *Notomys alexis* (MacMillen and Lee 1970) and *Rattus villosimus* (Collins and Bradshaw 1973). It has also been described in the smallest African gerbil, *Gerbillus pusillus* (13 g), a rodent that does not maintain precise homeothermy (Buffenstein and Jarvis 1985). The thermal neutral zone for this species lies between air temperatures of 31.4 and 38.0 °C. The body temperature starts rising at 34 °C (Fig. 4.6), and body heat is stored with this increase in body temperature. Heat dissipated by evaporative water loss amounted to 29% of metabolic heat pro-

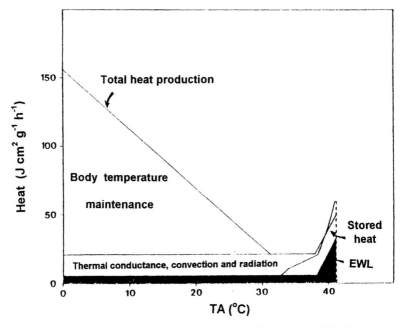

Fig. 4.6. The effect of air temperature (*TA*) on the proportion of heat produced, dissipated and stored in *Gerbillus pusillus*. All values are calculated from oxygen consumption and evaporative heat loss, assuming that 1 ml oxygen used in oxidative metabolism yields 20.46 kJ, the latent heat of vapourization dissipates 2.426 kJ g^{-1}, and the specific heat of mammalian tissue is 3.431 J g^{-1} $°C^{-1}$ (*EWL* evaporative water loss). (Buffenstein and Jarvis 1985)

duction below thermoneutrality. Evaporative water loss increased rapidly above an air temperature of 38.5 and at 41 °C (the lethal air temperature) it dissipated a maximum of 63% of the metabolic heat produced.

Larger body temperature fluctuations have been reported in active ground squirrels shuttling between sun and shade. When active in summer, the desert antelope ground squirrel (*Ammospermophilus leucurus*) shows a body temperature range of 7.8 °C, between a low of 35.8 °C in a cool burrow to a high of 43.6 °C above ground in the sun. These changes are relatively rapid and allow the animal to be active during the summer midday (Chappell and Bartholomew 1981a, b; Hainsworth 1995).

4.3.1 Torpor

Torpor, generally regarded as an extension of sleep (Harris et al. 1984), is a regulated reduction in body temperature as a strategy to reduce energy requirements for thermoregulatory purposes. It has been recorded in a number of eutherian mammals, including the orders Rodentia, Insectivora, Chiroptera, Carnivora and Primata, as well as in many marsupials (Geiser 1994). Among heteromyid rodents, torpor has been found to be well developed in all species of *Microdipodops, Chaetodipus* and *Perognathus* studied thus far (French 1993).

Two types of strategies have been described in the use of torpor (French 1993; Geiser 1994). The first is short-term, shallow daily torpor on a circadian and /or circannual basis when animals forage all year round. Episodes of daily torpor are controlled by the light:dark regime. Animals that exhibit this lower their body temperatures to 10–30 °C (Fig. 4.7) for fluctuations of less than 24 h duration (Fig. 4.8), and metabolic rates fall to about 10–60% of basal metabolic rates (BMR). Such shallow torpor is a response to food restriction and/or low

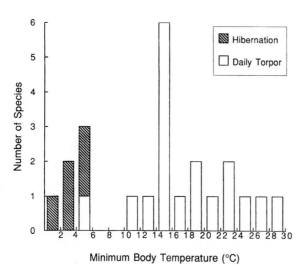

Fig. 4.7. Frequency distribution of the minimum body temperature (T_b) of 23 marsupial species. All but one displaying daily torpor has minimum T_bs higher than 11 °C, whereas all species displaying prolonged torpor (hibernation) had minimum T_bs lower than 6 °C. (Geiser 1994)

air temperatures. It is mainly used by mammals that inhabit regions with a relatively mild climate. Shallow torpor during winter has been well documented and is well developed in many heteromyid rodents such as *Chaetodipus californicus* (Tucker 1965), *C. fallax* (Nishimoto 1980, quoted by French 1993), *C. hispidus* (Wang and Hudson 1970) and *C. baileyi* (Hayden and Lindberg 1970). However, it is not well developed in all heteromyids. It is poorly developed in *Dipodomys*. Shallow torpor has been reported only during starvation in *D. merriami* (Carpenter 1966; Dawson 1955), *D. panamintinus* (Dawson 1955) and *D. deserti* (MacMillen 1983) and has not been reported in the tropical genera, *Heteromys* and *Liomys* (Fleming 1977; MacMillen 1983). Shallow torpor has also been reported in other small mammals such as in the cricetids *Calomys musculinus* (Bozinovic and Rosenmann 1988) and *Saccostomus campestris* (Ellison 1993), in the genus *Peromyscus* (Tucker 1965, 1966), in shrews (McNab 1991) and in small marsupials (Morton and Lee 1978; Geiser 1994).

The second type of torpor is long-term torpor. During this, animals can remain underground for several months. Body temperatures may be reduced to as low as 1 °C (Fig. 4.7; Geiser 1994), and metabolic rates fall to about 2–6% of BMR. Animals that exhibit long-term torpor are true hibernators. Such torpor occurs even when food is available. In the winter it occurs during periods of low air temperature and is known as hibernation and in the summer, at relatively high air temperatures, when it is known as aestivation (Schmidt-Nielsen 1964).

There are variations in these two extremes. For example, heteromyids such as *Microdipodops pallidus* and *Perognathus flavus* undergo torpor of several

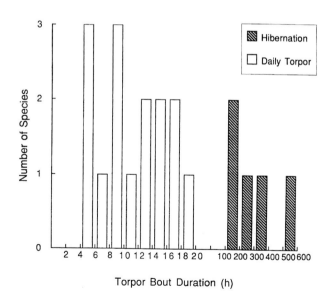

Fig. 4.8. Frequency distribution of the duration of the longest torpor bout of 20 marsupial species. All species displaying daily torpor had bouts shorter than 20 h, wheras all species displaying prolonged torpor (hibernation) had torpor bouts longer than 100 h. (Geiser 1994)

days' duration, but individuals forage above ground throughout the winter (Brown and Bartholomew 1969; Wolff and Bateman 1978; French 1989).

Extensive studies on long-term torpor have been reported in the heteromyid genus *Perognathus*. Photoperiod, air temperature, food availability and their effect on torpor have been examined in a number of studies on *P. longimembris*, a small (8 g) heteromyid inhabiting desert regions. This nocturnal rodent remains dormant underground all winter (about 5 months), but individuals can remain underground for as long as 10 months. Energy requirements during this underground period are provided by stored seeds collected prior to hibernation. This mouse is incapable of storing enough body reserves to last a long time. The mice cease their above-ground activity and go underground once reproduction has been completed and enough seeds accumulated. They do so whether or not food is available on the surface (French 1977a). When reproductive activity is prolonged, dormancy is delayed, and reproductively quiescent mice will forage actively above ground if not permitted to store enough food underground (French 1993). A relatively large store of food, which gradually diminishes, is available at the start of the dormant period. The cue to come above ground is provided by the warming of the top layers of the soil to nearly 30 °C.

The mice do not remain torpid throughout this time. That is, they are not hypothermic continually, but wake periodically and attain normal body temperatures. Bouts of torpor can last less than a day as well as several days. *P. longimembris* maintained at 8 °C showed bouts of torpor lasting 24, 48, 72, and 96 h – that is, discrete bouts of 1 day, 2 days' etc. The duration of torpor was found to be inversely related to air temperature: the maximum duration of torpor at 8 °C was 112 h. This was 2.5 times greater than the duration of 45 h at 18 °C. Interestingly, the periodic rhythmicity persisted even in continuous darkness, indicating its endogenous origin (French 1977b).

When given a choice, *P. longimembris* selects a temperature near to its thermal neutral zone (about 31 °C). It does not enter torpor and has an average metabolic rate of $1.8 \, \text{kJ} \, \text{g}^{-1} \text{day}^{-1}$ (French 1976). Energy used by this heteromyid over a 150-day dormancy period, therefore, amounted to $269.9 \, \text{kJ} \, \text{g}^{-1}$ (64.5 kcal g^{-1}). With restricted feed at air temperatures of 8 and 18 °C, *P. longimembris* consumed similar amounts of energy (256.8 and $261.5 \, \text{kJ} \, \text{g}^{-1}$, respectively). This was accomplished by spending more time in torpor at 8 than at 18 °C. With food available ad libitum at these air temperatures, the time spent in torpor was reduced, and daily energy intakes increased to 468.9 and $377.8 \, \text{kJ} \, \text{g}^{-1}$, respectively. Two of four rodents at 8 °C and two of eight at 18 °C did not enter torpor during a 10-month period (Table 4.1).

P. longimembris always selects the warmest temperatures and minimizes its torpor during winter hibernation. Indeed, in the winter the deeper layers of the soil are warmer than the top layers, and these mice occupy nests as deep as 2 m (Kenagy 1973a). Furthermore, the amount of time *P. longimembris* spends torpid is inversely related to food intake and therefore to daily metabolism (Fig. 4.9). This has also been reported in the heteromyids *Perognathus californicus* (Tucker 1966) and *Microdipodops pallidus* (Brown and Bartholomew

Table 4.1. Torpor and energy utilization during hibernation in *Perognathus longimembris* offered either restricted or ad libitum food and at an air temperature of either 8 or 18 °C. The 5-month period following the first recorded bouts of torpor was divided into 15 ten-day segments. The percentage of time in torpor was calculated for each segment, converted to energy metabolized, and then summed over the hibernation season. The four mice that did not become torpid were monitored for 10-months. (French 1976)

Restricted Food						Ad libitum food					
8 °C			18 °C			8 °C			18 °C		
Animal no.	Time torpid (%)	Energy used (kcal g^{-1})	Animal no.	Time torpid (%)	Energy used (kcal g^{-1})	Animal no.	Time torpid (%)	Energy used (kcal g^{-1})	Animal no.	Time torpid (%)	Energy used (kcal g^{-1})
1	57.8	62.69	5	33.7	72.60	13	32.7	95.73	17	11.2	101.19
2	64.0	54.07	6	32.3	70.97	14	56.5	63.95	18	29.2	76.88
3	55.5	65.73	7	31.0	74.32	15	0	144.30	19	30.4	74.08
4	57.3	62.98	8	49.7	53.02	16	0	144.30	20	26.5	81.40
			9	60.1	46.22				21	39.0	66.58
			10	47.3	59.10				22	22.9	84.38
			11	47.7	61.79				23	0	118.95
			12	41.9	61.89				24	0	118.95
x̄	58.7	61.37	x̄	42.2	62.50	x̄	22.3	112.07	x̄	19.9	90.30
(SE)	(1.9)	(2.53)	(SE)	(3.6)	(3.49)	(SE)	(13.8)	(19.71)	(SE)	(5.1)	(7.17)

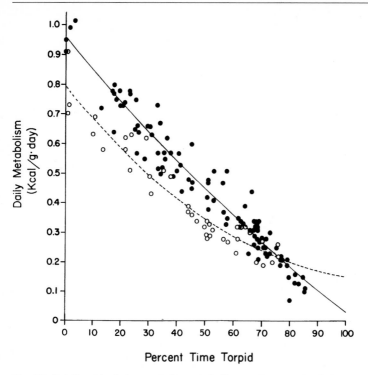

Fig. 4.9. Relationship between daily metabolism and amount of time spent in torpor in *Perognathus longimembris*. Measurements on 12 mice at 8 °C (*filled circles*) and 10 mice at 18 °C (*open circles*) were made for periods of 10–25 days. The best fit regression lines are described by: $MR_8 °C = 0.952 - 0.01(T) + 0.00001 (T^2)$; $MR_{18} °C = 0.794 - 0.0115(T) + 0.00005(T^2)$, where MR is daily mass specific metabolism (in kcal) and T is percent time torpid; 1 kcal = 4.184 kJ. (French 1976)

1969), indicating that these rodents prefer to remain euthermic, food availability permitting. In fact, when excessive food was made available to *M. pallidus,* this kangaroo mouse did not enter torpor even at temperatures as low as 6 °C (Trial 7, Fig. 4.10; French 1989). Thus, although maintaining normal body temperature can be energy costly for a small mammal, it may nevertheless be important since the "maximization of the time of eutheria reduces the chances of freezing during hibernation and enhances the animal's ability to escape from predators" (French 1976).

Energetic cost of torpor = (hours of euthermy) (cost/h) + (hours of torpor) (cost/h) + (number of arousals) (cost/arousal). The saving in energy can be calculated from the difference between (hours of hibernation) (euthermy cost/h) and the energetic cost of hibernation. It can be assessed similarly for the torpor period only.

Oxygen uptake is reduced considerably during torpor when compared with the uptake of euthermic non-torpid animals. For example, in the pouched

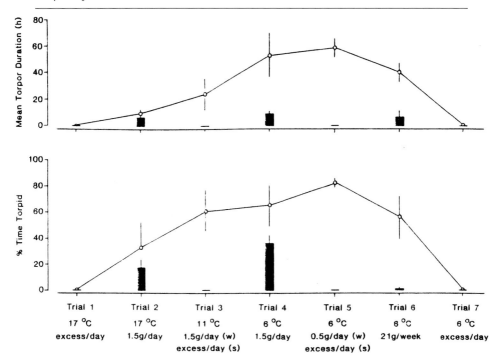

Fig. 4.10. The torpor responses (mean ± SD) of *Microdipodops pallidus* to various combinations of air temperatures and seed quantities. Data from animals caught in winter were obtained from Brown and Bartholomew (1969) and serve as a comparison for mice captured in spring and summer (*histogram*). Both the mean duration torpor and the percentage of time torpid were significantly different ($P < 0.05$) between mice captured in the winter and summer (trials 4 and 6). (French 1989)

mouse *(Saccostomus campestris)*, the reduction is greater than 50%. However, during arousal to eutheria, there is an abrupt increase in the rate of O_2 uptake which overshoots non-torpid values (Fig. 4.11; Ellison 1993). The energy saving per bout of torpor, during the duration of the torpor and its arousal, averages 25% with a range of 15–45%. The minimum rate of O_2 is not correlated with energy saving during torpor, however. The overshoot cost is positively correlated with the duration of the overshoot, and both the energy costs and duration of the overshoot are negatively correlated with energy saving. Energy saving in these cricetids is therefore dependent upon the number and duration of arousal bouts, "which suggests that the cost of regaining homeothermy can limit the energetic efficiency of short bouts of torpor" (Ellison 1993).

The Djungarian hamster *(Phodopus sungorus)* lowers its body temperature between 14 and 31.3 °C for 0.3–9.2 h daily when maintained at an air temperature of 5 °C (Ruf and Heldmaier 1992). The duration of episodes of torpor is inversely related to the minimum body temperature. With daily torpor in excess of 4 h, a 20.2% saving in energy is achieved. Energy saving is directly re-

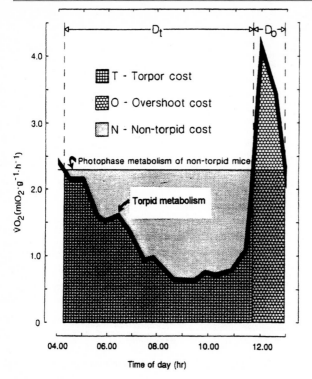

Fig. 4.11. Calculating the duration of torpor (D_t), the duration of overshoot (D_o), the cost of overshoot (O) and the equivalent cost of non-torpid pouched mice during the same period (N) by comparing the oxygen consumption of a torpid individual (Gam 1503) to the mean O_2 consumption of non-torpid animals from the same locality recorded during photophase. (Ellison 1993)

lated to torpor duration and inversely related to minimal body temperature (Fig. 4.12). Unlike other mammals in torpor (for example, pouched mice; Ellison 1993), there appears to be no overshoot in energy expenditure costs during arousal (Fig. 4.13).

Torpor on a daily basis is widespread among crocidurines, but among soricine shrews, it has been reported only in the North American desert shrew *Notiosorex crawfordi*. McNab (1983) suggested that torpor is related to BMR among small endotherms. Small endotherms with a relatively low BMR have a greater tendency to enter torpor than do species with a higher BMR. This point has been illustrated with examples of BMR and torpor in shrews and in small dasyurid marsupials (Fig. 4.14; McNab 1991). Dasyuridae, the marsupial equivalents of shrews, have the lowest BMR and enter torpor. Among the shrews, the Crocidurinae have a relatively low BMR and enter torpor, whereas the Soricidae have a relatively high BMR and do not. Crocidurines are mainly found in Africa and Southeast Asia, inhabiting either the tropics or deserts. The soricines which live in areas with the coldest winters of all the shrew habitats generally do not enter torpor.

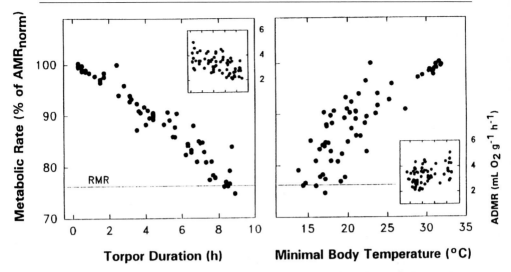

Fig. 4.12. Percentage energy expenditure (ADMR/ADMR$_{norm}$ × 100) as a function of torpor duration (*left*) and minimal T$_b$ during hypothermia (*right*). Extended torpor episodes lower energy requirements to the level of resting metabolic rate (RMR) during normothermia (*dotted lines*). *Inset graphs* show the same relations for original ADMRs measured at T$_a$s between 0 and 20 °C. (Ruf and Heldmaier 1992)

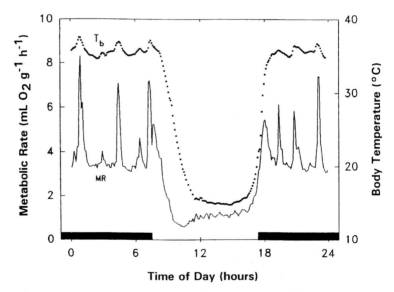

Fig. 4.13. Rate of oxygen consumption (*solid line*) and body temperature (*dots*) in a Djungarian hamster exposed to an air temperature of 5 °C. Body temperature and metabolic rate were recorded at 6-min intervals. *Black bars on the abscissa* indicate hours of darkness. (Ruf and Heldmaier 1992)

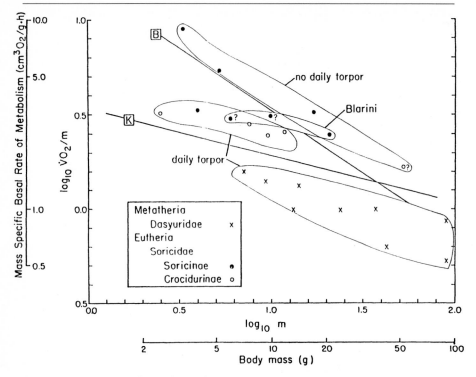

Fig. 4.14. Log mass-specific basal rate of oxygen consumption as a function of log body mass in shrews and in small dasyurid marsupials (data taken from MacMillen and Nelson 1969; Dawson and Hulbert 1970). *Question marks* indicate uncertainty as to whether a particular species enters torpor at all, or on a regular basis. (McNab 1991)

Differences between these two groups were well illustrated in a study comparing co-existing *Sorex coronatus* (Soricinae) and *Crocidura russala* (Crocidurinae) during winter (Genoud 1985). *S. coronatus* reduced its energy requirements by use of the Dehnel phenomenon, that is, a decrease in body mass, whereas *C. russala* reduced its energy requirements through torpor, nest huddling and selection of warm resting sites. *Notiosorex crawfordii* (4 g), a soricine occurring in three hot North American deserts, appears to be an exception. McNab (1991) suggests that in that desert soricine, torpor may be due to an "important ecological component". He questions whether *S. merriami* and *S. preblei,* two small soricine shrews that inhabit arid areas in North America, and the soricine shrews of the Eurasian deserts also enter daily torpor. When energy supplies are low, *N. crawfordii* drops its body temperature to 28 °C. This reduces its metabolic rate during the time of torpor by 50% at an air temperature of 20 °C and 20% at an air temperature of 25 °C. However, the coefficient of heat transfer (thermal conductance) remains the same whether the shrews maintain a body temperature of 38 or 28 °C, as the slopes of the lines describing specific metabolic rates at air temperatures below the lower critical

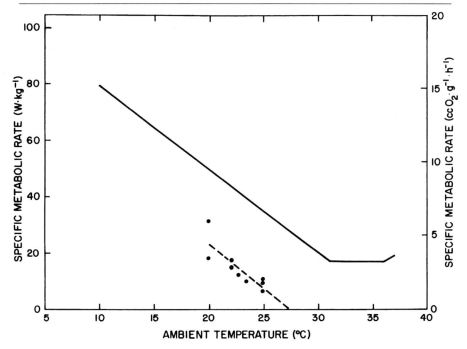

Fig. 4.15. Relation between specific metabolic rate and air temperature for desert shrews. *Solid line* (euthermia) was compiled from 79 individual experiments on 11 animals (from Lindstedt 1980a). *Dashed line* (hypothermia) is the result of 9 experiments on 3 shrews. The slopes of the lines are nearly identical, and their intercepts vary by some 96 mW. (Lindstedt 1980b)

temperature are nearly identical (Fig. 4.15; Lindstedt 1980b). Shallow torpor at a body temperature of 28 °C allowed quick entrance and exit from hyphermia and was described as a compromise between saving body energy and at the same time maintaining a fair degree of alertness (Lindstedt 1989b).

The argument for torpor being related to small body size, desert life and relatively low BMR is well supported by the arvicolid and heteromyid rodents. None of the arvicolids studied thus far enters torpor; in fact, all maintain rigid endothermy. No adult arvicolids are small (the smallest, *Clethrionomys californicus,* being over 18 g), they have relatively high metabolic rates, and they generally inhabit cool to cold environments (McNab 1992a). In contrast, heteromyid adults weigh less than 10 g and fulfill the other criteria related to torpor.

Gerbillinae also fit the criteria related to torpor and include small species. However, except for an account for torpor in *Gerbillus gerbillus* (Petter 1952, quoted by Schmidt-Nielsen 1964), in which the rectal temperature dropped from the normal 36 to 21.5–22 °C, there appears to be no other report available on torpor in these granivores. In fact, torpor was not noted in either *G. henleyi* or *Meriones crassus* when food intake was restricted (unpubl. data). *G. henleyi* is the smallest of the gerbils (8–12 g), inhabits some of the most extreme deserts (Shenbrot et al. 1994a), and should therefore be a prime candidate for entering torpor. *M. crassus* is much larger (50–110 g) and co-exists with *G. henleyi* over much of their ranges. Both species lost body mass on restricted diets, but neither entered torpor. The body temperature dropped to about 25 °C in *G. henleyi*, and without help, this gerbil was unable to recover euthermy. This hypothermic state was taken to be pathological and not a strategy to conserve energy. The body temperature of *M. crassus* fell to about 36 °C, and torpor was not evident.

4.4
Overall Thermal Conductance

Overall thermal conductance of an animal is a measure of the rate of heat loss per unit of thermal gradient and is presented in the units of watts per °C. It is usually estimated from the slope of the regression equation relating metabolic rate to air temperature between lower critical and lethal temperatures, if the curve extrapolates to the body temperature of the animal when the rate of metabolism equals zero (McNab 1980b). Minimal thermal conductance (C; °C^{-1}) can be estimated from the metabolic rate (BMR; W) at or below the T_{lc} and body (T_b; °C) and air (T_a; °C) temperatures using the equation (Bradley and Deavers 1980):

$$C = BMR/(T_b - T_a).$$

Bradley and Deavers (1980) found that for mammals ranging in mass from 10 g to over 100 kg, $C = 0.22\ m_b^{0.57}$. Furthermore, allometric equations relating overall thermal conductance to body mass have slopes which lie between 0.45 and 0.65, indicating that the rate of heat loss increases more slowly with body size than the metabolic rate does (Peters 1983). Therefore, the increase in metabolic rate at low air temperatures is proportionately greater for small animals than for large animals. In consequence, smaller animals are less tolerant of low air temperatures than larger animals. In addition, the body temperature is independent of body size, and the T_{lc} is therefore lower in larger animals than in smaller animals.

Low thermal conductance values reflect good insulation and an efficient use of physical thermoregulation (Aschoff 1981). This limits the rate of heat loss at low air temperatures, as well as the heat gain from the environment in the thermal neutral zone. Lower than predicted thermal conductance would consequently be advantageous for small, arid, nocturnal mammals exposed to low night air temperatures. Hinds and MacMillen (1985) found that the thermal conductance of nine arid-adapted heteromyid species was 26% lower than that predicted for mammals of their body masses. Similarly, lower than expected thermal conductance was found in several desert species of *Gerbillurus* (Buffenstein and Jarvis 1985; Downs and Perrin 1990) as well as in the big-eared desert mouse (*Malacothrix typica*; Knight and Skinner 1981). In contrast to results reported from desert rodents, tropical and subtropical heteromyid species which are rarely exposed to cold have thermal conductances that are close to or above that predicted for mammals of their size (Hinds and MacMillen 1985; French 1993).

High thermal conductances can also be useful in the dissipation of heat to small mammals exposed to high air temperatures. Diurnal desert species should theoretically fall into this category, and indeed, that is often the case. The spiny mice *Acomys russatus* and *A. cahirinus* are sympatric over much of their desert ranges, but *A. russatus* is diurnal, whereas *A. cahirinus* is nocturnal, and *A. russatus* inhabits more arid areas than *A. cahirinus* does. The thermal conductance of the diurnal species is higher than its co-generic nocturnal species (Haim and Izhaki 1993). Thermal conductance of the striped mouse (*Rhabdomys pumilio*), a diurnal murid, is also higher than that predicted for a mammal of the same body mass (Haim and Fairall 1986). Furthermore, the increased conductance in this species is more pronounced in a population inhabiting an arid area than in a population inhabiting a mesic area. Thermal conductance higher than that predicted has also been reported for two diurnal squirrels, *Xerus princeps* and *X. inauris*, species well adapted to a hot and arid environment (Haim et al. 1987).

High thermal conductance can also be useful to subterranean fossorial animals because their burrows are nearly saturated with water vapour, which prevents the use of evaporative cooling. Fossorial animals differ from other animals that inhabit burrows in that they usually plug the opening of the burrow and do their foraging underground, whereas other burrowing animals do not usually seal their burrows and forage above ground. The desert golden mole (*Eremitalpa granti namibensis*) has a thermal conductance which is 158% of that predicted for a mammal of its body mass (Fielden et al. 1990a). The round-eared elephant shrew (*Macroscelides proboscideus*) inhabits burrows characterized by high relative humidity in semi-arid to arid areas of south western Africa. It has a thermal conductance which is 365% of that predicted for a mammal of its mass.

It has been further suggested that it would be advantageous for small mammals inhabiting rocky crevices to have a lower thermal conductance than small mammals inhabiting burrows. The microenvironments of animals in rocky areas are less stable than those in burrows. These animals are therefore exposed to more extreme temperatures (Buffenstein 1984). In a comparsion of the thermal conductance of rodents differing in habitat, the rock rat (*Aethomys namaquensis*) which inhabits rocky outcrops in the desert has a thermal conductance 26% below that predicted for a mammal of its body mass, whereas the burrow-dwelling pigmy gerbil (*Gerbillurus paeba*) has a thermal conductance 7.2% above that predicted (Buffenstein 1984).

However, relationships between habitat, heat production and conductance have not been unequivocal. In general, desert rodents have metabolic rates and conductances lower than those expected. As a consequence, a significant positive correlation has been found between resting metabolic rate and thermal conductance (Haim and Izhaki 1993). Furthermore, in a comparison of thermal conductance between desert and non-desert rodents (Goyal and Ghosh 1983), it was concluded that the thermal conductance of desert rodents was lower than that of non-desert rodents. Nevertheless, some of the highest values were reported among the desert rodents (Fig. 4.16).

These inconsistencies could be the result of opposing responses within the same animal. For example, a desert, nocturnal, burrowing rodent should have a low conductance to conserve heat during the cool evenings. At the same time, because of the high relative humidity within its burrow, it may have no other way to dissipate heat than via high thermal conductance. Seasonal changes within species could account for some of the differences reported. Thermal conductance in the pouched mouse is 12% above the value predicted for an eutherian mammal in summer-adapted rodents. These rodents have significantly greater conductance values than do winter-adapted rodents. Furthermore, rock-dwelling rodents are more exposed to the environment than the burrow-dwelling rodents and should therefore have a lower conductance. However, the bushy-tailed jird (*Sekeetamys calurus*), a gerbillid inhabiting rocky areas, has a higher thermal conductance than most psammophile gerbils (Haim and Izhaki 1993).

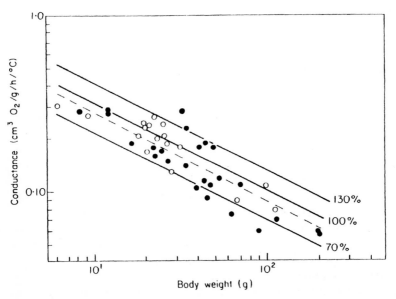

Fig. 4.16. Relationship between minimal thermal conductance and body mass in burrowing desert and non-desert rodent species. *Solid curves* labelled 100, 130 and 70% represent Herreid and Kessel's (1967) standard curve and 130 and 70% of that value, respectively. *Broken curve* is from Goyal and Ghosh (1983). ○ Non-desert species; ● desert species. For the solid curve of 100%, $C = 1.02 \, W^{-0.505}$, and for the broken curve, $C = 0.89 \, W^{-0.499}$. (Goyal and Ghosh 1983)

Overall conductance has also been calculated at air temperatures above the T_{uc}. At these temperatures, conductance increases rapidly as a function of air temperature. Increased conductance is facilitated by increased blood flow to the periphery. Vasodilation enhances the transport of body heat to the surface, where it is dissipated to the environment. The slope of the regression line relating heat production to air temperature above the animal's T_{uc} has been termed the coefficient of heat stress (Weathers 1981). The units of this term are also in W $°C^{-1}$, but it is not a measure of heat flux per unit of temperature gradient and, consequently, is not a measure of conductance. The increase in metabolic rate at high ambient temperatures is associated with the added cost of dissipating the heat load.

Dry Thermal Overall Conductance. The dry thermal conductance (C_d; W $°C^{-1}$) excludes heat loss through evaporative water loss (EWL; W) and is the rate of dry heat transfer per unit surface area (cm^2) to and from an animal per $°C$ ($T_b - T_a$; $°C$). It can be expressed as :

$$C_d = (MR - EWL)/A(T_b - T_a).$$

The dry thermal conductance is dependent upon insulation by hair and subcutaneous fat and by vasomotor changes. A decrease in metabolic rate, as is common in desert animals compared with their non-desert counterparts, results in a higher T_{lc}, while an increase in metabolic rate, as occurs during exercise and food digestion, results in a lower T_{lc}.

4.4.1
Pelage and Seasonal Changes in Thermal Conductance

Seasonal changes in the length and density of the hair fibres of homeotherms can be important adaptations in regulating thermal exchange under different climatic conditions. Both conductive and convective heat loss can be increased with a sparser and shorter pelage under hot conditions and can be reduced with a denser and longer pelage under cold conditions. The fur of small animals differs substantially from that of large animals in that it is much denser, shorter and shallower (Steudel et al. 1994). This is the case for small animals even in the summer. Therefore, they have little latitude in changing fur depth and hair density in the winter and are also limited by biomechanical constraints. In contrast, summer-adapted large mammals have very low fur densities and much latitude for increase in the winter. A comparsion of the pelage of eutherian mammals with different body masses bore this out. Seasonal pelage characteristics of deer mice (16.3–18.9 g) showed virtually no difference between summer and winter; lemmings (54.4–75.8 g) approximately doubled both their fur depth and density in the winter; and red deer (63–124 kg) showed a sixfold increase in fur depth and a more than fivefold increase in fur density in the winter. The winter fur density of the red deer is much lower than even the summer densities of the smaller species, but the hair length and fur depth of the winter pelage of the red deer are much greater. Based on these differences, which are related to body size, Steudal et al. (1994) concluded that seasonal changes in the pelage of small mammals can have only a modest effect on mass specific metabolic rate, whereas it can have a substantial effect on large animals. They further concluded that "Bergmann's rule seems more relevant to small animals than to large ones from a heat transfer perspective".

Nontheless, seasonal changes in the pelage of small mammals that can affect thermal heat exchange have been reported. The Djungarian hamster changes the colour of its fur from dark grey in summer to white in the winter (Fig. 3.4). It also increases the fur depth by 27%, from 8.05 to 10.20 mm, when living outdoors under conditions of natural photoperiod and by 22%, from 7.84 to 9.58 mm, while living indoors under conditions of natural photoperiod (Heldmaier and Steinlechner 1981).

The effect of pelage on solar heat gain was examined in the rock squirrel (*Spermophilus variegatus*) which inhabits arid, rocky canyons and slopes (Walsberg 1988; Walsberg and Schmidt 1989). Solar heat gain is a function of environmental properties such as wind speed and solar radiation intensity, as well as an animal's coat properties, including structure, insulation, short-wave reflectivity, hair optics and skin colouring. The coat of the rock squirrel consists of two layers: a dense inner layer of predominantly fine, dark hairs and a sparse, outer layer of predominantly coarse, lighter hairs. The inner layer comprises approximately 63% of the total coat depth and 76% of the total coat insulation, and its thermal resistance is 80% greater than that of the outer layer (Walsberg 1988). The ratio of these two layers minimizes solar heat gain. Furthermore, solar heat gain at low wind speeds is about 20% higher in the winter than in the summer. "This change is in an apparently adaptive direction and is predicted to have a major effect on the animal's heat balance in the nature" (Walsberg and Schmidt 1989). Both the minimization of solar heat gain as well as seasonal differences indicate that fur structure "may represent an effective means of adjusting solar heat gain independent of coat insulation and surface coloration" (Walsberg 1988).

Overall heat transfer (thermal conductance) was measured in black-tailed jack-rabbits (*Lepus californicus*) in both winter and summer at different wind velocities (Harris et al. 1985). These jack-rabbits are well adapted to the arid regions of southwest USA. At baseline (wind speed 0.0 m s^{-1}) the thermal conductance of the summer pelage was 48.9% higher than that of the winter pelage. At wind velocities of up to 2.8 m s^{-1}, there was only a slight increase in the thermal conductance of the pelage. Above a wind velocity of 2.8 m s^{-1}, thermal conductance increased in both seasons, but at a faster rate in summer so that with a wind velocity of 11 m s^{-1}, the summer pelage had a thermal conductance which was 113.7% higher than that of the winter pelage (Fig. 4.17).

Most of the difference in heat transfer between seasons was a consequence of convection. The boundary layer of air trapped in the fur of the jack-rabbit provides a source of insulation. Wind speeds in excess of 2.8 m s^{-1} disrupt the fur physically, reducing the boundary layer and increasing convective heat loss. Convective heat transfer has a greater effect on the overall heat transfer on the summer pelage than on the winter pelage.

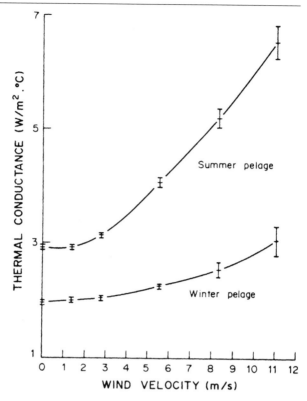

Fig. 4.17. Relationship between the thermal conductance of winter and summer pelage of *Lepus californicus* and ambient wind velocity. The mean and 95% confidence intervals are shown for each wind velocity. (Harris et al. 1985)

4.4.2
Thermal Exchange and Effective Surface Area

Thermal exchange is dependent upon the effective surface area (ESA) for heat exchange and the surface temperature of the animal. In order for heat exchange to take place, a thermal gradient must exist between the surface and the air. ESA, therefore, reflects surfaces whose temperatures deviate from the air temperature over at least part of the air temperature range. Maximum ESA (ESA_{max}) is assumed to occur during the summer and the minimum ESA (ESA_{min}), during the winter. ESA_{max} and ESA_{min} were determined for 29 mammal species ranging in body mass from the 0.02 kg white-footed mouse (*Peromyscus leucopus*) to the 4000 kg African elephant (*Loxodonta africana*) using infrared scans (Phillips and Heath 1995). Measurements of ESA, metabolic heat production and thermal limits of the animals were then used to calculate the index of vasomotion (VMI) as:

$$VMI = (SMR)/[(SA \times ESA_{max})/(T_b - T_{lc})](\Delta\ ESA),$$

where SMR is standard metabolic rate in W; SA is surface area in m^2; T_b and T_{lc} are body and lower critical temperatures, respectively.

The VMI indicates the extent to which an animal controls its surface temperature as a consequence of changes in the exposed surface area, fur or fat, and/or vasomotion, and its consequent dependence upon that ability (Phillips and Health 1995). Higher VMI values indicate a greater degree of control and a higher level of dependency to regulate surface temperature. VMI scales positively with an animal's body mass (Fig. 4.18), which suggests that the ability

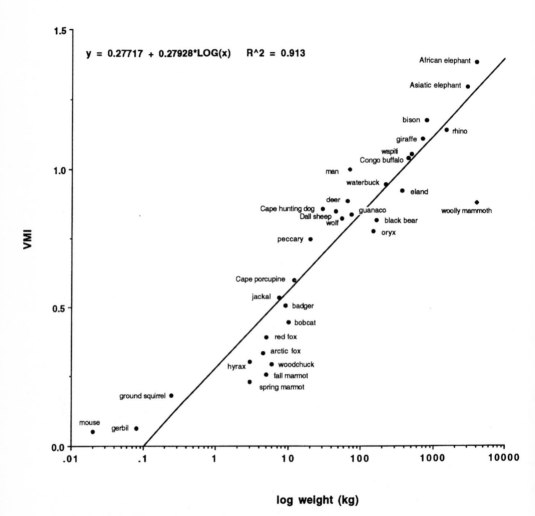

Fig. 4.18. The relationship between vasomotor index (VMI) and body mass in eutherian mammal species. The ability to control surface temperature, as described by VMI, scales positively with body mass. Large animals exert more control over their surface temperature than do small animals. VMI values for man and the woolly mammoth are included for references and are not used in the calculation of the regression line. (Phillips and Heath 1995)

to control the surface temperature becomes increasingly more important with an increase in the ratio of surface area to volume.

This indicates that smaller animals must use different strategies for the control of heat loss than larger animals. Phillips and Heath (1995) summed up these differences as "small animals can be described as metabolic specialists, relying on changes in metabolic rate to maintain body temperature. Large animals do not change metabolic rate. Rather, they are able to regulate heat exchange as the ambient conditions warrant by controlling surface temperature." This is achieved mainly by seasonal changes in fur cover and internal fat. Phillips and Heath (1995) also calculated the functional conductance for the 29 mammalian species, that is, thermal conductance per unit of ESA. Mean functional conductance was $3.20\ W\ m^{-2}\ {}^{\circ}C^{-1}$ with a range of 1.45 to $6.15\ W\ m^{-2}\ {}^{\circ}C^{-1}$. No correlation was found between functional conductance and body mass (Fig. 4.19) nor with habitat, fur depth or taxonomy. This indicates that the functional conductance rate from bare skin is about the same in all mammals.

4.5
Shivering and Non-Shivering Thermogenesis

Heat production for thermoregulatory purposes is derived from shivering and non-shivering thermogenesis (NST). As the temperature declines below the thermoneutral zone, shivering due to muscular contractions produces metabolic heat. This is the primary source of heat generation in larger mammals (Hemingway 1963, Bell et al. 1974). In addition, metabolic heat is produced by freeing chemical energy without involving muscular activity by NST.

NST is of particular importance in newborn animals, where it can increase the rate of heat production by two to three times the BMR. It declines in importance as the animal grows older (Mount 1979). This method of heat production is important in that it can be activated in a very short space of time (Wunder 1985) and is considered to be the major thermoregulatory heat production pathway of small mammals (Heldmaier et al. 1986). NST is also the major source of heat production for arousal from hibernation (Smith and Horwitz 1969). Fat is the main fuel for NST. The principal site of NST production is brown adipose tissue (BAT), and brown fat cells contain many mitochondria, indicating a high level of metabolic activity. Apparently, it also occurs in other tissues such as skeletal muscle (Mount 1979), but this has not been reported in some small mammals (Wickler 1981).

The process of NST is controlled by the sympathetic nervous system. It can therefore be activated by noradrenaline injections. Thermoregulatory NST is usually measured as the maximal oxygen consumption (V_{O_2na}) and body temperature (T_{bna}) in response to noradrenaline injection (NST_{NA}). NST capacity is calculated from the ratio between V_{O_2na} and V_{O_2min}, the latter measured at or close to the T_{lc} prior to noradrenaline administration (Jansky 1973).

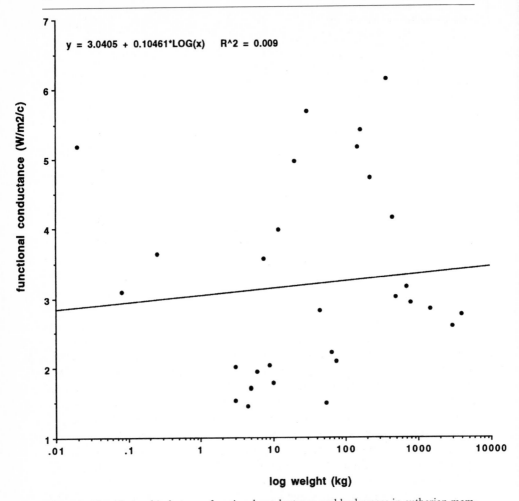

Fig. 4.19. The relationship between functional conductance and body mass in eutherian mammal species. Unlike conductance measured in other ways, their is no apparent correlation between functional conductance and the size of an animal. Functional conductance describes the heat flux across only those surface areas examined for heat exchange. Since this eliminates heavily furred areas, it is essentially a measure of heat flux across bare skin, which should be the same for all animals. (Phillips and Heath 1995)

NST is generally associated with small mammals acclimated to cold and long scotophase (Jansky 1973; Heldmaier et al. 1989). Winter-acclimatized rodents have increased BAT mass and metabolism and increased oxidative enzymatic activity compared with summer-acclimatized rodents (Rabi and Cassuto 1976; Wickler 1981). A combination of cold acclimation and long scotophase further increases NST_{NA} over that in some just cold-acclimated rodents such as the white-footed mouse (Lynch et al. 1978), but not in others such as the Alaskan red-backed vole (*Clethrionomys rutilus*; Feist and Feist 1986).

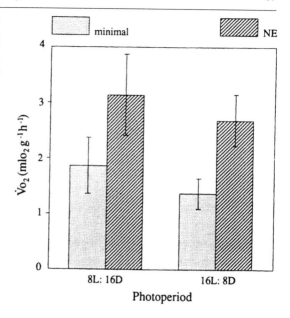

Fig. 4.20. Minimal oxygen consumption measured below the lower critical temperature, and maximum oxygen consumption measured as the maximal response to norepinephrine injection (NE) in long scotophase-acclimated voles (8L:16D) and in long photophase-acclimated Levant voles (16L:8D). Values are means ± SD, n = 6. (Banin et al. 1994)

An increase in NST in response to a short daily photoperiod could be important in preparing small rodents for winter conditions, but research results have been equivocal. Golden (*Mesocricetus auratus*) and Djungarian hamsters (Heldmaier and Hoffmann 1974), white-footed mice (Lynch et al. 1978) and fat jirds (*Meriones crassus*; Haim and Levi 1990) increased BAT and NST_{NA}. In constrast, no such NST_{NA} response to a long scotophase was noted in white rats (Hagelstein and Folk 1978), Alaskan red-backed voles (Feist and Morrison 1981) and Levant voles (*Microtus guentheri*; Banin et al. 1994). In the last, the NST responses of voles acclimated to short and long photoperiods did not differ from each other (Fig. 4.20), and the authors concluded "that changes in photoperiod are not a cue for seasonal acclimatization of NST", and that "the relatively high values (of NST) recorded for LP-voles (long photophase) might be explained as a compensation for their reduced RMR in a thermally fluctuating habitat".

Heat-acclimated golden hamsters showed a decline in BAT mass and metabolism, and increased oxidative potential (Cassuto 1968). This would appear to indicate that desert rodents have a relatively low NST and NST capacity, but apparently the opposite is true (Haim and Izhaki 1993). Regression analysis of the relation between NST and body mass in rodent species showed that habitat and temporal activity explained 66% of the variation found in NST. In these, the resting metabolic rate was lower in arid than in mesic species, as generally found in comparisons, but NST and NST capacity did not follow this trend. NST capacity was higher in arid than in mesic species (3.7, n = 10 vs 2.5, n = 13) and in nocturnal than in diurnal species (4.2, n = 7 vs 2.5, n = 13). No difference in NST capacity was found between subterranean and non-subterranean mesic species (Table 4.2). Haim and Izhaki (1993) concluded that arid,

Table 4.2. Various metabolic and thermoregulatory variables measured for different rodent

Species	Habitat[a]	Habit[b]	Body mass (g)	RMR (ml O_2 $g^{-1}h^{-1}$)	Conductance[d]	T_{lc} (°C)[e]
Sciuromorpha Sciuridae						
Xeris inauris	A	DE	542	0.602	0.071	29
Xerus princeps	A	DE	602	0.565	0.048	22
Hystricomorpha						
Bathyergidae						
Cryptomys hottentotus	M	– S	102	0.681	0.144	31
Myomorpha						
Cricetidae						
Cricetulus migratorius	M	NE	30.7	1.43	0.255	33
Microtidae						
Microtus guentheri	M	NE	43.8	1.83	0.26	30
Gerbillidae						
Gerbillus gerbillus	A	NE	29.7	1.43	0.233	32
Gerbillus pyramidum	A	NE	53	1.0	0.166	32
Gerbillus allenbyi	M	NE	35.3	1.1	0.126	28
Gerbillus nanus	A	NE	28.4	0.78	0.146	33
Gerbillus dasyurus	A	NE	27.6	1.06	0.176	32
Sekeetamys calurus	A	NE	56.9	0.78	0.18	34
Meriones tristrami	M	NE	112	0.88	0.120	30
Muridae						
Otomys irroratus	M	NE	102	0.832	0.08	24
Lemniscomys griselda	M	DE	47.5	1.213	0.157	30
Rhabdomys pumilio	A	DE	39.6	0.810	0.176	32
Praomys natalensis	M	NE	41.5	0.79	0.121	32
Acomys russatus (S)	A	NE	60	0.75	0.18	33
Acomys russatus (EG)	A	NE	60	0.75	0.18	33
Acomys cahirinus	A	NE	48	1.13	0.12	27
Apodemus sylvaticus	M	NE	23.9	1.81	0.187	29
Apodemus mystacinus	M	NE	40.4	1.387	0.134	25
Spalacidae						
Spalax ehrenbergi (2n = 52)	M	– S	138	0.86	0.124	28
Spalax ehrenbergi (2n = 54)	M	– S	134	0.76	0.136	30
Spalax ehrenbergi (2n = 58)	M	– S	135	0.85	0.142	30
Spalax ehrenbergi (2n = 60)	M	– S	133.6	0.62	0.14	30

[a]Habitat: A, semi-arid-arid; M, mesic.
[b]Habit: D, diurnal; N, nocturnal; (–), unknown; S, subterranean; E, non-subterranean.
[c]RMR: resting metabolic rate, oxygen consumption at the thermoneutral zone.
[d]overal minimal thermal conductance measured by the formula $C = (M(\text{metabolism}))/(T_b - T_a)$, M measured as RMR ($V_{02}$) below T_{lc}.
[e]Lower critical point.

species. (Haim and Izhaki 1993)

T_b (°C)[f]	T_{bmin} (°C)[g]	T_{bNA} (°C)[h]	NST capacity[i]	VO_2min (ml O_2 g^{-1}h^{-1})	V_{O_2NA} (ml O_2 g^{-1}h^{-1})	Reference
36.8	32.8	39.3	4.8	0.43	1.99	Haim et al. (1986)
37.6	33.2	37.3	4.6	0.37	1.66	Haim et al. (1986)
35.8	32.9	38.3	4.5	0.77	3.46	Haim and Fairall (1986)
38.1	38.1	38.6	3.3	1.56	4.16	Haim and Martinez (unpub.)
38.3	37.2	37.4	1.9	1.64	3.06	Banin (1990)
37.2	37.8	39.3	2.9	1.42	4.15	Haim and Harari (1992)
37.6	36.2	38.3	2.6	1.28	3.26	Haim and Harari (1992)
36.3	33.6	36.3	2.6	0.16	1.51	Haim (1984)
38.8	34.2	36.7	2.6	0.70	1.74	Haim (1984)
38.6	37.7	39.9	3.1	0.82	2.50	Haim (1987)
37.5	35.7	38.7	2.9	0.78	2.26	Haim and Borut (1986) Haim (unpub.)
36.5	34.6	37.0	2.6	0.70	1.82	Haim (unpub.)
37.6	36.3	35.8	3.1	0.77	2.40	Haim and Fairall (1987)
36.9	34.9	39.2	3.3	0.85	2.81	Haim (1981)
37.0	34.6	38.8	5.2	0.72	4.00	Haim and Le Fourie (1980a)
38	35.1	38.3	2.7	0.79	2.13	Haim and Le Fourie(1980)
37.5	35.0	39.4	5.8	0.70	4.10	Borut et al. (1978) Haim and Boret(1981)
37.5	35.2	35.8	2.7	0.74	2.00	Borut et al. (1978) Haim and Borut(1981)
37.5	36.3	37.1	2.4	1.03	2.49	Weissenberg (1977) Haim (unpub.)
36.7	35.7	38.3	2.1	1.62	3.40	Haim et al.(1986)
35.5	32.2	31.2	1.4	0.69	0.95	Haim et al. (1986) Yahav et al.(1982)
34.9	35.4	35.2	1.8	0.52	0.84	Nevo and Shkolnik (1974) Haim et al. (1984)
35.8	35.0	35.8	2.1	0.53	1.08	Nevo and Shkolnik (1974) Haim et al. (1984)
36	33.8	32.7	1.0	0.39	0.40	Nevo and Shkolnik (1974) Haim et al. (1984)
35.5	34.1	36.2	3.3	0.28	0.92	Nevo and Shkolnik (1974) Haim et al. (1974)

[f]Body rectal temperature at T_{lc}.

[g]V_{O_2min}, T_{bmin}: minimal oxygen consumption and body (rectal) temperature.

[h]V_{O_2NA}, T_{bNA}: maximal oxygen consumption and body (rectal) temperature as a response to noradrenaline injection.

[i]The ratio between V_{O_2NA}/V_{O_2min} = NST capacity.

diurnal species are characterized by high NST capacity in comparison with mesic, nocturnal species, while maintaining a low resting metabolic rate (RMR). The reduced RMR and high NST capacity allows low metabolic heat production for desert rodents during normal activity but provides the opportunity to increase heat production and body temperature quickly. This thermogenic mechanism could be important for a small mammal that may be exposed to large nychthemeral and seasonal changes in air temperature and in which NST capacity may differ significantly from season to season.

This latter point was well illustrated by the kangaroo rat, *Dipodomys ordii* (Gettinger et al. 1986). The NST capacity for animals captured in winter was twice that of animals captured in summer. Furthermore, the NST capacity could be enhanced by manipulation of the photoperiod and air temperature. This would suggest that the proportion of brown fat may vary widely between seasons, which may explain the seemingly contradictory results reported by Cassuto (1968) and Haim and Izhaki (1993).

CHAPTER 5

Behavioural Adaptations

"Most terrestrial vertebrates weigh less than 100 g, and few arthropods weigh more than a gram. For these creatures the physical environment consists of cracks and crevices, sheltered nooks and tunnels. For them, distances are measured in meters, not kilometers, and the difference between sun and shade or surface and burrow can mean the difference between life and death. A terrestrial animal can pick and choose among the different environmental conditions that occur from place to place and time to time in its habitat. By its behaviour it can assemble its own environment. This behaviorally generated physical microenvironment is the one with which the organism's physiological capacities can cope."

Bartholomew (1987)

Most studies on small desert mammals have been carried out in North America, especially on the kangaroo rat *Dipodomys* and kangaroo mouse *Microdipodops* (Heteromyidae) of the Sonoran Desert (Genoways and Brown 1993). In general, these rodents are nocturnal, burrow-dwelling, highly granivorous and bipedal. They are well-adapted to deserts and their ability to survive on a diet of seeds without drinking water has become legendary (Schmidt-Nielsen 1964). As a result, the ecophysiological responses, temporal activity, type of refuge, trophic status and mode of locomotion of these rodents have been assumed to represent the normal suite of adaptations of small desert mammals. However, although many characteristics are found in small mammals of other deserts that are similar to those described in the Heteromyidae, they are not at all representative of small desert mammals in general (Morton 1979; Kerley 1992a; Mares 1993; Shenbrot et al. 1994b).

5.1
Temporal Patterns

The temporal behaviour patterns of small desert mammals can play a vital role in their thermal balance. Most small desert mammals are nocturnal and thereby avoid the desert heat during the day. However, several species are diurnally active and have adapted to cope with daytime desert conditions.

5.1.1
Small Nocturnal Mammals

Little information is available about the nocturnal activity of small desert mammals, doubtless because of the difficulty in observing them. Crepuscular activity has been observed in some nocturnal rodents, at least in heteromyids

(Reichman and Price 1993). In a study of Heteromyidae, Kenagy (1973a) found that *Dipodomys microps, D. merriami* and *Perognathus longimembris* spend most of their lives solitarily underground and forage for as little as an hour nightly. About 1 h of above-ground activity was also reported for *D. ordii*, most of which was spent foraging (Langford 1983). In this latter heteromyid, activity is greatest during the first hour after sunset and then, but to a lesser extent, several hours before dawn. This bimodal activity behaviour is common in a number of rodent species. Air temperatures in the winter are highest at these peaks of nocturnal activity, and perhaps these rodents stay in their burrows at other times for thermoregulatory purposes. Smaller species indulge in foraging activity at dusk before larger species emerge (Kenagy 1973a), and it has been suggested that this may be a consequence of smaller rodents being more vulnerable to low air temperatures (Reichman and Price 1993).

5.1.2
Small Diurnal Mammals

Day-active desert rodents are relatively few in number, but include species from all the world's deserts: for example, the fat sand rat (*Psammomys obesus*; Ilan and Yom-Tov 1990) and Egyptian spiny mouse *Acomys russatus* (Shkolnik 1971a) in North Africa, the great gerbil (*Rhombodys opimus*) and ground squirrels (for example, *Spermophilopsis leptodactylus*) in the Kyzylkum desert (Shenbrot et al. 1994b), Brants whistling rat (*Parotomys brantsii*), the Karoo rat (*Otomys unisulcatus*; du Plessis et al. 1992) and Cape ground squirrel (*Xerus inauris*; van Heerden and Dauth 1987) in South Africa, the Indian desert gerbil (*Meriones hurrianae*) and bush rat (*Golunda ellioti gujerati*) in India (Ghosh et al.1979), and ground squirrels (*Spermophilus tereticaudatus*; Drabek 1973; *Spermophilus elegans*; Byman 1985) in North America. Among other small desert mammals that are diurnal are lagomorphs and hyraxes. The bat-eared fox (*Otocyon megalotis*) of the South African deserts forages nightly in the summer when midday soil temperatures reach 70 °C, and diurnally in the winter when air temperatures drop to − 10 °C during the night (Lourens and Nel 1990; Nel 1990).

Two main activity patterns have been described for diurnal rodents. In the first, exemplified by the fat sand rat and Indian desert gerbil, the animals manage to avoid extreme heat in the summer by being mainly crepuscular. In the second pattern, exemplified by ground squirrels, the rodents are active all day but reduce their activity when air termperatures are high. Both of these patterns limit the above-ground activity time of the animals.

Fat sand rats exhibit crepuscular activity during the summer but not in the winter (Ilan and Yom-Tov 1990). In both seasons, these gerbils are usually away from their burrows when air temperatures vary between 14 and 36 °C, being active less than 1% of the time outside this temperature range (Fig. 5.1). In the summer, they limit their above-ground activities to the early morning and late afternoon and, during this time, remain mainly in the shade of bushes. In the

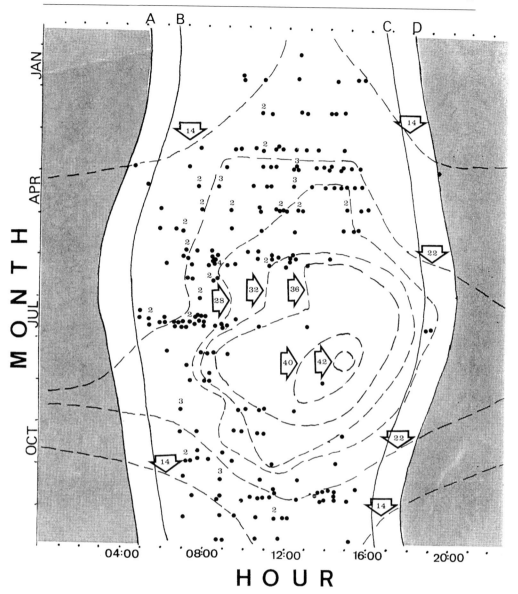

Fig. 5.1. Timing of activity of *Psammomys obesus* in the Negev Desert, Israel. Captures of sand rats indicated by *solid circles*; accompanying *numerals* indicate number of individuals captured simultaneously. A Early morning first light, the sun 18° below the horizon; B sunrise; C sunset; D latest evening twilight, the sun 18° below the horizon. *Dashed line* indicates temperature isotherm at 10 cm above ground. *Arrows* show °C. (Ilan and Yom-Tov 1990)

winter, they are inactive during rain, and activity starts later in the day. Sunbathing is common on first emergence from the burrow. Thus, a bimodal pattern of activity is exhibited in the summer and a unimodal pattern in the winter.

A bimodal activity pattern in summer has also been reported for the rock hyrax (*Procavia capensis*). Feeding occurs primarily in the cooler hours of the day, that is, in the morning from 07:30 to 09:30 h and in the evening from 15:30 to 18:30 h (Sale 1965). Between feeding bouts, hyraxes usually rest in the sun or shade, depending on the air temperatures, and only at extreme temperatures do they retreat into their refuges. In the winter, they bask in the sun to aid thermoregulation (Rubsamen et al. 1982). Gundis (*Ctenodactylus gundi*), diurnal Ctenodactylidae that also inhabit rocky areas, appear to use similar behavioural adaptations (Happold 1984).

In contrast, the Cape ground squirrel (*Xerus inauris*) is active throughout the day all year round, but reduces the time spent outside its burrows at temperatures above 27.5–29 °C and then seeks shade (van Heerden and Dauth 1987). A similar pattern of activity has been described in the diurnal striped mouse (*Rhabdomys pumilio*), which also limits its activities to shaded places at high air temperatures (Kerley et al. 1990). In squirrels, part of the shade is provided by a raised parasol tail which reduces the environmental thermal load (Bennett et al. 1984). *Ammospermophilus leucurus* can lower the operative temperature (T_e) by 3 °C and its standard operative environmental temperature (T_{es}) by 6–8 °C during the hot part of the day by shading its back with its tail and by using optimum postures and orientation (Chappell and Bartholomew 1981b). Similarly, *Xerus inauris* can reduce T_e by 5–8% and in this way increase its foraging time (Bennett et al. 1984).

In addition to using its tail to provide shade during the hot part of the day, Cape ground squirrels minimize contact with the hot soil surface by standing on their toes while feeding, and also minimize exposure to direct sunlight by orienting their bodies according to the direction of the sun. These behavioural traits were observed early in the day when there was no wind, and none of them appeared when conditions were overcast (van Heerden and Dauth 1987).

5.1.3
Shuttling

Shuttling has been described in both desert vertebrates and invertebrates. The most extensive studies on small desert mammals have been performed on ground squirrels, which typically shuttle between foraging sites and cooling sites. Foraging sites provide the squirrels with their energy sources; cooling sites such as burrows or shade allow them to regain normal body temperatures. The body temperature of the squirrel sets a limit on the time that it can spend travelling to and foraging at a site, and signals the time at which it must return to the non-resource cooling site.

Bennett et al. (1984) reported that *Xerus inauris* moves frequently to and from its burrow when temperatures are high and retreats into its burrow when

T_e exceeds 40 °C. Van Heerden and Dauth (1987) did not notice this frequent movement, but reported that the squirrels spent more time inside their burrows at high temperatures. Similarly, Byman (1985) reported that bouts of aboveground activity of the Wyoming ground squirrel (*Spermophilus elegans*) decreased with an increase in T_{es}: generally, bouts were not time-limited at operative temperatures below 31 °C, but were less than 10 min when operative temperatures reached or exceeded 47 °C. Furthermore, he demonstrated that the T_{es} the squirrel experiences is dependent upon posture. The animal is in a prostrate position at the highest T_e, in a standing position at intermediate T_e, and in a picket pin position at the lowest T_e.

Vispo and Bakken (1993) reported maximum above-ground activity in 13-lined ground squirrels (*Spermophilus tridecemlineatus*) between T_{es} of 25 °C and 40 °C, with a decrease in activity at temperatures both below and above this range. At a T_{es} of about 55 °C, the squirrels were able to remain outside their burrows for about 50% of the time. They allowed their body temperature to increase, then shuttled to their burrows to allow it to decline, then re-emerged, and so on (Fig. 5.2).

Hainsworth (1995) used data from the literature on the desert antelope ground squirrel (*Ammospermophilus leucurus*) to examine and generate models of its body temperature changes during shuttling behaviour. The data described the environment to which the squirrels were exposed (Table 5.1) and their physiological responses. Minimum body temperature was 35.8 °C at 14:30 to 18:30 h and maximum body temperature was 43.1 °C at 10:30 h, but could increase to an upper extreme of 43.6 °C. Ground level T_{es} in the sun is above the

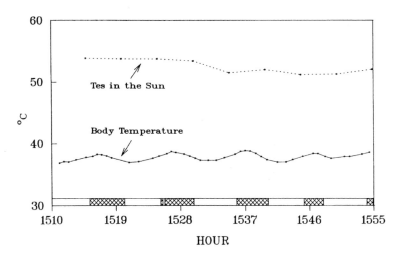

Fig. 5.2. Pattern of body temperature fluctuation and burrow use, illustrating shuttling. Body temperature and T_{es} in the sun are plotted against time for a 45-min period in late June. The *bar across the bottom* indicates the squirrel's location in the burrow (*cross hatched*) or on the surface in the sun (*open*). Body temperature rises while the squirrel is above ground and drops while it is in the burrow. (Vispo and Bakken 1993)

Table 5.1. Minimum, average and maximum measured (T_b) for antelope ground squirrels and the standard effective temperatures (T_{es}) in the sun at different times of the day in summer. (Hainsworth 1995)

Time of day (h)	Min T_b(°C)	Mean T_b(°C)	Max T_b(°C)	Sun T_{es}(°C)
0630	36.2	38.6	39.9	28
0830	37.8	40.0	41.9	43
1030	37.7	40.2	43.1	60
1230	37.5	38.7	40.0	69
1430	35.8	38.0	39.8	69
1630	35.8	39.2	42.1	65
1830	36.5	39.4	41.7	49

Data from Chappell and Bartholomew (1981b).

thermal neutral zone (30–38 °C) of *A. leucurus* by 08:00 h and reaches nearly 70 °C between 12:30 and 14:30 h. In the shade, peak T_{es} reaches 42 °C at midday, and in the burrow at a depth exceeding 40 cm, T_{es} remains constant at 32 °C (Chappell and Bartholomew 1981a; b). Standard operative temperatures were then used to calculate changes in body temperature over time with Newton's Law of Cooling as:

$$dT_b/dt = (C/K)(T_b - T_{es}),$$

where T_b is average body temperature (°C) of the ground squirrel; T_{es} is standard operative temperature (°C); C is measured thermal conductance of *A. leucurus* (J g^{-1} min^{-1} °C^{-1}) and K is specific heat of *A. leucurus* (J g^{-1} °C^{-1}).

Integration of the above equation results in:

$$\log(T_b - T_{es}) = \log E + (C(t)/K(\log e),$$

where E is value of (T_b-T_{es}) when t (time in min) = 0, and e is base of the natural logarithm.

For thermal conductance, Hainsworth (1995) used values for a 100-g *A. leucurus* as reported by Chappell and Bartholomew (1981a): 0.1 W °C^{-1} for a surface-active squirrel and 0.55 W °C^{-1} for a squirrel in a burrow. With a specific heat value for *A. leucurus* of 3.431 J g^{-1}°C^{-1}, C/K equals 0.0175/min and 0.0962/min for squirrels active on the surface and remaining in their burrows, respectively.

Body temperatures of the squirrels were calculated using metabolic heat production and evaporative heat loss. Basal and field metabolic rates were taken to be 0.55 and 1.7 W, respectively (Hudson 1962), of which 30 and 33% respectively, of the heat production were lost by evaporation. The metabolic heat produced and not lost by evaporation was assumed to be stored as body heat, resulting in an increase in body temperature. To increment the value for E in the above equation, the non-evaporative heat rate from metabolism was divided by the specific heat of *A. leucurus* and multiplied by the time spent heating or cooling. These equations took the forms for heating while travelling and foraging as:

$$T_b = T_{es} - [E \text{-} 0.1962(t)](e)^{-0.0175(t)},$$

and for cooling in the burrow (T_{es}) as:

$$T_b = 32 + [E + 0.0673(t)](e)^{-0.0962(t)}.$$

Rates of heating and cooling when *A. leucurus* enters the burrow with a body temperature of 43 °C and exits with a body temperature of 34 °C are presented in Fig. 5.3. The time required to heat *A. leucurus* from a body temperature of 34 to 43 °C at a T_{es} of 70 °C is about the same as to cool from 43 to 34 °C in a burrow at T_{es} of 32 °C. At a T_{es} of 50 °C, it takes about twice as long to heat to 43 °C as it does at a T_{es} of 70 °C.

The rates of heating and cooling of *A. leucurus* are not linear with time; the rate of heating above ground decreases as the body temperature is higher, and the rate of cooling in the burrow decreases as the body temperature is lower. Therefore, to maximize the time spent above ground (travelling plus foraging) "predicts that ground squirrels should quickly dart in and out of their burrows with burrow exit T_b (body temperature) high and only infinitesimally less than burrow entry T_b," (Hainsworth 1995). However, this does not occur. Although quick darting in and out of the burrow would allow maximum time above ground, it would not lead to a maximal foraging time. When subtracting a travelling time of 4 min and assuming a heating T_{es} of 50 °C, a cooling T_{es} of 32 °C and a return T_b of 43 °C, the optimum burrow exit temperature to maximize foraging time is 37 °C (Fig. 5.4). Under these conditions, but at a return T_b of 43.6 °C and a travelling time of 2 min, maximum foraging time is also achieved at an exit T_b of 37 °C. But, with a heating T_{es} of 70 °C, maximum foraging is achieved at a T_b of 38.5 °C (Fig. 5.5).

Thus, travelling time has a major effect on the length of time ground squirrels can spend above ground. An option for the ground squirrel is to cool at a

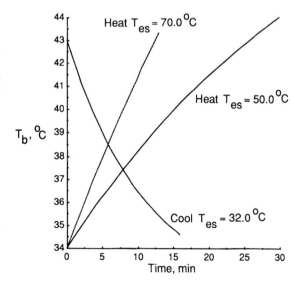

Fig. 5.3. Heating and cooling functions for antelope ground squirrels. The two heating functions are for an initial T_b of 34.0 °C, for a T_{es} of 50.0 (*lower curve*) and 70.0 °C (*upper curve*), each assuming a thermal conductance of 0.1 W °C^{-1} with a metabolic rate of 1.7 W and 33% of heat production lost by evaporation. The cooling function is for an initial T_b of 43.0 °C and a T_{es} of 32.0 C, assuming a thermal conductance of 0.55 W °C^{-1} with a metabolic rate of 0.55 W and 30% of heat production lost by evaporation. (Hainsworth 1995)

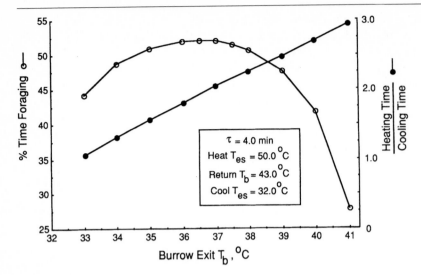

Fig. 5.4. Percent time foraging (○) in the vicinity of the optimum burrow exit T_b of 37.0 °C and heating time/cooling time (●) as functions of burrow exit T_b for a travel time of 4-min, a heating T_{es} of 50 °C, a cooling T_{es} of 32 °C and a return T_b of 43 °C. (Hainsworth 1995)

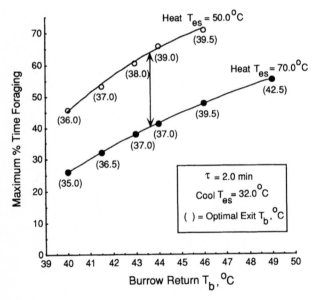

Fig. 5.5. Maximum percent time foraging as a function of burrow return T_b for a travel time of 2 min, and a cooling T_{es} of 32°C, for a heating T_{es} of 50 °C (○) and 70 °C (●). *Arrow* shows the maximum observed T_b for *A. leucurus* of 43.6 °C. Optimal exit T_b (in °C) to maximize percent foraging times for each return T_b are shown in *parentheses*. (Hainsworth 1995)

site other than its burrow. T_{es} in the shade is 35 °C rather than 32 °C in the burrow. Cooling would therefore be slower in the shade, but this could be a trade-off for the reduction in time of travelling and heating.

Antelope ground squirrels (Hainsworth 1995) and 13-lined ground squirrels (Vispo and Bakken 1993) neither leave their burrows at the same body temperature nor cease foraging and enter their burrows at the same body temperature on every occasion. Maximum return T_b at midday is about 40 °C, well below the maximum T_b, and the lowest return T_b recorded during the day. The authors suggested that this may be related to the high rate of heating and might be a precaution against the risk of overheating.

5.1.4
Nocturnal vs Diurnal Activity for Small Desert Mammals

Crepuscular activity and/or shuttling allow small mammals to be diurnal and yet not be exposed to extreme heat over extended periods of time. Furthermore, solar radiation in the winter could be used to advantage by the animal for thermoregulatory purposes and to economise its energy requirements. So, why are not more small mammals diurnal? A number of possible factors come to mind that may influence temporal activity. These include body size, dietary habits, water requirements and energy needs, as well as interaction among these factors.

Small mammals have a low thermal inertia, and this severely limits the length of time that they can be exposed to high air temperatures and yet maintain homeothermy within physiological limits. The smallest diurnal desert mammals, *Acomys russatus and Rhabdomys pumilio*, have a mean adult body mass of about 50 g, and this mass may be the lower limit which would allow crepuscular mammals to remain active for any reasonable length of time. Shuttling would not be a practical option for a very small mammal as its body temperature increases very quickly when above ground during the hot part of the day. As a result, the animal would spend most of its time moving to and from its burrow. The smallest shuttling mammal weighs about 100 g.

None of the diurnal mammals are granivores. Granivory may be related to body size and energy needs since smaller animals are forced to select diets that are more energy-rich (Demment and van Soest 1985). Moreover, obligatory granivores may be forced into nocturnal foraging due to considerations of water balance. Some seeds and green vegetation are hydroscopic, and preformed water can increase dramatically at night. In fact, Taylor (1970) has shown that some large desert mammals shift foraging from day to night and thereby satisfy their water needs.

The foraging time allotted by diurnal activity is perhaps insufficient to meet the energy requirements of small mammals. This reasoning, however, seems unlikely. Degen et al. (1986) showed that activity was more costly in terms of energy requirements for the nocturnal *Acomys cahirinus* than for the diurnal *A. russatus,* yet the nocturnal niche appears to be preferred by both species. In fact, *A. russatus* becomes nocturnal when *A. cahirinus* is removed from the habitat, suggesting that *A. cahirinus* may force *A. russatus* to become diurnal where both species occur together (Shkolnik 1971a). It may be that factors other than those discussed above, such as predation pressure, favour nocturnal over diurnal activity.

5.2
Refuge Microhabitat

Refuges of small desert mammals can provide a thermally comfortable, mesic micro-environment within a harsh, xeric macro-environment. The most easily recognized of these refugia are the burrows of desert rodents. However, crevices in rocks and under boulders, roosts, dens and above-ground nests also provide protection against extreme conditions. These refugia dampen the large daily and seasonal fluctuations of temperature and relative humidity in deserts. This is important as soil surfaces often attain temperatures above 70 °C, with daily fluctuations greater than 45 °C (Bennett et al. 1988) and seasonal fluctuations greater than 80 °C (Schmidt-Nielsen 1964).

5.2.1
Burrows

Burrows are usually excavated near the base of bushes, on sand plains and hammadas, and in *wadis* (Happold 1984). They protect the occupants against predators and extreme environmental conditions in the summer and winter. They also serve as food storage areas for several gerbil species (*Meriones persicus, M. libycus, M. meridianus* and *Rhombomys opinus*) in the deserts of Mongolia and Central Asia (Naumov and Lobachev 1975) and for some heteromyids in North America (Eisenberg 1975), but apparently not for gerbils (*Meriones hurrianae* and *Tatera indica*) in the Thar desert of India (Goyal and Ghosh 1993). Burrows also provide areas for brooding and, depending on the burrowing species, can either house individuals, families and colonies or act merely as temporary gathering places for individuals (Kucheruk 1983).

Burrows differ greatly from species to species, ranging from simple systems such as the "Y" shaped burrow of *Tatera indica* to very extensive complicated systems, as found in *Meriones hurrianae* (Goyal and Ghosh 1993). They can also vary greatly among members of one species, as reported for *Gerbillurus setzeri* (Fig. 5.6; Downs and Perrin 1989a). In addition, the number of surface openings may also vary greatly among and within species. For example, burrows of *M. hurriance* can have between 4 and 30 openings (Goyal and Ghosh 1993).

Burrows may be non-chambered, chambered without nests, contain nests without chambers, nests with chambers, and nests with a system of branched tunnels near the surface (Kucheruk 1983). They may contain several nests which

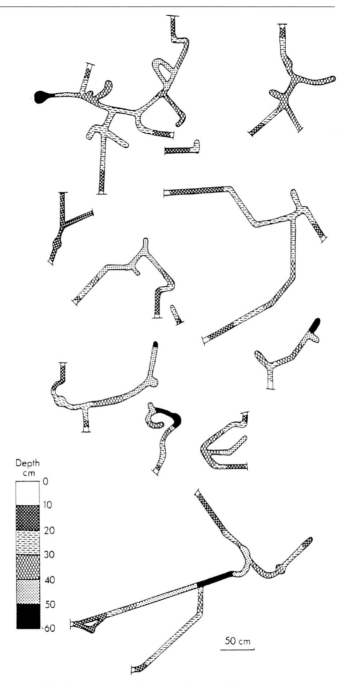

Fig. 5.6. Burrow structures of *Gerbillurus setzeri*. (Downs and Perrin 1989a)

differ in temperature due to differences in depth, and the animal can then choose the desired temperature. This is very evident in the African rodent mole (*Cryptomys damrensis*) which has an extensive, complex burrow system to a depth of 2.5 m. This mole is sensitive to high air temperatures and, when summer temperatures are extreme, it retreats deeply into the foraging burrow (Bennett et al. 1988).

Although some small, fossorial desert mammals have burrows of over 2 m in depth, most rodent burrows reach a depth of only 20–70 cm (Table 5.2; Goyal and Ghosh 1993). However, differences in the depths of burrows, varying according to season, have been reported for some species. For example, in the Sahara Desert, the burrow of the jerboa (*Jaculus jaculus*) is approximately 25 cm deep in winter and 70 cm deep in summer (Ghobrial and Hodieb 1973). Similar depths, varying according to season, have also been reported for the burrows of *Pygeretmus pumilio* in the deserts of Turkestan (Shenbrot et al. 1992). The deeper summer burrows provide a cooler environment and the shallower winter burrows, a warmer one.

Temperature fluctuations within burrows are dependent on their depth. Diurnal fluctuations decrease with the depth of soil (Happold 1984; Bennett et al. 1988; Downs and Perrin 1989a) and, in general, are only about 3 °C at 20 cm and negligible below 80–100 cm (Schmidt-Nielsen 1964; Happold 1984). Annual fluctuations in air temperature at 1 m are about 12 °C. Fluctuations in burrow air temperatures are also dependent on soil characteristics and ground cover. For example, temperature fluctuations are smaller in very hard soils than in wind-blown soil and fine grain sands. They are also smaller when there is less ground cover (Bennett et al. 1988).

Summer studies in the deserts of Arizona revealed that an air temperature of less than 31 °C was maintained within the burrows of banner-tailed kangaroo rats (*Dipodomys spectabilis*) and less than 29 °C in the deep burrow nests of round-tailed ground squirrels (*Citellus tereticaudus*) (Schmidt-Nielsen 1964).

Table 5.2. Burrow depth of various rodent species. (Goyal and Ghosh 1993)

Species	Burrow depth (cm)	Reference
Dipodomys merriami	60	Schmidt-Nielsen and Schmidt-Nielsen (1951)
D. merriami	20–30	Kenagy and Smith (1973)
D. microps	20–30	Kenagy and Smith (1973)
Geomys bursarius	65	Kennerley (1964)
Parotomys brantsii	30	Graaf and Nel (1965)
Desmodilus auricularis	30–60	Nel (1967)
Neotoma albigula	20–30	Kenagy and Smith (1973)
Perognathus longimembris	20–30	Kenagy and Smith (1973)
Desert gerbils and jerboas of former USSR	30–70	Naumov and Lobachev (1975)
Meriones hurrianae	45	Goyal and Ghosh (1993)
Tatera indica	35–50	Goyal and Ghosh (1993)

Measurements in the burrow of the fat jird (*Meriones crassus*) in Israel showed that the burrow temperature was relatively constant and under 30 °C while the outside air temperature fluctuated between 14 and 38 °C and the soil temperature fluctuated between 14 and 61 °C. Relative humidity within the burrow was also relatively constant at about 55%, while the outside relative humidity fluctuated between 18 and 72% (Fig. 5.7; Shkolnik 1971b). In southern Africa, mean burrow temperatures of four desert *Gerbillurus* species ranged between 22.9 and 31.1 °C, and the maximum reached was less than 33 °C. There was little daily temperature fluctuation in the burrow (Downs and Perrin 1989a).

Fewer data are available on the winter burrow temperatures of desert mammals, as most studies have concentrated on hot summer conditions. In *Paratomys brantsii* of the Karoo, the mean winter burrow temperature at a depth of 15 cm was about 12 °C, with a range of 8.4 to 16.6 °C, when the outside air temperature ranged between 3.4 and 23.2 °C (du Plessis et al. 1992). In the rodent mole *Cryptomys damarensis* of Namibia, the mean burrow temperature at a depth of 13 cm was 18.6 °C, with a range of 15.0 to 22.8 °C. The mean summer burrow temperature was 13 °C higher (Bennett et al.1988).

It is apparent that burrow temperatures in the summer are generally within or close to the thermoneutral zone of the animal; however, in the winter this does not appear to be the case since the burrow temperatures are considerably lower. The main concern among researchers has been to explain how desert rodents survive the hot summer and much less attention has been devoted to how

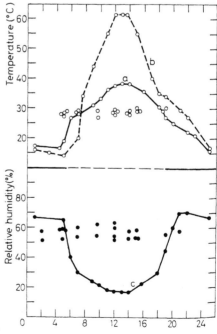

Fig. 5.7. Temperature of air (*a, solid line*) and soil (*b, dashed line*) and relative humidity (*c, below*) during August in the Ramon cirque near burrows of the fat jird (*Meriones crassus*). *Single open circles* are temperatures and *single solid circles,* relative humidities measured in the burrows of the fat jird. The *full line in the middle of the drawing* indicates the time of activity of the animals outside their burrows, the *dashed line* the time the animals are inactive inside their burrows. (Shkolnik 1971b)

they survive low air temperatures. For example, Schmidt-Nielsen (1964) referred to a study of burrow temperatures of *Dipodomys spectabilis* in which the lowest temperature was 9 °C in the winter and the highest, 31 °C in the summer. Schmidt-Nielsen (1964) then stated that these results were "widely quoted and form the basis for the statement that burrow temperatures never reach levels that require active defences against an undue rise in body temperature." No mention, however, was made of the low burrow temperature of 9 °C which, presumably, was well below the animals thermoneutral zone. It is possible that the nest within a burrow provides insulation and/or that by huddling together the animals improve their thermal conditions.

5.2.2
Rock Crevices

Little has been written about temperature fluctuations in rocky habitats. Happold (1984), however, surmised that rock domiciles "provide a cool non-fluctuating microclimate similar to that of burrows. All species obtain excellent insulation from the outside climatic conditions in their burrows and rock crevices." Indeed, this premise was supported by temperature measurements in rock crevices used as refuges by rock hyraxes (*Procavia capensis*). Temperature fluctuations 2 m inside a hyrax shelter did not exceed 4 °C while the diurnal air temperature outside the shelter fluctuated 23 °C (Sale 1966). Furthermore, in areas with strong winds, the entrance of the hyrax shelter was either protected from or positioned against prevailing winds (Rubsamen et al. 1982).

5.2.3
Nests

Temperature measurements were made in above-ground nests of *Otomys unisulcatus*, a herbivore inhabiting the Karoo that builds nests of sticks under shrubs (du Plessis et al. 1992). These nests average 0.45 m in height and have a mean of nine openings. Temperatures measured in the nests were compared with warren temperatures of *Parotomys brantsii*, a sympatric herbivore. Fluctuations were much larger in the winter than in the summer in the refuges of both species. Interestingly, the nest of *O.unisulcatus* showed less temperature variation in the winter and lower temperatures in the summer and was less affected by outside air temperature than was the warren temperature of *P. brantsii*, indicating its effectiveness (Fig. 5.8). Relative humidities inside the nest of *O. unisulcatus* ranged between 64 and 74% in the summer and between 56 and 83% in the winter; in the warrens of *P. brantsii*, relative humidities ranged between 51 and 60% in the summer and between 56 and 98% in the winter.

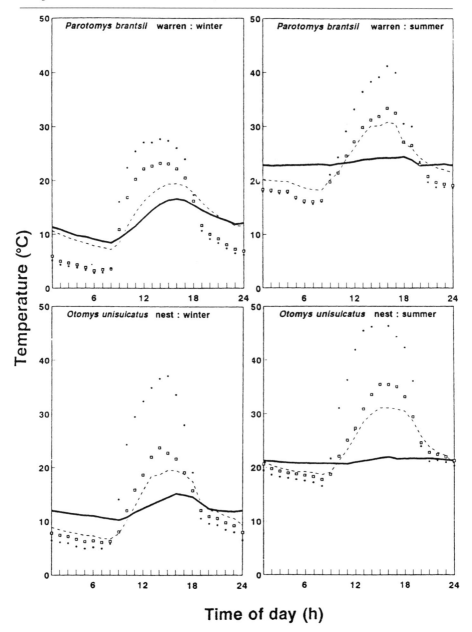

Fig. 5.8. Daily temperature profiles in *Parotomys brantsii's* warren and *Otomys unisulcatus'* nest in winter and summer. Means (n = 5) are indicated, where *large points* are ambient temperature at the warren or nest, and *small points* are black bulb temperatures. *Broken lines* indicate superficial refuge temperatures and *solid lines*, interior temperatures. (du Plessis et al. 1992)

5.2.4
Geothermally Heated Mines

Perhaps the most stable diurnal and seasonal air temperatures occur in geo-
thermally heated abandoned mines. These are used as roosts in the Colorado
Desert by the bat *Macrotus californicus* (Phyllostomidae). With outside air temperatures ranging between 5 and 35 °C, temperature at the entrance to the mine
ranges between 19.5 and 25.8 °C but, deep in the mine where the bats roost, it
varies only between 28.2 and 29.3 °C. In fact, between December and March,
the temperature at a distance of 50 m from the entrance to the mines varied
less than 0.5 °C, and the relative humidity remained within $22 \pm 1\%$ range (Bell
et al. 1986).

5.3
Dietary Habits

Classifying mammals by their dietary habitats has proven to be extremely
difficult. In general, descriptive terms such as "granivore" (seed-eater), "herbivore" (vegetation-eater), "omnivore", "carnivore" (vertebrate-eater), "insectivore" (invertebrate-eater), "folivore" (leaf-eater), "nectarivore" (nectar or
pollen-eater) and "frugivore" (fruit-eater) are used. Kerley and Whitford (1994)
suggested that the "terms granivore, insectivore and herbivore should refer to
animals whose diet is dominated ($> 50\%$) by one of these dietary categories,
while omnivore should refer to those species whose diets are not dominated by
any particular category." Other terms such as "mixed feeders", "grazers",
"browsers" and "concentrate selectors" are used as well to describe feeding behaviour.

Only a few small mammal species consume a single type of diet, thus making their classification straightforward. Such is the case with the herbivorus fat
sand rat which subsists wholly on chenopods and also with certain shrews (e.g.
Notiosorex spp. and *Crocidura* spp.) and bats *(Dasyurus* spp.), which consume
insects only. However, the majority of mammals consume a variety of items in
their diet. At times, these can be lumped together as, for example, "grasses and
forbs" or "seeds and nuts". More often, the dietary habits of a mammal are described as "primarily", such as primarily granivorous or primarily insectivorous.

Season and habitat can affect the dietary habits of animals, and this is mainly
the result of differences in feed availability. In addition, animals exhibit dietary
shifts due to changes in physiological condition, as for example changes in reproductive state. These differences have led to much confusion in the catego-
rization of animals according to diet. For example, the pigmy gerbil, *Gerbillurus
paeba*, a species inhabiting the sand dunes of the Namib and Kalahari deserts,
has been described as omnivorous (Kerley 1989; Perrin et al. 1992), herbivorous (Kerley et al. 1990; Kerley 1992a), granivorous (Nel 1978) and insectivo-

rous (Boyer 1987, quoted in Kerley 1992a). Furthermore, many authors have classified a large number of mammals as omnivorous, but there are still wide dietary differences within this category. These can range from a mainly vegetarian diet to a mainly carnivorous diet. Indeed, McNab (1992b) pointed out some of these problems and stated that omnivory "is unlikely to be a uniform category and should be excluded (in the classification of animals)" and that "errors and judgements (in categorizing mammals) influence statistical calculations and may lead to erroneous conclusions".

5.3.1
Granivory

The best-known, small, granivorous mammals are the heteromyids (kangaroo rats, kangaroo mice and pocket mice) of the North American deserts. These rodents have a number of adaptations suited for granivory, including specialized cheek pouches for the collection and transport of seeds (Morton et al.1980; Randall 1993) and scratch-digging behaviour for caching seeds over extended periods (Morgan and Price 1992). Gerbillidae in the North African and southwestern Asian deserts have also been described as granivorous (Bar et al.1984). This has led to much discussion on convergence in dietary habits and in other characteristics between the Old World desert Gerbillidae and New World desert Heteromyidae (Abramsky 1983; Abramsky et al. 1985a).

Mammals are the main granivorous taxa in the North American, North African and Negev deserts (Mares and Rosenzweig 1978; Abramsky 1983; Mares 1993). At least in North America, rodents remove more than 75% of the seed bank (Chew and Chew 1970; Soholt 1973; Nelson and Chew 1977; Brown et al. 1979). This, however, is not the case in the deserts of South America (Mares and Rosenzweig 1978), Australia (Morton 1985) and southern Africa (Kerley 1991; Kerley and Whitford 1994). Here, ants appear to be the main granivores not mammals (Fig. 5.9). Seed consumption by small mammals at Tierberg, in the Karoo Desert, was estimated to be only about 0.5% of seed production (Kerley 1992a). No obligatory granivores are known to occur in the South American and Madagascaran deserts (Mares 1993). In a comparison of North American and Australian deserts, Morton (1979) listed 109 desert-dwelling mammals in North America of which 33 were considered to be granivores, and 73 desert-dwelling mammals in Australia of which only 12 were granivores. The Australian desert species were mainly members of the genus *Notomys*, which are bipedal murids, and the genus *Pseudomys*, which comprises quadrupedal murids (Morton 1979). Furthermore, examination of the dietary habits of the Australian granivores (Morton and Baynes 1985; Murray and Dickman 1994a) revealed that they consumed a much more varied diet than the heteromyids.

Data on granivory in the above-mentioned deserts were collected by measuring the rate of seed removal from trays by different taxa, and the common methodology in these studies has allowed comparisons to be made among deserts. This method has not been employed in the Asian deserts. However, a

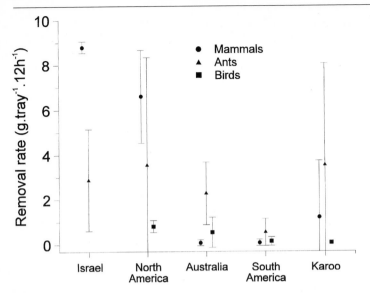

Fig. 5.9. Mean seed removal (*vertical bars* are SD) of experimental seed baits in Israeli (Abramsky 1983), North and South American (Mares and Rosenzweig 1978), Australian (Morton 1985) and the South African Karoo deserts (Kerley 1991). Data for the first four localities redrawn from Morton (1985). (Kerley and Whitford 1994)

description of the food habits of rodents from Asian deserts (Shenbrot et al. 1994b) suggests that granivory is much less important to mammals in these deserts than in North American (Chihuahuan) and Middle Eastern (Negev) deserts and is similar to the situation in the deserts of South America, Australia and South Africa (Fig. 5.10). Only 1 (*Salpingotus heptneri*) of 26 rodent species in the Kyzylkum Desert of Uzbekistan, 4 (*Cardiocranius paradoxus, Salpingotus crassicauda* and two *Phodopus* spp.) of 25 rodent species in the Gobi Desert of Mongolia and 2 (*Gerbillus* spp.) of 12 rodent species in the Thar Desert of India were categorized as granivores. In contrast, 9 (*Perognathus, Chaetodipus* and *Dipodomys* spp.) of 28 rodent species in the Chihuahua Desert and 6 (all *Gerbillus* spp.) of 18 rodent species in the Negev Desert are granivores.

The dietary classification of two species of Australian desert murids, *Notomys alexis* and *Pseudomys hermannsburgensis*, that have been considered to be granivorous (Morton 1979) has recently been questioned. Seeds are a major component of the diet of both species but, in the autumn, invertebrates comprise nearly 50% of stomach contents in *N. alexis* and 55–60% in *P. hermannsburgensis* (Fig. 5.11; Murray and Dickman 1994a). Furthermore, when offered a choice of dietary items in the laboratory, these species preferred invertebrates to seeds (Fig. 5.12; Murray and Dickman 1994b). Murray and Dickman (1994a) concluded that both these species should be classified as omnivores, and cautioned against interpreting ecological studies of desert rodents based solely on a granivorous diet.

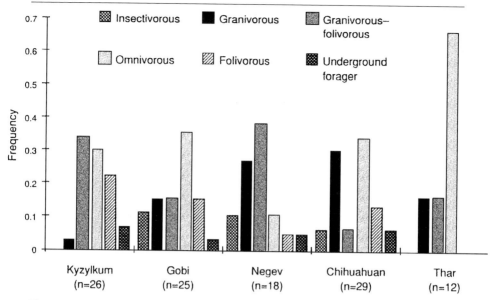

Fig. 5.10. Distribution of rodent feeding habits in various deserts (n = number of species). (Shenbrot et al. 1994b)

Examination of the dietary habits of the granivorous Negev gerbils also revealed that their diet was somewhat varied and included a considerable fraction of green vegetation and insects (Bar et al. 1984). For example, on a dry matter basis, *Gerbillus pyramidum* consumed only 63% seeds over a year, *G. allenbyi* 46%, *G. gerbillus* 69% and *Meriones sacramenti* 52%. Furthermore, none of these Gerbillidae consumed more than 90% seeds in any one season (Table 5.3).

Of the dry matter intake of *G. gerbillus* in winter, 63% consisted of seeds; that is, at a time when seeds were relatively scarce. This percentage of seeds was considerably higher than that of the other *Gerbillus* spp. (Bar et al.1984), which is of particular interest since, of these species, *G. gerbillus* inhabits areas with the least vegetation and lowest seed availability. It appears that *G. pyramidum* may exclude *G. gerbillus* from areas in which seeds are more abundant (Abramsky et al.1985b). In contrast, in its diet during the winter, *G. allenbyi* consumes only 11% seeds.

This would suggest that *G. gerbillus* is either more efficient at foraging for seeds, caches seeds, and/or is more dependent on seeds in its diet. The last alternative is a strong possibility on the basis of body size differences: *G. gerbillus* is smaller than the other three species (21.7 g vs 26.2–53.3 g; Abramsky et al. 1985a). This would indicate that it is least able to rely on forage for its maintenance energy requirements (Demment and Van Soest 1985) and, as a result, is the one most dependent on a concentrated energy source. Indeed, Yom-Tov (1991) found a general increase in seed consumption with a decrease in body mass in Gerbillidae. A relationship between the ratio of seed to vegetation in-

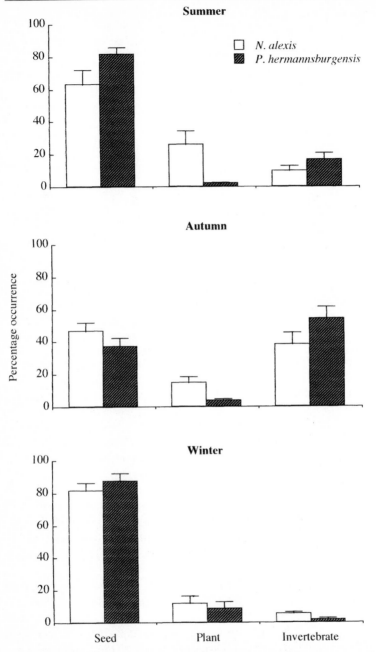

Fig. 5.11. Percentage seasonal occurrence of each of the three broad categories of food in the diets of *Notomys alexis* (*open columns*) and *Pseudomys hermannsburgensis* (*solid columns*). For each season, values are means ± SE of the stomach contents of 10 individuals of each species. Occurrences were obtained from counts of 120 food fragments (*N. alexis*) and 135 fragments (*P. hermannsburgensis*) for all individuals. (Murray and Dickman 1994a)

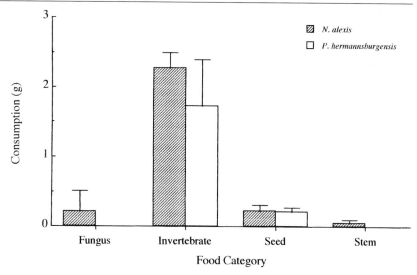

Fig. 5.12. Consumption of food items (mean ± SD) by six *Notomys alexis* and six *Pseudomys her-mannsburgensis* in broad food-category experiments. (Murray and Dickman 1994b)

Table 5.3. Percent composition by dry mass of seeds, greenery, and insects (± SE) in the diets of gerbilline rodents. (Bar et al. 1984)

	N1	N2	Seeds	Greenery	Insects
Gerbillus pyramidum					
Winter	36	35	40.0 ± 6.1	55.0 ± 5.8	4.9 ± 2.1
Spring	83	44	85.3 ± 2.7	8.7 ± 1.7	6.0 ± 1.8
Summer	50	41	59.3 ± 4.8	30.6 ± 4.5	10.0 ± 1.8
Total/mean	169	120	63.2 ± 3.1	29.7 ± 2.9	7.1 ± 1.3
Gerbillus allenbyi					
Winter	83	48	11.3 ± 3.4	81.8 ± 3.6	6.9 ± 2.2
Spring	60	38	89.6 ± 1.9	7.4 ± 1.3	2.9 ± 1.5
Summer	75	50	48.7 ± 5.0	39.7 ± 4.9	11.6 ± 2.9
Total/mean	218	136	46.3 ± 3.5	45.5 ± 3.4	7.5 ± 1.4
Gerbillus gerbillus					
Winter	10	9	63.3 ± 12.7	35.4 ± 12.7	1.2 ± 0.7
Spring	5	3;	82.1 ± 11.5	1.7 ± 0.5	16.2 ± 12.3
Summer	6	3;	73.5 ± 26.4	21.2 ± 22.0	5.2 ± 5.3
Total/mean	21	15	69.1 ± 9.0	25.6 ± 8.9	5.0 ± 2.7
Meriones sacramenti					
Winter	12	12	1.5 ± 0.5	97.1 ± 1.8	1.4 ± 0.9
Spring	14	14	56.9 ± 1.0	41.5 ± 7.3	1.6 ± 1.2
Summer	19	18	25.5 ± 7.5	69.7 ± 7.9	4.8 ± 4.3
Total/mean	45	44	29.0 ± 5.0	68.2 ± 5.1	2.8 ± 1.8
All species All seasons	453	315	51.7 ± 1.9	41.7 ± 2.1	6.6 ± 0.8

N1 number of individuals captured; *N2* number of stomach contents analysed.

take and body mass is also apparent within seasons in a New World and in Old World desert granivores; however, the ratio differs between seasons (Fig. 5.13).

The foraging efficiency of *G. gerbillus* has yet to be examined. It is possible that *G. gerbillus* is more efficient at harvesting seeds at low density (when they are scarce) than are the other species. Such was the case when *G. allenbyi* was compared with *G. pyramidum* (Kotler and Brown 1990; Kotler et al. 1993).

Small desert mammals are often limited in their food choices and are forced to select more digestible, energy-rich diets (such as seeds) than large mammals. Gut capacity in mammals is related linearly to body mass, whereas the energy requirement is related exponentially. As a result, smaller animals require more energy per unit gut volume than do larger mammals (Demment and van Soest 1985). The smallest gerbil, the 8–12 g *G. henleyi* of the Negev Desert, is highly granivorous (Shenbrot et al. 1994a), even in the winter when many gerbils shift their diet to include more vegetation (unpubl. data). Small, highly granivorous rodents of approximately 10 g also inhabit other deserts, for example, *G. nanus indus* in the Thar desert, *Salpingotus crassicauda* in the Gobi Desert, *Perognathus longimembris* in the Sonoran Desert and *P. flavus* in the Chihuahuan Desert (Shenbrot et al. 1994a).

Among mammals, granivory appears to be exclusive to rodents, in particular the smaller species. However, the Sechuran or Peruvian desert fox, *Dusicyon sechurae*, joins this exclusive group at certain times of the year. This primarily nocturnal fox mainly inhabits flat, sandy terrain of the Sechuran Desert in northwestern Peru. Its diet before the El Niño rains consists of approximately 99% seeds and seed pods, and while on this dietary regime, it can survive "with limited or no access to drinking water". *D. sechurae*, however, is an opportunistic feeder, and after the El Niño rains, its diets shifts dramatically to also include grasshoppers (42%) and rodents (20%; Asa and Wallace 1990).

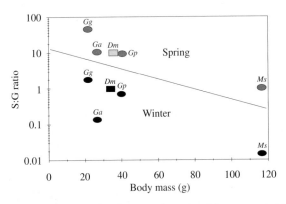

Fig. 5.13. Relationship between the ratio of dry matter intake of seeds to vegetation (*S:G*) and body mass in a free-living New World (*squares*; calculated from Nagy and Gruchacz 1994) and in Old World desert rodents (*circles*; calculated from Bar et al. 1984) in spring and winter. *Ms, Meriones sacramenti; Mc, M. crassus; Gp, Gerbillus pyramidum; Ga, G. allenbyi; Gg, G. gerbillus; Dm, Dipodomys merriami.* (Kam, Khokhlova and Degen, unpubl.)

Divergence in Granivory Among Deserts. What are the reasons for the apparent divergence in granivory among desert rodents? Ants and birds do not exclude rodents from granivory, at least not in Australia (Morton 1985). In addition, the numbers of seeds in the Karoo (Kerley 1992a) and Australian (Morton 1985) deserts are approximately equal to that of the North American deserts. Morton (1985) suggests that the difference in granivory among desert rodents may lie in the rate of seed renewal. In the North American deserts, seed production is relatively reliable owing to a comparatively dependable precipitation regime, adequate soil nutrients, and apparently rapid recycling of nutrients. In contrast, precipitation in the Australian deserts is more unpredictable, and the sandy soils are poor in nutrients, making seed production unstable. Such a situation would be unreliable for supporting granivorous, sedentary homeotherms, such as rodents, that require a steady energy source. However, it could be exploited by poikilotherms such as ants, a taxon which does not require a steady energy intake to maintain body temperature, and also by highly mobile homeotherms, such as birds, which can forage for seeds over vast areas.

The extinction of small granivorous mammals from some deserts has also been suggested as a possible reason for the differences in mammalian granivory among deserts. For example, a diverse family of marsupials, the Argyrolagidae, used to be present in the South American deserts but became extinct some 1 to 2 million years ago. According to fossil remains, this family showed extensive skeletal convergence with the Heteromyidae, which may suggest that they were granivorous. Since their extinction, there has not been sufficient time for the evolution of another group of small mammals with the requisite physiological and morphological adaptations to be successful desert granivores (Mares and Rosenzweig 1978). The lack of certain trophic strategies as a consequence of extinction has also been proposed for the Australian deserts (Morton and Baynes 1985). However, this cannot explain the lack of granivorous species in the Karoo, where there is no evidence for the extinction of other small mammals (Kerley 1991; Kerley and Whitford 1994).

5.3.2
Herbivory

Strict herbivory is characteristic of a large number of medium and large mammals. Where it occurs, it is usually confined to the larger of the small mammals. This is so because small mammals simply cannot digest enough herbage to satisfy their energy demands, especially in the dry seasons when forage is often of poor nutritive value. Small mammals require more energy in proportion to their body mass than do large mammals (see Sect. 5.3.1). Even among ruminants this is evident: choice in the quality of diet can be very varied among herbivores. Most of the smaller ruminants, such as the dik-dik and suni, consume diets of high quality, which are easily digestible and rich in accessible plant cell contents (Hofmann 1989).

Nonetheless, the number of herbivores among small mammals is relatively numerous. Morton (1979) compared herbivory in the Australian and North American deserts and found the absolute numbers to be similar. He listed 18 species of desert herbivores in Australia, mainly marsupials (Macropodidae), and 19 species in North America, some of which are ruminants (Rumentia). Four rodent species were cited in Australia, all murids. There are more rodent herbivores in the North American deserts, and these include Geomyidae and a number of Cricetidae (*Neotoma* spp. and *Sigmodon* spp.). Morton (1979) also included as a herbivore *Dipodomys microps*, a heteromyid that can survive while only consuming the chenopod *Atriplex confertifolia* (Kenagy 1973b), but which is often classified as an omnivore.

Among rodents, herbivores are well represented in Dipodidae (bipedal rodents common in Asian deserts, particularly Turkestan, China and Mongolia), Cavidae in the South American deserts and Arvicolidae – voles common in a number of deserts – and Cricetidae (*Hypogeomys* and *Macrotarsomys* spp.) in Madagascar. *Psammomys obesus*, a gerbil of North Africa, only consumes chenopods such as *Atriplex halimus*, a plant high in salt content and low in energy yield. Larger herbivorous rodents are represented by the springhare *Pedestes* in South African deserts and by the porcupines *Erethizon* in North America and *Hystrix* in North Africa and Asia.

Fossorial rodent herbivores consume underground parts of plants and are well represented in world deserts: the gophers (Geomyidae) in North America, Ctenomyidae in South America, moles (Spalacidae) in North Africa, and rodent moles (Bathyergidae) in South Africa. All these rodent species have an adult body mass of over 100 g.

Thus, in general, about 50 g adult body mass is the lowest weight for a herbivore. A notable exception is *Malacothrix typica*, a small cricetid of 7–13 g (Mares 1993) or 22 g (Knight and Skinner 1981) that inhabits the Karoo Desert. How this mouse can survive on such a diet is unknown, but it is possible that it is highly selective in its green matter intake, or that it may feed on seeds as well (Knight and Skinner 1981).

Except among the Carnivora, which are small-sized mammals but larger than rodents, mammalian orders usually show a high proportion of herbivores. For example, most Lagomorpha weigh between 1.5 and 3.0 kg and are strictly herbivorous. These mammals are widely distributed, with Leporidae occurring in most deserts. In addition, Hyracoidea, which are found in Africa, weigh between 3 and 5 kg and consume only green vegetation.

5.3.3
Insectivory and Carnivory

Insects and meat are highly digestible and provide a high yield of metabolizable energy per unit intake. As a result, this dietary source is very desirable for small mammals, and indeed this trophic status is practised by shrews and bats that weigh only a few grams.

The greatest abundance of insectivorous species occurs in the Australian deserts. In a comparison between Australian and North American deserts, Morton (1979) listed 25 insectivores from 15 genera in Australia, but only 8 species from 6 genera in North America. Marsupials (Dasyuridae) were the most common, there being 7 desert-dwelling insectivorous genera. The North American insectivorous species are represented, among others, by two genera of Soricidae and the grasshopper mouse, *Onychomys* spp.

Morton (1979) offers several suggestions as to why this large difference in the number of insectivores exists between the two deserts. First, the relatively mild Australian winters provide a reliable supply of insects throughout the year, which is not the case in the much colder North American deserts. This contrasts with the prevalence of seed availability and granivory in these deserts. Secondly, North American deserts almost totally lack the order Insectivora. Except for two genera, shrews have not been able to penetrate desert regions in any part of the world.

Insectivora, however, are well represented in Africa and Asia by two species of hedgehogs (Erinaceidae) and three species of shrews (Soricidae) which inhabit the deserts of these continents. Moreover, two genera of Tenrecidae are endemic to the deserts of Madagascar (Stephenson and Racey 1993a,b). Desert foxes are carnivorous, although they are opportunistic and consume most food items available. Two species have been described as highly insectivorous, the 1 kg Blanford's fox, *Vulpes cana* (Ilany 1983; Mendelssohn et al. 1987), and the 4 kg bat-eared fox (Nel and Mackie 1990). Blanford's fox, however, also includes fruit in its diet (Geffen et al.1992a,b).

5.3.4
Omnivory

Small, omnivorous mammals are well represented by the New World desert Cricetidae and the Old World Muridae. In addition, the omnivorous Sciuridae are common in most deserts. For example, the antelope ground squirrel *Ammospermophilus leucurus* has a varied diet consisting of seeds, vegetation and insects. The proportions of the dietary items shift seasonally and are dependent upon nutritional requirements and availability (Karasov 1985). In particular, the content of nitrogen and water are important constraints on the diet consumed by these ground squirrels. Furthermore, in order to satisfy their energy demands, these squirrels must consume a diet that has a dry matter digestibility exceeding about 50%. It may also be that many more small mammals are omnivorous than was once thought (Murray and Dickman 1994a).

5.3.5
Dietary Preference and Selection Indices

Dietary selection can be divided at three levels: (1) selection among dietary items; for example, among plant species, between plants and invertebrates or between seeds and vegetation; (2) selection of components within a dietary item, for example, leaves, stems or buds within a plant species; and (3) selection within a component. The first level is usually determined by cafeteria trials in laboratory studies and by stomach analysis in field studies. The second and third levels are less common, in particular for small mammals, because of technical problems involved in doing the measurements.

Most measures of dietary preference depend on food availability (Cock 1978; Krebs 1989). The relative abundance (RA) of either a plant species or component of a plant (RA_i) can be calculated as the ratio between the dry matter offered of either the plant species or a component (DMO_i) and the total dry matter of plants offered (DMO_p) as follows:

$$RA_i = DMO_i/DMO_p.$$

The dietary ration (DR_i) consumed of either a plant species or a component of a plant can be defined as the ratio between its dry matter intake (DMI_i) and the total dry matter intake of plants (DMI_p). This ratio can be used as a measure of food preference and may be expressed as:

$$DR_i = DMI_i/DMI_p.$$

Several indices for selectivity have been suggested in which selectivity was defined as a function of RA and DR. This assumes, therefore, that selectivity is affected by the abundance of food items. The values of the indices do not necessarily indicate the proportion of the actual dietary fraction intake of a specific item, as the dietary item may be highly preferred but may comprise only a minor part of the total diet.

In comparing the suitability of different preference indices, Cock (1978) suggested that three criteria should be considered: (1) scale of the index: "It is preferable for both negative and positive preference to have finite scales, symmetrical about zero"; (2) adaptability of the index: "The adaptability of the method to include more than two prey types"; (3) range of the index: " Whether maximum index values are attainable at all combinations of prey densities." The most familiar and frequently applied index of relative selectivity is the ratio between DR_i and RA_i. This index, initially suggested by Savage (1931) as a measure of preference, is known as the forage ratio (FR; Jacobs 1974) and can be expressed as:

$$FR_i = DR_i/RA_i.$$

Values of FR are all positive when negative selection is represented by $0 < FR < 1$, positive selection by $1 < FR < \infty$, and the absence of selection by the value 1. The main shortcoming of this index is its asymmetry, and therefore criterion 1 is not fulfilled. Moreover, it is sensitive to the relative densities of food types (Jacobs 1974) and, therefore, also violates criterion 3.

To introduce a symmetrical distribution for positive and negative selection, Ivlev (1961) transformed the FR to produce symmetrical values for negative and positive selection about zero. Ivlev's selectivity index (Fig. 5.14), called the electivity index (E), is expressed as:

$$E_i = (DR_i - RA_i)/(DR_i + RA_i).$$

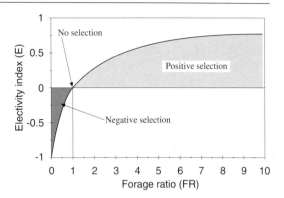

Fig. 5.14. Relationship between Ivlev's electivity index (*E*) and forage ratio (*FR*) in which the regions of positive and negative selection are indicated

Although this index is commonly used, it also violates criterion 3; that is, it is sensitive to the relative densities of the dietary items (Jacobs 1974).

These two indices cannot therefore be used to study the correlation between relative abundance of food and food selection. Jacobs (1974) modified them and used the index Q (presented as its logarithm) instead of FR as follows:

$$Q_i = [DR_i(1 - RA_i)]/[RA_i)1 - DR_i)],$$

and the index D instead of Ivlev's index as:

$$D_i = (DR_i - RA_i)/(RA_i + DR_i - 2RA_iDR_i).$$

Both these indices meet all three criteria; however, they differ in that log Q is more sensitive to changes at levels of high selection while D is more sensitive to changes at levels of low selection (Fig. 5.15.).

Dietary Plant Selection. Plant selection and its components were examined in three co-existing, desert gerbilline rodents that differ in body size, *Gerbillus henleyi* (8–12 g), *G. dasyurus* (20–33 g) and *Meriones crassus* (50–110 g). The three species were offered husked millet seeds of approximately 40 and 80% mainte-

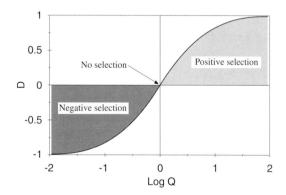

Fig. 5.15. Relationship between Jacobs' selectivity indices (*E*) D and log Q in which the regions of positive and negative selection are indicated

nance energy requirements and six plant species ad libitum. These plant species were dominant in the natural habitat of the rodents and consisted of *Atriplex halimus, Moricandia nitens, Zygophyllum dumosum, Lycium shawii, Anabasis articulata* and *Thymelaea hirsuta*.

All three rodent species showed a positive selection for *At. halimus* and *M. nitens*, a negative selection for *An. articulata* and *T. hirsuta*, and no selection for either *Z. dumosum* or *L. shawii* (Fig. 5.16). Within a plant, all rodent species preferred buds over all other components. Buds comprised approximately 3.5 and 5.6% of DM content for *Z. dumosum* and *M. nitens*, and the preference exhibited was 12 times their relative abundance in the diet of *G. dasyurus*. However, buds in total comprised less than 6% of the total DM of the plants offered, and less than 10% of the DM consumed. They could not therefore contribute substantially to the energy intake of the rodents.

Selection of different plant components within each plant species differed among rodent species. For example, the leaf:stem (L:S) intake ratio on a DM basis for *At. halimus* was 4.8, but 7.4, 4.0 and 1.7 for *M. crassus, G. dasyurus* and *G. henleyi*, respectively. These values indicate selection of leaves by *M. crassus*, no selection by *G. dasyurus* and selection of stems by *G. henleyi*.

Although the rodent species were offered the same plant species, gross energy of the consumed plants differed among them: 15.15 ± 0.02, 15.91 ± 0.04 and 16.55 ± 0.05 kJ g^{-1} DM for *M. crassus, G. dasyurus* and *G. henleyi*, respectively. This indicates that the smaller gerbil species select parts of the plant that have a higher energy content than do larger species. Furthermore, in all gerbil species, dry matter digestibility and digestible energy of the consumed diet decreased with an increase in level of plants consumed; the greatest decrease was in *M. crassus*, the least in *G. henleyi*, and it was intermediate in *G. dasyurus* (Fig. 5.17).

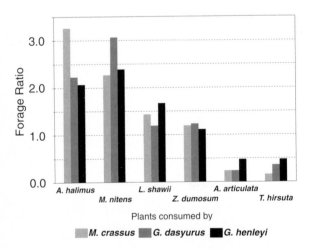

Fig. 5.16. The forage ratio of six plant species (*Atriplex halimus, Moricandia nitens, Zygophyllum dumosum, Lycium shawii, Anabasis articulata* and *Thymelaea hirsuta*) consumed by *Meriones crassus, Gerbillus dasyurus* and *G. henleyi*. (Kam, Khokhlova and Degen, unpubl.)

The difference among species in the energy consumed could be a consequence of plant manipulation. *M. crassus* consumes all the parts of the plants in toto with no noticeable manipulation, while *G. henleyi* seems to select among the components. For example, *At. halimus* stems not eaten by *M. crassus* are complete in contrast to the shredded stems left by *G. henleyi*. Similar shredding behaviour has been reported in *G. henleyi* when consuming *Hammada salicornica* (Khokhlova, Kam and Degen, unpubl.). In this latter study, *G. henleyi* also selected more energy-rich components than *M. crassus* did.

A high degree of manipulative dietary behaviour has been reported in the New World desert heteromyid *Dipodomys microps* and the Old World desert gerbilline *Psammomys obesus*. Both these rodents strip the salt-laden epidermis off *Atriplex* leaves before ingesting them (see Sec. 6.2.5). Reichman and Price (1993) suggested that the manipulation of plants before their consumption has a generic basis. This is supported by the study on the gerbils in that their manipulative behaviour differed among the species but was not affected by the level of seed offered.

5.4
Locomotion

Bipedal locomotion allows rodents to travel much faster than with quadrupedal locomotion. It has evolved in four families: Heteromyidae in North America,

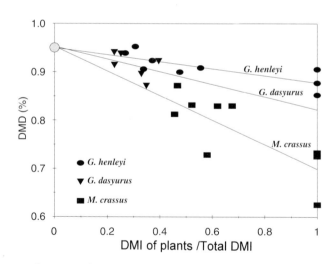

Fig. 5.17. Relationship between dry matter digestibility (DMD) and the proportion of dry matter intake (DMI) of plants of the total DMI (DMI$_p$) in *Gerbillus henleyi* (*circles*). *G. dasyurus* (*triangles*) and *Meriones crassus* (*squares*). For *G. henleyi*: DMD (%) = 0.95–0.07 DMI$_p$ ($n = 9$; SE$_a$ = 0.01; SE$_b$ = 0.02; SE$_{y\cdot x}$ = 0.02; r^2 = 0.26; P < 0.02); *G. dasyurus*: DMD (%) = 0.97–0.19 DMI$_p$ ($n = 6$; SE$_a$ = 0.05; SE$_b$ = 0.16; SE$_{y\cdot x}$ = 0.02; r^2 = 0.26; p < 0.29); and *Meriones crassus*: DMD (%) = 0.62–0.26 DMI$_p$ ($n = 9$; SE$_a$ = 0.06; SE$_b$ = 0.08; SE$_{y\cdot x}$ = 0.05; r^2 = 0.61; P < 0.02) (Kam, Khokhlova and Degen, unpubl.)

Pedetidae in southern Africa, Dipodidae in Europe, Asia and northern Africa and Muridae in Australia (Berman 1985). A generalized summary of small, bipedal mammals and their food habits in the deserts of the world is presented in Table 5.4. Quadrupedal species of small desert mammals are much more common than bipedal species. Most of the world's desert granivores are quadrupedal; however, most small, quadrupedal desert mammals are herbivorous or omnivorous (Table 5.5; Mares 1993).

Table 5.4. Bipedal small mammals and their general food habits in the deserts and semi-deserts of the world. (Mares 1993)

Desert	Bipedal			
	Granivore	Herbivore	Insectivore	Omnivore
North America	*Microdipodops, Dipodomys*	None	None	None
South America	Argyrolagids[a]	None	None	None
Australia	*Notomys*	Various marsupials	*Sminthopsis*	None
Sahara	*Jaculus*	*Allactaga*	*Elephantulus*	None
Southern Africa	None	*Pedetes*	*Elephantulus, Macroscelides*	None
Madagascar	None	None	None	None
Middle East-Thar	*Jaculus*	Various dipodids	None	*Salpingotus*
Turkestan	*Cardiocranius*	Various dipodids	None	*Pygeretmus, Salpingotus, Dipus*
Chinese/Mongolian	*Cardiocranius*	Various dipodids	None	*Salpingotus, Dipus*

[a]Extinct.

Table 5.5. Quadrupedal small mammals and their general food habits in the deserts and semi-deserts of the world. (Mares 1993)

Desert	Smaller quadrupedal			
	Granivore	Herbivore	Insectivore	Omnivore
North America	Perognathines	Various rodents	*Notiosorex, Onychomys*	Various rodents
South America	None	Various rodents	*Marmosa, Lestodelphis*	Various rodents
Australia	*Pseudomys*	Various rodents	Various marsupials	None
Sahara	Gerbillines	*Psammomys*	*Crocidura*	Various rodents
Southern Africa	Gerbillines, *Saccostomus*	Various rodents	*Crocidura*	Various rodents
Madagascar	None	*Hypogeomys, Macrotarsomys*	*Geogale*	*Echinops*
Middle East-Thar	Gerbillines, *Cricetulus*	*Rhombomys*	Soricids, erinaceids	Various rodents
Turkestan	*Meriones, Cricetulus, Phodopus*	Various rodents	*Crocidura, Diplomesodon, Hemiechinus*	None
Chinese/ Mongolian	*Meriones, Cricetulus, Phodopus*	Various rodents	*Hemiechinus, Crocidura*	*Spermophilus*

Much of what is known about bipedal locomotion is based on studies of the subfamily Dipodomyinae (*Dipodomys* and *Microdidops*) of the Heteromyidae (Brylski 1993; Hafner 1993), although this group differs, both in morphological features and in locomotion, from other bipedal rodents (K.A. Rogovin, pers. comm.). Heteromyidae are not highly specialized for bipedal running compared with the other groups; the Old World jerboas are most specialized and appear to be the most advanced of bipedal rodents (Fokin 1978; Berman 1985). Furthermore, within the bipedal species, jerboas are faster than kangaroo rats (K.A. Rogovin, pers. comm.). Differences in home range among the families have also been noted. Bipedal heteromyids have relatively small home ranges which are not correlated with their body sizes. In contrast, bipedal jerboas have extensive home ranges which are correlated with their body sizes, and they also have overlapping territories. This is made possible due to their specialization in fast locomotion (K.A. Rogovin, unpubl. data).

A number of hypotheses have been proposed for the evolution of bipedal locomotion, at least among the Heteromyidae, mostly based on functional and causal explanations. The "functional explanations explain morphology in terms of its purpose to the animal and disregard prior states, whereas causal explanations focus on prior morphological states from earlier ones" (Hafner 1993). The main hypotheses proposed for the evolution of bipedal locomotion are as follows: (1) As a means to avoid predators. Long ricochetal, erratic jumping is an effective means of escaping predators; however, quadrupedal pocket mice (subfamily Perognathinae) rely on the same response. Results, thus far, have been inconclusive as to whether bipedal animals suffer lower rates of predation than do quadrupedal animals (Hafner 1993; Kotler et al. 1994). (2) As a means to conserve energy during locomotion. It has been proposed that the elasticity of the muscles and tendons of the long hind legs would permit energy to be stored in one stride and released in the following. However, studies have shown that little or no energy is stored in the tendons. Oxygen uptake increases linearly with running speed in both *Dipodomys merriami* and *D. deserti* even when they are running bipedally, and their costs of locomotion were similar to that of other small quadrupedal mammals (Thompson et al. 1980). Furthermore, no difference in energy expenditure at any speed has been noted in kangaroo rats when they were using either bipedal or quadrupedal gaits (Thompson 1985), and the energy expenditure of bipedal heteromyids is similar to that predicted for mammals of their body size (Taylor et al. 1970). (3) As a means to facilitate the handling of and digging for food items, in particular seeds. Food gathering and handling are similar, however, in bipedal kangaroo rats and quadrupedal pocket mice (Hafner 1993). Furthermore, the cost of digging may be even greater for bipeds than for quadrupeds (Brown et al. 1988), and bipeds are poorer diggers than quadrupeds due to the mechanical disadvantage of their shorter lever-arms (Kotler et al. 1994). (4) An adaptation for harvesting food when distributed in patches in desert environments. It has been proposed that bipedal mammals mainly use open areas and that quadrupedal mammals use sheltered areas. However, recent studies have indicated that bipedal heteromyids use space more evenly than do quadrupedal desert rodent, implying that these rodents forage equally in sheltered and open areas.

Therefore, the hypotheses proposed regarding bipedal locomotion have not been substantiated. This led Hafner (1993) to question the functional and causal explanations of bipedal locomotion. Perhaps the answer was best summed up by Kotler et al. (1994) who stated that, "the advantages and disadvantages of bipedal locomotion discussed suggest that bipedal locomotion is just another characteristic with its various costs and benefits; it is not necessarily the ultimate morphology for desert rodents. Bipedal locomotion may infer superior ability to move rapidly between patches and superior ability to dodge predators." Unfortunately, few studies have been carried out on rodents other than the heteromyids, and as a result, our understanding of bipedal locomotion is largely based on these studies. Studies on more specialized bipedal desert species, such as those from the Old World, may show different responses that could be important in the understanding of bipedal locomotion.

CHAPTER 6

Water Requirements and Water Balance

"The major climatic factors that shape the hot deserts of the world are high temperatures and low and irregular rainfall. The primary physiological problems of life in such areas are directly related to heat and lack of water; other aspects of animal survival, e.g. food resources, predators, cold, reproduction, etc. do not differ in principle from similar problems in non-deserts areas."

(Schmidt-Nielsen 1975b)

Animals are composed mainly of water, it comprises about 60–80% of their body mass and over 98% of all body molecules (Macfarlane and Howard 1972). The large percentage is a consequence of the small size of water molecules compared with lipid, protein and carbohydrate molecules. Water is essential to the body as it acts as a solvent and is involved in a large number of body functions including temperature regulation, transport of metabolic products, hydrolytic reactions, excretion of waste material, and transport of sound and light within the ear and eye, respectively (Robinson 1957; Robbins 1983).

Most small desert mammals, in particular rodents, obtain their water requirements from the food that they consume and do not need to drink in addition (MacMillen and Christopher 1975). They have been classed roughly into three groups on the basis of the water content of their diets: (1) granivores, (2) omnivores and (3) herbivores/insectivores/carnivores. Granivorous rodents require the least amount of water of these three groups and have evolved efficient water-conserving mechanisms, notably high urine concentration, dry faeces and low evaporative water loss, that allow them to survive in deserts on seeds only (Schmidt-Nielsen 1964). In contrast, herbivorous, insectivorous and carnivorous rodents can afford to be more liberal with their water losses as they obtain proportionally more water from their food. They require drinking water in order to survive when they only consume seeds, at least under laboratory conditions. Omnivorous rodents are in an intermediate position as regards their water needs, but they too generally require water when they are only offered seeds (MacMillen 1983).

While most granivorous desert rodents can maintain a constant body mass on a diet of seeds alone without drinking water, other rodents cannot. When offered a diet of seeds only, the rate of body mass loss in rodents is related directly to their normal total water intake or the moisture content of their natural diet. Rodents that normally consume a moister diet lose body mass at a faster rate than do rodents that normally consume a drier diet (Schmidt-Nielsen 1964; MacMillen 1983). This has been reported both in New World (Fig. 6.1) and in Old World (Fig. 6.2) desert rodents. In addition, when deprived

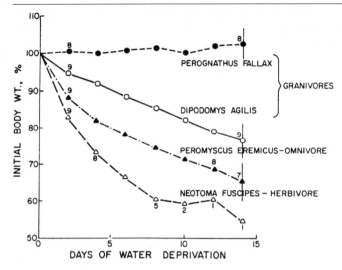

Fig. 6.1. Response of body mass to water deprivation in four species of co-existing southern Californian rodents that differ in their food habits. *Lines* connect the means of sample sizes as indicated by the numerals; *numerals* also indicate the number of animals surviving; *vertical lines* encompassing the means on day 14 are 95% confidence intervals. (MacMillen 1983)

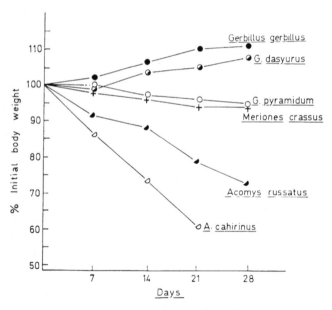

Fig. 6.2. Body mass changes of various Old World desert rodents living for 4 weeks on dry barley only. The animals were maintained at 30 °C and 30% RH. (Shkolnik and Borut 1969)

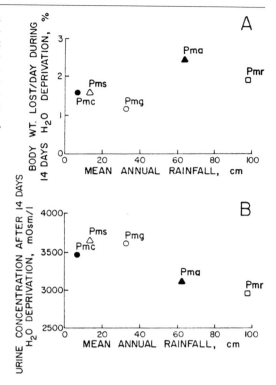

Fig. 6.3. A Relationship between body mass loss during water deprivation and annual precipitation at collecting sites for five Western subspecies of *Peromyscus maniculatus*. The *symbols* represent mean values from 10–15 animals. **B** Relationship between urine concentration during water deprivation and annual precipitation in five Western subspecies of *Peromyscus maniculatus*. The *symbols* represent mean values from 10–14 animals; subspecies, *a, austerus*; *c, cooledgei*; *g, gambelii*; *r, rubidus*; *s, sonoriensis*. (MacMillen 1983)

of drinking water and offered only a dry diet, subspecies of *Peromyscus* that inhabit xeric areas lose body mass at a slower rate than do those from more mesic areas (MacMillen 1983; Fig. 6.3).

6.1
Water Input

Dietary water available to an animal comes in the form of performed water (free water in food) and metabolic water (water produced upon the oxidation of hydrogen). In general, granivores obtain relatively little water in the form of preformed water, and most of their water is obtained from metabolic water. The opposite is true for non-granivores (Table 6.1).

6.1.1
Water for Granivores

Preformed water in desiccated or air-dried seeds is generally low, as the moisture content of seeds can only average about 3% of the fresh mass (Morton and MacMillen 1982; Nagy and Gruchasz 1994). Surface seeds usually have a content of under 10%, although higher percentages have been recorded after rain

Table 6.1. Water balance in a desert granivorous, omnivorous and herbivorous rodent consuming a near natural diet. An omnivorous rodent (*Ammospermophilus leucurus*) is included. Water was not available to the animals

Species	*Dipodomys merriami*[a]	*Acomys cahirinus*[b]	*Ammospermophilus leucurus*[c]	*Psammomys obesus*[d]
Dietary habit	Granivore	Omnivore	Omnivore	Herbivore
Food type in study	Millet seeds	Millet seeds, snails	Crickets	*Atriplex halimus*
Body mass (g)	35	49	88	135
Dry matter intake (g day^{-1})	3.57	3.14	6.55	12.11
Water influx (ml day^{-1})	2.13	4.99	14.21	46.47
Metabolic water (ml day^{-1})	1.91	1.29	2.42	3.45
(% water influx)	89.7	25.9	17.0	7.4
Preformed water (ml day^{-1})	0.22	3.70	11.79	42.84
(% water influx)	10.3	74.1	83.0	92.2
(% fresh matter)	5.8	59.9	68.5	71.4
Water efflux (ml day^{-1})	2.13	4.99	14.21	46.47
Faecal water (ml day^{-1})	0.07	0.44	1.30	7.57
(% water efflux)	3.2	8.8	9.2	16.3
Urine water (ml day^{-1})	0.49	2.11	6.73	18.90
(% water efflux)	23.0	42.3	47.4	41.6
Evaporative water loss (ml day^{-1})	1.57	2.44	6.26	20.0
(% water efflux)	73.7	48.9	44.0	43.1

Data calculated/taken from: [a]Schmidt-Nielsen (1964). [b]Kam and Degen (1991). [c]Karasov (1983). [d]Kam and Degen (1988).

(Morton and MacMillen 1982). However, seeds are hygroscopic and, therefore, increase their moisture content when stored in a humid environment. Such is the case with many seeds that are collected by rodents and stored in underground burrows where moisture gradients may increase to 100% relative humidity (Studier and Baca 1968; Kay and Whitford 1978). The hygroscopic characteristic of seeds could be used to advantage by desert mammals in maintaining their water balance. Reichman et al. (1986) reported that the kangaroo rat *Dipodomys spectabilis* can distinguish differences in relative humidity and select areas of high relative humidity in which to store their seeds.

Morton and MacMillen (1982) examined the effect of relative humidity on the moisture content of 10 species of seeds. The desiccated mass of these seeds averaged 2.9% of the fresh mass (Table 6.2). After 54 days of incubation at different relative humidities, the moisture content of the seeds increased to 7.2% when maintained at a relative humidity of 6%, to 10.0% at a relative humidity of 45% and to 15.5% at a relative humidity of 76% (Table 6.2). The study also revealed that: (1) hygroscopic water uptake was rapid and in most cases reached equilibrium within 10 days, and (2) larger seeds took up proportionately less water than did smaller seeds. Therefore, granivores that cache their seeds for a week or so can consume seeds with a moisture content of 15 to 20%, or even higher, when the ambient relative humidity approaches 100%.

Table 6.2. The moisture content of test seeds when dessicated, and their total water content after 54 days exposure to three levels of humidity. Values of the latter are means from 10 and 25 °C. (Morton and MacMillen 1982)

Species and reference no.	No. of seeds used per test	Average dry mass of individual seed (mg)	Moisture content of desiccated seeds (% wet mass)	Moisture contents at experimental humidities (% final wet mass)		
				6%	45%	76%
1. *Descurainia pinnata*	100	0.1	6.3	12.1	14.2	20.7
2. *Amaranthus blitoides*	70	0.2	3.9	11.0	14.2	19.5
3. *Salvia columbariae*	40	0.7	2.4	5.5	8.0	11.7
4. *Phacelia distans*	30	0.8	2.2	7.3	10.6	17.9
5. *Viguiera deltoidea*	25	0.6	4.0	7.9	8.8	15.3
6. *Chanenactis fremontii*	20	0.7	3.5	11.0	11.4	17.2
7. *Bromus rubens*	15	2.1	2.5	6.7	12.2	17.8
8. *Opuntia* sp.	10	21.5	0.7	2.8	7.1	11.1
9. *Isomeris arborea*	10	29.1	1.7	4.1	6.8	13.8
10. *Yucca schidigera*	5	112.7	1.7	3.1	6.3	10.0
mean ± SD	–	–	2.9 ± 1.6	7.2 ± 3.4	10.0 ± 3.0	15.5 ± 3.7

Can rodents distinguish between seeds containing different amounts of pre-formed water, and if so, do they choose seeds on the basis of water content? Furthermore, does the water requirements of the rodent influence the selection of seed with different preformed water content? Frank (1988a, b) asked these questions when examining kangaroo rats (*Dipodomys* spp.). He offered *D. merriami* husked barley seeds with different moisture contents. Moisture content differences were obtained by incubating seeds at 35 °C and at 100% relative humidity over different periods of time. The seeds were then colour coded according to moisture content and offered to the rodents. Energy yield of the dry matter and nutritional composition of control and incubated seeds were similar. Therefore, the selection of seeds by the rodents must be on the basis of water content. In all cases, the rodents consumed more dry matter of the seeds with higher preformed water content and were able to respond "to a difference in water content that was at most 10.28%" (Frank 1988a). Before consumption, the rodents would gather all the seeds in one pile. Selection was made from this pile, which suggests that discrimination in water content was made on a seed-by-seed basis: thus the ability to sense differences in water content between individual seeds was involved.

In a study of *D. spectabilis*, Frank (1988b) found that this species could not maintain body mass when consuming husked barley seeds with a preformed water content of 9.63%, but could maintain body mass when consuming seeds with one of 10.59%. He then concluded that *D. spectabilis* requires a diet of barley seeds with 10–11% water content, at least under laboratory conditions, to satisfy its water requirements. These rodents were presented with husked barley seeds containing different amounts of preformed water which had been prepared as described above. The moisture contents of the seeds were above and below those required for water balance. The studies showed that "these rodents can sense a 6.05% difference in water content and always prefer the moistest seeds available, even when selecting among seeds containing over twice the minimal water necessary for a positive balance. These results indicate that kangaroo rats do not ingest only enough preformed water to meet their requirements but instead maximize water intake through their seed preferences" (Frank 1988b).

Studies on two granivorous Australian desert rodents, the spinifex hopping-mouse (*Notomys alexis*) and the sandy inland mouse (*Pseudomys hermannsburgensis*) "provided some indication that the water content of food items may underlie diet selection" when only different seed species were offered, but the results were not conclusive (Murray and Dickman 1994b). However, when these granivores were offered invertebrates as well as fungi, in addition to seeds they preferred the invertebrates. Murray and Dickman (1994b) concluded that "water and protein may be important dietary components" in diet selection, and their flexible omnivorous diet was advantageous to these rodents in allowing them to survive in Australian deserts.

Metabolic water gained by granivores is dependent upon the composition of the seeds they consume and the relative humidity of the ambient air. At low relative humidity, carbohydrate oxidation results in a net metabolic water gain, lipid oxidation results in a net water loss, and protein oxidation results in a still larger net water loss, mainly as a consequence of urinary water loss. At high relative humidity, carbohydrate oxidation still produces a net metabolic water gain, but lipid metabolism produces the largest net metabolic water gain. Protein oxidation still results in a large net metabolic water loss, but slightly less than at low relative humidity (Frank 1988c; Table 6.3).

Therefore, to maximize net metabolic water gain under conditions of low relative humidity, carbohydrate intake should be maximized, while protein and lipids should be minimized, only enough being ingested to meet nutritional and energy needs. Under conditions of high relative humidity, a high lipid in-

Table 6.3. Net amounts of metabolic water produced by the oxidation of various nutrients by *Dipodomys spectabilis* breathing dry (RH = 20%) or humid (RH = 66%) air at 25 °C. All units are expressed per gram of nutrient oxidized. (Frank 1988c)

Nutrient class	Water produced by oxidation[a] (g/g)	Oxygen required for oxidation[a] (ml/g)	Dry air		Humid air	
			Evaporative water lost (g/g)	Net metabolic water yield (g/g)	Evaporative water lost (g/g)	Net metabolic water yield (g/g)
Starch	0.556	828	0.472	0.084	0.270	0.286
Hexose	0.600	746	0.425	0.175	0.243	0.357
Lipid	1.071	2019	1.151	− 0.080	0.658	0.413
Protein	0.396	967	0.551	− 1.613[b]	0.315	− 1.377[b]

[a] From Peters and VanSlyke (1946), Schmidt-Nielsen (1964).
[b] Including 1.458 g of water lost through urination (Schmidt-Nielsen 1964).

take should be maximized, and protein intake should just meet nutritional demands. The importance of the protein content of seeds in determining the water balance of rodents was shown by *Dipodomys spectabilis* when offered a diet of sunflower seeds (Frank 1988c). These seeds are characterized by their high protein and lipid contents. When the rodents were maintained at 25 °C and a relative humidity of 20% and offered these seeds (moisture content of 4.78%), they lost body mass. However, when maintained at an air temperature of 25 °C and a relative humidity of 65% and offered seeds of 19.0% moisture content, the rodents gained body mass. This demonstrated that the loss of body mass in dry air was due to a negative water balance. However, it failed to show whether the increased net metabolic water gain under the conditions of higher relative humidity and/or the increased preformed water was sufficient for a positive water balance and allowed a gain in body mass.

Studies were made on a number of Heteromyidae to examine the efect of different environmental conditions (mainly air temperature and relative humidity) and different diets (mainly high and low protein content) on net metabolic water balance, and to examine whether desert rodents select dry diets on the basis of metabolic water production. Hulbert and MacMillen (1988) offered three, different-sized, co-existing, heteromyid rodents, *Dipodomys merriami* (39 g), *Perognathus fallax* (23 g) and *P. longimembris* (9 g), either wheat or hulled sunflower seeds at an air temperature of 15, 20, 25 or 30 °C. The net metabolic water gain with wheat was much higher than with the high-protein sunflower seeds. A change in body mass was used to indicate the water balance of the rodents. These authors found that: (1) all three species were able to maintain a more positive water balance on wheat than on sunflower seeds; (2) for all species there was an air temperature above which body mass could not be maintained, this air temperature being higher with the wheat diet than with sunflower seeds; and (3) the water regulatory efficiency was inversely related to body mass.

In further studies, the three heteromyid species were offered both wheat and sunflower seeds at an air temperature of 15, 25 or 30 °C. *D. merriami*, the largest one maintained its body mass at 15 °C but lost body mass at 25 and 30 °C. It showed no preference for either type of seed at any of the air temperatures, regardless of whether it was gaining or losing body mass, and consumed about 50% of each one. Both *Perognathus* spp. were able to maintain their body mass at all air temperatures. *P. fallax* consumed about 50% of each seed type at 15 °C but increased its intake of wheat to 56–57% of the total seed intake at both 25 and 30 °C. *P. longimembris*, the smallest of the three species, always consumed a greater proportion of sunflower seeds at all air temperatures, but there was a significant decrease in sunflower consumption at 25 °C relative to 15 and 30 °C.

Frank (1988c) also demonstrated that *Dipodomys spectabilis* selects dry diets that result in the highest net water gain. Under conditions of low relative humidity, hydrated *D. spectabilis* preferred a diet consisting of high lipid and carbohydrate and intermediate protein. When water-stressed, this species preferred a low protein diet, and its preference for high lipid diets decreased. At high relative humidities, *D. spectabilis* always preferred diets high in lipid content. When normally hydrated, these rodents preferred a diet of 15% protein over 10 and 20% protein in one study, and there was no difference in selection between 10 and 20% protein in another study. However, when water-stressed, they preferred a diet of 10 and 15% protein over 20% and a diet of 10% protein over 15% (Fig. 6.4).

6.1.2
Water for Non-Granivores

Omnivores must adjust their diet to fulfill their water, energy and nutrient (mainly protein) requirements. Extensive studies have been undertaken on the antelope ground squirrel (*Ammospermophilus leucurus*) in the Mohave Desert to demonstrate seasonal shifts in diet to optimize intake (Karasov 1983, 1985). Water influx parallels the moisture content of the plants consumed by the squirrels. Minimum water requirements of outdoor caged squirrels, as calculated by the linear regression equation of change in body mass on water influx, was 68 ml kg^{-1} day^{-1}. To satisfy water needs these squirrels must consume plants with a moisture content of at least 45% (0.9 mlg^{-1} dry mass) preformed water. Such is the case in spring when the squirrels are basically herbivores. However, in autumn and winter, the moisture content of the plants generally declines to below 40%, at which time *A. leucurus* includes arthropods in its diet. These play an essential role in water balance. Following winter rains and the emergence of green vegetation, the diets of these squirrels include 1–2% arthropods. This percentage increases to 20–30% in the autumn, prior to the winter rains.

Changes in body mass related to preformed and total water intake were examined in two co-existing omnivorous desert rodents, *Acomys cahirinus* and *A. russatus* (Kam and Degen 1993b). These murids eat seeds, green vegetation and animal material, in particular snails. In the study, the rodents were offered dry pellets ad libitum and different quantities of snails. Regression analyses for

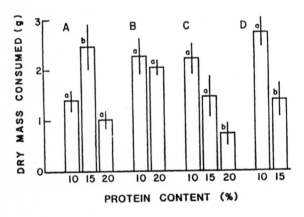

Fig. 6.4. Histograms indicating the mean ($\pm$ SE) intake of each diet in protein preference experiments. *Abscissa* indicates diets that are controls (10% protein), medium protein (15%) and high protein (20%). *Bars with different letters* within an experiment are significantly different ($P < 0.05$) from each other. (Frank 1988c)

A. cahirinus showed that in order to maintain body mass this species required 2.1 ml day^{-1} (4.7% body mass) preformed water and a diet that contained 46.5% water content. Regression analyses of change in body mass on preformed and dietary water intakes for *A. russatus* were not significant, but it was estimated that this species required 1.0 ml day^{-1} (2.3% body mass) and a diet that contained approximately 27% water content. To fulfill water needs, *A. russatus* requires about three snails and *A. cahirinus* about six snails per day. When given a choice, these murids adjust their diets as necessary to maintain their water balance.

6.2
Water Output

Water is lost from an animal via evaporation from the skin and respiratory surfaces, urine, faeces, and saliva (Table 6.1; Fig. 6.5). During lactation, females also lose water via milk (Fig. 8.5). Small desert mammals, in particular granivorous rodents, are known for their ability to reduce water losses through these avenues.

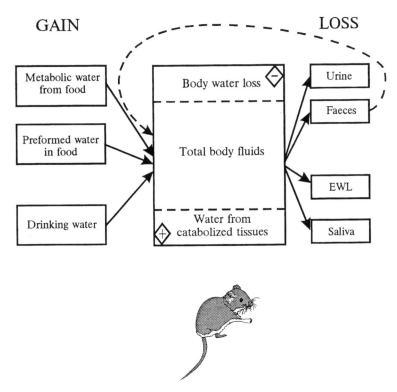

Fig. 6.5. Water balance in a mammal depicting sources of water input (gain) and water output (loss). *EWL* Evaporative water loss

6.2.1
Evaporative Water Loss

Small desert mammals cannot remain exposed to hot summer conditions for lengthy periods without resorting to evaporative water cooling for heat dissipation. Their relatively large ratio of surface area to body mass compared with larger animals would force them to use large volumes of water which would quickly cause dehydration. They do not possess sweat glands, and cutaneous water loss occurs by diffusion through tissues. Emergency thermoregulation is achieved by salivating on the neck and chest (Sect. 6.3).

Small desert mammals are characterized by low evaporative water loss (EWL) compared with mesic species. This is the result of a reduction in both cutaneous and respiratory water losses. Respiratory evaporation, which can be a major route of water loss, is a function of: (1) the difference between the water content of expired and inspired air; (2) the volume of air passed over the respiratory surfaces; and (3) the area of respiratory surfaces. However, much of the exhaled water is recovered as a result of a counter-current heat exchange in the nasal passages (Schmidt-Nielsen 1964). The nasal cavity features a series of thin, often convoluted sheets of bone, the turbinals. The dorsal and posterior portions of the nasal passages divide the region into a series of blind recesses that are partially protected from respiratory currents. These turbinals greatly increase the surface area of the tissue and are covered mainly by olfactory epithelium. The turbinals in the anteroventral portion of the nasal cavity, the maxilloturbinals, are covered with non-olfactory, respiratory tissue and are located directly in line with the respiratory nasal cavity. These are "typically complex structures, consisting of elaborately scrolled, folded, or finely branched lamellae" (Hillenius 1992). The maxilloturbinals have been associated with the recovery of respiratory water by desert mammals. During inhalation, water evaporates from the moist nasal surfaces and cools the surrounding tissue. As it passes through the convoluted tubules of the maxilloturbinals, the exhaled air, saturated with water at lung temperature, is cooled to temperatures near to, or even slightly lower, than that of inhaled air. Water is thus exhaled below body temperature, and some of the water vapour condenses on the epithelial surface and is recovered during inspiration. Depending on the air temperature and relative humidity, about 0.67–0.75 of the water vapour added to the inspired air is recovered in the nasal passages (Collins et al. 1971).

Heteromyids and other desert mammals use this mechanism to conserve body water (Table 6.4). However, complex turbinates are found in virtually all mammals (Hillenius 1992), and all mammals show a distinct temperature gradient decreasing from the proximal cribiform plate to the distal openings of the external nares (Schmid 1976). Furthermore, the turbinates substantially reduce respiratory water loss in a number of non-desert mammal species as well (Table 6.4; Hillenius 1992). For example, the laboratory rat is just as efficient in cooling exhaled air as are desert heteromyids (K. Schmidt-Nielsen and Schmidt-Nielsen 1952). Hillenius (1992) concluded that "turbinates did not evolve

Table 6.4. Mammalian orders and families in which reduced expired air temperatures have been recorded. The number of species studied in each family in parentheses. (Hillenius 1992)

Order	Family	Sources
A. Desert animals		
Rodentia	Heteromyidae (2)	Jackson and Schmidt-Nielsen (1964); Schmidt-Nielsen et al. (1970); Collins et al. (1971)
	Muridae (1)	Withers et al. (1979b)
Artiodactyla	Giraffidae (1)	Langman et al. (1979)
	Bovidae (6)	Langman et al. (1979); Kamau et al. (1984)
	Camelidae (1)	Schmidt-Nielsen et al. (1981)
Perissodactyla	Equidae (1)	Langman et al. (1979)
B. Non-desert animals		
Marsupialia	Didelphidae (1)	Hill (1978); Hillenius (1992)
Insectivora	Soricidae (2)	Schmid (1976)
Chiroptera	Vespertilionidae (3)	Schmid (1976)
Rodentia	Muridae (10)	Jackson and Schmidt-Nielsen (1964); Getz (1968); Schmid (1976); Bintz and Roesbery (1978); Withers et al. (1979a); Edwards and Haines (1978); Hillenius (1992)
	Sciuridae (10)	Schmid (1976); Bintz and Roesbery (1978); Withers et al. (1979a) Welch (1984); Hillenius (1992)
Lagomorpha	Leporidae (2)	Schmid (1976); Caputa (1979); Hillenius (1992)
Carnivora	Mustelidae (3)	Schmid (1976); Withers et al. (1979a); Hillenius (1992)
	Canidae (2)	Verzar et al. (1953); Schmid (1976)
	Phocidae (1)	Folkow and Blix (1987)
	Otariidae (1)	Huntley et al. (1984)
Artiodactyla	Cervidae (1)	Blix and Johnsen (1983); Langman (1985)
Primates	Hominidae (1)	Cole (1953); Proctor et al. (1977)

primarily as an adaptation to particular environmental conditions, but in relation to high ventilation rates, typical of all mammals". The volume of respiratory water loss, therefore, is related to the volume of air passing over the respiratory system. As the metabolic rate of xeric mammals is lower than that of mesic species and assuming that the efficiency of oxygen extraction is similar for the two groups, less air must pass over the respiratory surfaces, and less respiratory water is lost (French 1993).

Few studies are available that have measured the contribution of respiratory and cutaneous water to total evaporative water loss. In laboratory rats (Tennant 1945) and *Peromyscus maniculatus sonoriensis* (Chew 1955), cutaneous water loss amounts to approximately 50% of the total evaporative water loss. However, in the heteromyid *Dipodomys merriami*, only 16% of evaporative water loss is the result of cutaneous water loss (Chew and Dammann 1961).

MacMillen and Hinds (1983) developed a model based on the ratio of metabolic water production (MWP) to EWL to describe *water regulatory efficiency* in granivores. Both EWL and MWP are related to air temperature and respiratory function. Below thermal neutral conditions, MWP increases and thus is negatively related to air temperature, whereas EWL is either positively related to, or independent of, air temperature. It was therefore predicted that, for each rodent species, an air temperature exists in which MWP exceeds EWL. This would allow these granivorous rodents to be independent of drinking water. The air temperature at which MWP = EWL was taken as indicating water regulatory efficiency. Species with a high air temperature at equality are more efficient than those with a lower air temperature at equality. The composition of the diet (protein:lipid:carbohydrate ratio) is important in determining this air temperature since carbohydrates provide the maximum MWP to total water losses, and proteins provide the minimum.

Thirteen species of Heteromyidae consuming millet seeds were used to test the model. The semi-logarithmic relationship between MWP/EWL and air temperature showed that a ratio of 1 was attained at a higher air temperature with a decreasing body mass (Fig. 6.6A). The regression line of air temperature where MWP = EWL (T_a at MWP = EWL) on body mass (m_b) took the form (Fig. 6.6B):

$$T_a \text{ at } (MWP = EWL) = 29.682 \, m_b^{-0.137}.$$

This equation indicates that water regulatory efficiency increases with decreasing body mass. However differences were found in that the smaller *Perognathus* (n = 5; body mass = 8–29 g) conforms to the equation, whereas the larger *Dipodomys* (n = 4; 64–105 g) does not. In the latter, water regulatory efficiency was independent of body mass and averaged 18.1 °C, which is a value similar to that of the largest *Perognathus*, about 30 g (Fig. 6.7). From their study, MacMillen and Hinds (1983) concluded that "initially selection favored a decrease in body mass with a concomitant increase in water regulatory efficiency and reduction in absolute energy needs in heteromyids, tracking progressive aridity during the Tertiary. In *Dipodomys*, the option of bipedality was adopted, apparently freeing them from energetic constraints imposed strictly by mass, coupled with an intermediate and fixed level of water regulatory efficiency that dictates use of seeds with high metabolic water yields. The quadrupedal *Perognathus* have retained a mass-specific water regulatory efficiency, ensuring maintenance of both water and energy balance on a broad array (with respect to protein:lipid:carbohydrate and metabolic water yield) of seeds, when their more-limited locomotor powers are consistent with seed availability; torpor is a trade-off, enhancing survival during energetically demanding periods."

6.2.2
Salivary Water Loss

A number of small mammals neither sweat nor pant during heat stress. Instead, they spread saliva over vascularized surfaces which provides a means of heat loss through evaporative cooling.

"In the kangaroo rat it seems that salivation is an emergency reaction which occurs when the body temperature approaches the lethal limit. A similar response occurs in ground squirrels and in the

Fig. 6.6. A model of water regulatory efficiency in heteromyid rodents. **A** The hypothetical semi-logarithmic relationship between metabolic water production/evaporative water loss (MWP/EWL) and ambient temperature (T_a) for a size range of heteromyids all oxidizing the same kinds of seeds. *Horizontal line* designates the level at which MWP = EWL; the *intercept between the horizontal line and the diagonal line* for each species represents the relative value of water regulatory efficiency for each species (T_a at MWP = EWL), with the highest values being the most efficient. **B** The hypothetical double logarithmic relationship between water regulatory efficiency (T_a at MWP = EWL) and body mass in heteromyid rodents while oxidizing the same kinds of seeds. Each value represents one heteromyid species, and the *diagonal line* connects those values. (MacMillen and Hinds 1983)

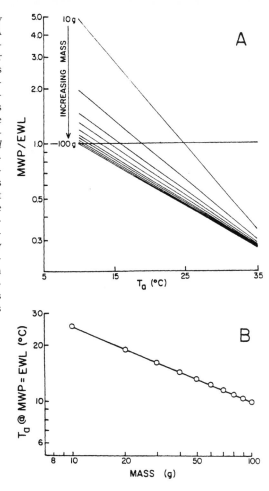

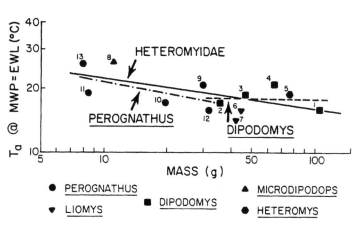

Fig. 6.7. The double logarithmic relationships between water regulatory efficiency (T_a at MWP = EWL) and body mass in five genera of heteromyid rodents. Regression lines are provided for all species (—) and for *Perognathus* spp. (–·–) and both have significant slopes. *Horizontal line* (---) represents the average value for each species; the regression equations are: T_a at MWP = EWL) = 29.682 $m_b^{-0.137}$ in all 13 heteromyid species, T_a at (MWP = EWL) = 31.078 $m_b^{-0.166}$ in *Perognathus* spp. The *geometric symbols* represent mean values for each species. (MacMillen and Hinds 1983)

Egyptian rodent *Dipus*. It can only give a temporary defence against the temperature rise, but its biological significance seems evident. If an animal temporarily is unable to retreat to its burrow, it may be essential to survival that it has some means of resisting a temperature rise. Its water resources are not sufficient to carry on active heat regulation by evaporation for any length of time, but an emergency reaction of this type may still be the difference between life and death."

(Schmidt-Nielsen 1964)

Saliva spreading and its effect on thermoregulation has been described best in rats (Hainsworth 1967, 1968; Hainsworth et al. 1968; Hainsworth and Stricker 1970; 1972), but is has also been observed in other rodents, several marsupial species and bats, particularly, Megachiroptera (see discussion by Hainsworth and Stricker 1970).

In rats, salivary water loss, as a proportion of total evaporative water loss, increases with a rise in air temperature. It comprises approximately 50% of total water evaporated at an air temperature of 36 °C, and 65% at an air temperature of 44 °C (Hainsworth and Stricker 1970).

Saliva from the submaxillary gland, but not from the parotid gland, is crucial to the thermoregulatory responses of male and female rats when exposed to heat stress. Rats lose their ability to regulate body temperature following surgical removal of the submaxillary glands, but not following surgical removal of the parotid glands. This was attributed to the greater volume of saliva secreted by the submaxillary glands than by the parotid glands once salivary secretion increases during heat stress (Hainsworth and Stricker 1972).

Golden hamsters (*Mesocricettus aureatus*), widely distributed in deserts and steppes of Asia Minor and Caucasus, also depend on saliva spreading for cooling during heat stress (Rodland and Hainsworth 1974). At 30 °C, saliva secretion in these hamsters was negligible, and of the EWL, 27% was respiratory water and 73% cutaneous. At 40 °C, there was a fivefold increase in evaporative water loss when compared with the loss at 30 °C, and this was mainly a result of increased salivary water spreading. Of the total evaporative water loss, 70% was comprised of salivary water, 22% was respiratory water, and only 8% was cutaneous water.

Volumes of saliva secretion were similar in either submaxillary or parotid (partially) desalivated hamsters, and they were able to produce 74–83% of the saliva volume of controls. At 40 °C, normal hamsters could dissipate more than 100% of their total heat production through evaporative water loss, and partially desalivated hamsters could dissipate approximately 90% of their total heat production, but for completely desalivated hamsters only 25% was possible.

Saliva secretion in submaxillary and parotid desalivated hamsters differed from that reported in rats where removal of the submaxillary glands greatly reduced EWL, but the effect of parotid desalivation was negligible. This may be a consequence of the ionic composition of the saliva in the two glands. In the rat, submaxillary saliva has a low sodium concentration and is hypotonic to plasma, whereas parotid saliva is nearly isotonic. In the hamster, saliva from both glands is hypotonic, with a low sodium concentration. Rodland and Hainsworth (1974) suggested that the sodium concentration "may be of primary importance in determining the importance of individual salivary glands as thermoregulatory effectors. When both the submaxillary and parotid glands produce a low sodium saliva, as in the hamster, both glands seem to be freely interchangeable as thermoregulatory effectors. However, when saliva from one set of glands has a high sodium concentration those glands appear to have a negligible role in thermoregulation." Nonetheless, normal and submaxillary desalivated hamsters were able to regulate body temperature to a greater degree than completely and parotid desalivated hamsters.

The salivary secretion rate of non-acclimatized fat sand rats (*Psammomys obesus*) is approximately 2.5 times that of the rat (Horowitz and Mani 1978). During 28 days of exposure to 34 °C, fat sand rats always secreted more saliva than rats. Initial saliva secretion was 10 µl g^{-1} 15 min^{-1} for *P. obesus* and 4 µl g^{-1} 15 min^{-1} for rats. In fat sand rats, saliva volume decreased by 44% during the second day and then increased gradually to near, but still below, normal levels. Rats maintained a constant saliva volume throughout (Fig. 6.8). The saliva was isotonic to plasma fluid in the fat sand rat but hypotonic to plasma fluid in the rat.

Horowitz and Mani (1978) concluded that the ability of *P. obesus* to secrete large volumes of saliva is probably related to its diurnal habits. It is sporadically exposed to high daily air temperatures, and therefore "*P. obesus* has developed emergency cooling mechanisms to a greater extent that the rat. Also, this animal can give preference to water conservation, whereas *R. norvegicus* must give preference to emergency cooling." Similarly, the diurnal *Citellus leu-*

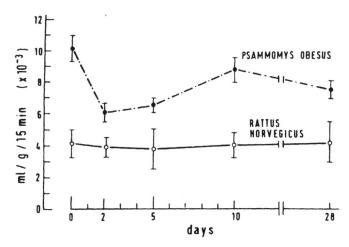

Fig. 6.8. Mean ($\pm$ SE) mixed saliva volume of *Psammomys obesus* (n = 9) and *Rattus norvegicus* (n = 10) during 28 days of exposure to heat of 34 $\pm$ 1 °C. (Horowitz and Mani 1978)

curus that inhabits the Mohave Desert has the ability to secrete large volumes of saliva compared with nocturnal desert rodents (Hudson 1962).

6.2.3
Kidney Function and Electrolyte Excretion

Mammalian kidneys are important in the regulation of body fluid volumes and electrolyte concentrations and in the elimination of nitrogenous wastes. The kidney is composed of an outer cortex and medulla; the medulla, in turn, is composed of the outer medulla, an outer zone, and the inner medulla. The outer medulla is subdivided into an outer stripe and inner stripe; the latter forms the papilla which protrudes into the pelvis.

The single unit within the kidney is called a nephron. It is composed of a Malpighian body, a proximal convoluted tubule, loop of Henle and a distal collecting tubule. The Malpighian bodies and the proximal and distal convoluted tubules are contained in the cortex of the kidney; the loop of Henle and the collecting ducts are contained in the medulla. The distal collecting tubule leads fluid to the collecting duct. A small artery runs into the Malpighian body and breaks up into a network of capillaries known as the glomerulus.

There are two types of nephrons which differ in the length of their loop of Henle: (1) short loops in which the renal capsule lies near the surface of the cortex (superficial nephrons), and the turn of the loop to the ascending limb lies in the outer medulla. Thin descending limbs are present but thin ascending limbs are not; and (2) long loops in which the renal corpuscle lies at the junction of the cortex and medulla (juxtamedullary nephrons), and the turn of the loop lies in the inner medulla. These nephrons have both thin descending limbs and thin ascending limbs. Short loops are generally more homogeneous in their length than are long loops; long loops turn back at different levels within the inner medulla. Kidneys of some species have only short loops and therefore have no inner medulla. Kidneys of this type are represented by mammal species such as the pig, beaver and muskrat, and all have poor urine-concentrating ability. Only long loops have been found in the cat and dog. Unlike what might be expected, their urine-concentrating abilities are no better than average. Most species have both short and long loops of Henle, and there does not appear to be any relationship between the ratio of short to long loops of Henle and the urine concentrating ability of an animal. In fact, some rodents with the ability to highly concentrate their urine have more short than long loops (Kriz 1981).

Vascular bundles, known as descending and ascending vasa recta, run parallel to the loops of Henle and are permeable to water and solutes. These vascular bundles can fuse in the inner stripe of the outer medulla to form secondary vascular bundles, as reported for the jerboa and *Meriones*, or even to form giant bundles, as in fat sand rats. In contrast, the bundles can be primary only and not fuse into secondary bundles, as in gundis (*Ctenodactylus vali*; Rouffignac et al. 1981). The nephrons may be surrounded by vasa recta, as is the case for

Psammomys obesus, or have no nephrons inside the vascular bundles, as in the dog (Schmidt-Nielsen 1979).

The formation and excretion of concentrated urine are achieved by two processes: (1) ultrafiltration of the plasma in the glomerulus, where substances of large molecular size such as proteins are filtered out; and (2) reabsorption of water as well as needed substances such as glucose, amino acids and vitamins, from the filtered fluid as it passes through the loop of Henle and the collecting duct. The kidney can process large volumes of fluids in this manner and reabsorb more than 99% of the filtered volume.

For ultrafiltration to occur, the blood pressure must be greater than the colloidal osmotic pressure of the blood, that is the osmotic pressure due to plasma proteins. The filtrate then enters the tubule, where it is modified both by tubular reabsorption and by secretion to form urine. Water, as well as many solutes such as salt and glucose, is absorbed in the proximal convoluted tubule, while the distal convoluted tubule continues to modify the tubular fluid.

In addition to the net pressure gradient, the glomerular filtration rate is dependent upon "glomerular plasma flow; the permeability of the glomerular capillaries; capillary surface area; and the number, activity and heterogenous nature of the glomeruli" (Yokota et al. 1985). The glomerular filtration rate of mammals is influenced by water influx and the rate of metabolism. It increases when there is a need to eliminate (1) excess water loads and (2) metabolic wastes such as nitrogenous wastes. In contrast, the rate decreases when there is a need to conserve (1) body water during dehydration and (2) salts, metabolites and energy. A short-term modification occurs by changing the glomerular plasma flow and/or glomerular capillary osmotic pressure. Furthermore, with an increase in dietary salt, there is an increase in glomerular filtration in the smaller superficial cortical nephrons and a decrease in the larger juxtamedullary nephrons; the opposite occurs with a low salt diet (Braun 1985).

The glomerular filtration rate (GFR; ml/h) has been found to be allometrically related to body mass in 41 mammals weighing 17 to 500×10^3 g. The equation had an r = 0.8 and took the form (Yokota et al. 1985):

$$GFR = 1.24 \, m_b^{0.765}.$$

Desert mammals usually have a reduced glomerular filtration rate under normal conditions. Two arid zone heteromyids, *Perognathus penicillatus* and *Dipodomys merriami*, have glomerular filtration rates that are significantly less than predicted for mammals of their body masses, whereas a third heteromyid, *D. spectabilis*, has a rate similar to that predicted (Yokota et al. 1985). Either when water-stressed or when acclimated to water shortage, desert mammals show further reduction in their glomerular filtration rate and urine output. Spiny mice (*Acomys cahirinus*) that were acclimated to chronic water shortage reduced their glomerular filtration rate by 55% and their urine output by 91% (Haines and Schmidt-Nielsen 1977). Similarly, mice that were offered one-fourth and one-eighth of ad libitum water intakes reduced their glomerular filtration rates by 45% and 47%, respectively (Haines and McKenna 1988).

Urine Concentration. Mammals are able to produce urine that is considerably more concentrated than plasma and, in this way, rid the body of excess osmolytes with a minimum of water. Small desert rodents are particularly noted for the ability to concentrate their urine highly. A urine osmolality of 9370 mOsmol/kg H$_2$O has been reported for the Australian desert murid rodent *Notomys alexis* (Table 6.5). This is the highest recorded value and is more than 24 times its plasma concentration (MacMillen and Lee 1967, 1969).

The concentrating ability of the kidney is dependent upon a counter-current multiplier system featuring the loop of Henle. The loop of Henle consists of a thick descending limb leading into a thin descending limb, a hairpin turn into a thin ascending limb leading into a thick ascending limb. Solutes, most probably Cl$^-$, are actively transported out of the thick ascending limb of the loop of Henle. Cl$^-$ is followed by Na$^+$ as the counter ion, and in this way NaCl is actively transported out of the loop. This creates a corticomedullary gradient of increasing osmolality in the interstitium of the renal medulla. The magnitude of the gradient between the tubule fluid and the interstitial medullary tissue is dependent upon the length of the loop of Henle. As a result, animals with longer loops of Henle and collecting ducts should be able to produce a more concentrated urine (Beuchat 1990a, b). Water is then withdrawn from the descending limb, thereby osmotically concentrating the tubular fluid making the turn. Chloride is pumped out from the ascending limb and "in this way the single effect of the ion pump is multiplied as more water is withdrawn and more and more concentrated fluid is presented to the ascending limb" (Schmidt-Nielsen 1979). The vasa recta are permeable to water and solutes, but not to protein, and act as a counter-current diffusion exchanger. The concentration of solutes within these capillaries increases in the descending limb, as does the tubular fluid and interstitium. The colloidal pressure due to the proteins results in the osmotic movement of water into the bloodstream. Tubular fluid remaining after the ascending thick limb is relatively dilute, but urea becomes concentrated at

Table 6.5. Maximum urine concentration ratio for rodents independent of exogenous water. (Buffenstein 1985)

Species		Urine Concentration (mOsm)	Source
Peromyscus critinus	N. America	3047	MacMillen and Christopher (1975)
Dipodomys spectabilis	N. America	4090	Schmidt-Nielsen (1964)
Gerbillus pusillus	N. Africa	4380	Buffenstein (1984)
Rhabdomys pumilio	Namibia	4554	Buffenstein (unpubl. observ.)
Notomys cervinus	Australia	4920	MacMillen and Lee (1967)
Dipodomys merriami	N. America	5540	Schmidt-Nielsen (1964)
Gerbillus gerbillus	N. Africa	5590	Burns (1956)
Gerbillurus tytonis	Namibia	6324	Buffenstein (unpubl. observ.)
Jaculus jaculus	N. Africa	6500	Schmidt-Nielsen (1964)
Notomys alexis	Australia	9370	MacMillen and Lee (1967)

this point and diffuses into the interstitium, increasing the osmotic pressure. Water is osmotically reabsorbed from the fluid in the collecting ducts because of the gradient in the interstitium as a result of the salts and urea, and leaves a concentrated urine to be excreted. The urea that remains contributes to more than 50% of the urine osmolytes of most desert rodents.

It is not only the length of the loop of Henle that determines the concentrating ability of the kidney. As pointed out by Beuchat (1990b), this also depends on "the organization of the medullary vascular supply (the vasa recta), the presence of extensions of the renal pelvis into the medulla (specialized pelvic fornices), and the degree of confluence of the collecting ducts in the inner medulla". This was well illustrated in the desert rodent, the gundi (*Ctenodactylus vali*). This animal has a long and well developed papilla, but the vascular bundles in the outer papilla are very numerous, simple and small. The glomerular filtration rate is low and similar to that of other desert rodents such as the chinchilla, fat sand rat, *Meriones* and jerboa, species that have better developed vasa recta and which can concentrate their urine highly. However, unlike the high urine concentration reported in the other desert rodents, the maximum urine concentration found in the dehydrated gundi was only 1361 mOsmol/kg (Rouffignac et al. 1981).

Kidney Size, Body Size and Urine Concentration. The kidney mass of mammals increases with body mass, and in 149 species takes the allometric form (Beuchat 1996):

$$\text{Kidney mass} = m_b^{0.88},$$

where kidney mass is the mass in grams of both kidneys, and m_b is body mass in kilograms. No difference in scaling of kidney mass with body mass was apparent among species from arid, mesic and fresh-water habitats.

However, Sperber (1944) noted that mammals from arid areas could concentrate their urine to a greater extent than species from mesic areas, and that they had thick medullae relative to kidney size. In contrast, mesic species are less able to concentrate their urine and possess kidneys with thinner medullae relative to kidney size. To quantify the relationship between medullary thickness and kidney size, he introduced the index known as the relative medullary thickness (RMT) which takes the form:

$$\text{RMT} = L/(H \times D \times W)^{0.33},$$

where L is absolute medullary thickness defined as the maximum length from the corticomedullary boundary to the tip of the papilla, and H, D and W are the absolute height, depth and width of the kidney. The RMT of mammals decreases with increasing body mass, that is, smaller species have a thicker medulla relative to the size of their kidneys than do larger species. In 165 species, this relationship takes the form (Fig. 6.9A; Beuchat 1996):

$$\text{RMT} = 5.32 \, m_b^{-0.11}.$$

The RMT is greatest in species inhabiting arid regions followed by those species from mesic and freshwater habitats (Fig. 6.9B).

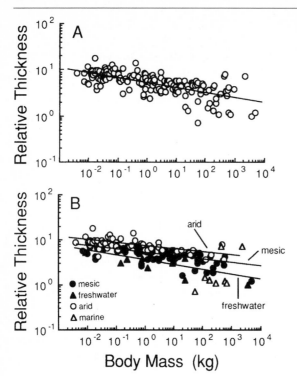

Fig. 6.9. Scaling of relative medullary thickness (**RMT**) and body mass (kg) in all mammal species (**A**) and in mammal species differentiated by habitat (**B**). **RMT** declines with increasing body mass, and the highest values are found in species inhabiting arid environments. (Beuchat 1996)

Brownfield and Wunder (1976) introduced a modified index called the relative medullary area (RMA) in which the areas of the medulla and cortex are used. The relationship takes the form:

RMA = medullary area/cortical area.

The maximum urine concentration in mammals is inversely correlated with body size. A large number of mammals weighing over 10 kg can concentrate their urine to 3000–4000 mOsmol/kg H_2O, but only the smallest mammals, weighing less than 400 g, can concentrate their urine well above this value. Two exceptions are the 475 g chinchilla (*Chincilla laniger*), which can produce urine of 7600 mOsmol/kg H_2O, and the 4.5 kg dikdik (*Madoqua kirkii*) which can produce urine of 4760 mOsmol/kg H_2O (Beuchat 1990a). Both these mammals inhabit arid regions. The allometric regression equation describing the relationship of maximum urine concentration (U_{osm}, mOsmol/kg H_2O; and body mass m_b, kg) for 169 species takes the form (Fig. 6.10; Beuchat 1996);

$$U_{osm} = 2667 \, m_b^{-0.09}.$$

This regression equation appears to contradict what is known about the relationship between the length of the loop of Henle and the concentrating ability of the kidney. Medullary thickness, a measure of the length of the loop of Henle,

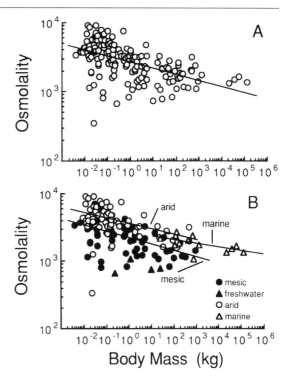

Fig. 6.10. Maximum recorded urine osmolality (mmol/kg H_2O) as a function of body mass (kg) in all mammal species (**A**) and in mammal species differentiated by habitat (**B**). The highest urine osmolalities are found in species from arid habitats, although some mesic species concentrate urine quite well, and some arid species have only modest concentrating abilities. Marine species concentrate urine about as well as would be expected for an arid species of the same body mass. (Beuchat 1996)

increases with a rise in body mass (Table 6.6). Small desert rodents that produce the most osmotically concentrated urine have short loops of Henle. In absolute terms, these usually range between 7 and 14 mm. Larger mammals, such as cows and horses, are able to produce urine of much lower concentration yet have loops of Henle that are much longer. For example, the loop of Henle in the horse is about 37 mm in length (Greenwald 1989). The allometric relationship between medullary thickness (MT; mm) and body mass (m_b; kg) for 152 mammalian species takes the form (Fig. 6.11; Beuchat 1996);

$$MT = 8.11 \, m_b^{0.14}.$$

The relationship between maximum urine concentration and medullary thickness in 55 mammalian species was not found to be significant (Beuchat 1990a). However, as pointed out by Sperber (1944) and Schmidt-Nielsen and O'Dell (1961), urine concentration is related to the relative medullary thickness, an index of the relative length of the loop of Henle rather than its absolute length. The allometric regression equation for this relationship takes the form (Beuchat 1990a):

$$U_{osm} = 341 \, RMT^{1.229}.$$

The slope of the equation does not differ significantly from 1.0, and therefore U_{osm} is directly proportional to the relative medullary thickness. The linear re-

Table 6.6. Maximum urine concentration for mammals in which measurements are also provided for medullary thickness (MT) and relative medullary thickness (RMT). (Adapted from Beuchat 1990a)

Order	Species	Body mass (kg)	MT (mm)	RMT	Osmolality (mosmol kg^{-1} H$_2$O)	Reference
Monotremata	*Tachyglossus aculeatus*	4.250	7.0	3.30	2300	Sperber (1944); Bentley and Schmidt-Nielsen (1967)
Marsupialia	*Didelphis virginiana*	5.000	13.0	5.70	1497	Sperber (1944) Plakke and Pfeiffer (1970)
	Setonix brachyurus	5.000	14.1	5.40	2188	Purohit (1974a)
	Macropus robustus	19.800	18.0	5.00	4054	Purohit (1974b)
Rodentia	*Pseudomys delicatula*	0.012	3.3	7.70	6930	Purohit (1974a, c)
	Pseudomys hermannsburgensis	0.013	3.8	7.50	8970	MacMillen and Lee (1967); Purohit (1947c)
	Notomys alexis	0.029	5.2	7.90	9370	MacMillen and Lee (1967); MacMillen and Lee (1969)
	Mus musculus	0.030	4.9	8.00	7000	Sperber (1944); Haines et al. (1973)
	Acomys cahirinus	0.033	6.4	9.40	6039	Purohit (1975)
	Dipodomys merriami	0.035	5.0	8.50	6382	Sperber (1944); Carpenter (1966)
	Jaculus jaculus	0.060	7.4	9.30	6500	Sperber (1944); Schmidt-Nielsen (1964)
	Meriones hurrianae	0.062	9.9	12.57	3731	Purohit (1975)
	Meriones unguicaulatus	0.070	6.5	7.51	5400	Edwards et al. (1983)
	Dipodomys spectabilis	0.100	7.7	8.50	4692	Munkacsi and Palkovits (1977)
	Mesocricetus auratus	0.105	8.7	8.01	5340	Munkacsi and Palkovits (1977); Trojan (1977)
	Psammomys obesus	0.150	13.9	10.90	6340	Sperber (1944); Schmidt-Nielsen (1964)
	Spermophilus lateralis	0.212	5.5	5.44	2425	Blake (1977); Munkacsi and Palkovits (1977)
	Neofiber alleni	0.250	3.0	3.60	658	Sperber (1944); Pfeiffer (1970)
	Chinchilla laniger	0.475	8.7	9.60	7599	Weisser et al. (1970)
	Aplodontia rufa	0.785	5.0	2.90	820	Sperber (1944); Dolph et al. (1962)
	Myocaster coypus	10.000	8.5	3.50	741	Sperber (1944); Pfeiffer (1970)
	Castor canadensis	25.000	17.7	1.30	537	Munkacsi and Palkovits (1977); Schmidt-Nielsen et al. (1948)
Chiroptera	*Eptesicus fuscus*	0.013	4.3	9.30	3675	Carpenter (1969); Geluso (1978)
	Leptonycteris sanborni	0.024	2.5	4.30	342	Carpenter (1969)
	Desmodus rotundus	0.026	8.0	7.40	6250	Carpenter (1969); McFarland and Wimsatt (1969)

Order	Species					Reference
Insectivora	*Hemiechinus auritus*	0.301	8.9	7.00	4010	Yaakobi and Shkolnik (1974)
	Paraechinus aethiopicus	0.352	10.9	7.50	3634	Yaakobi and Shkolnik (1974)
	Erinaceus europaeus	0.637	8.1	4.80	3062	Yaakobi and Shkolnik (1974)
Carnivora	*Fennecus zerda*	1.200	9.8	5.35	4022	Noll-Bonholzer (1979 b)
	Felis domesticus	5.000	14.0	4.80	3250	Sperber (1944); Schmidt-Nielsen (1964)
	Canis domesticus	15.000	17.0	4.30	2608	Sperber (1944); Schmidt-Nielsen et al. (1948)
	Mustela vison	106.000	10.5	4.60	2087	Sperber (1944); Eriksson et al. (1984)
Primates	*Galago s. senegalensis*	0.180	6.5	6.50	2020	Munkacsi and Palkovits (1977)
	Macaca mulatta	3.000	7.0	3.10	1607	Tisher (1971)
	Homo sapiens	70.000	19.0	3.00	1430	Schmidt-Nielsen (1964)
Artiodactyla	*Sus scrofa*	20.000	14.0	1.60	1075	Sperber (1944); Schmidt-Nielsen et al. (1948)
	Bos taurus	225.000	19.0	1.60	1160	Sperber (1944); Taylor and Lyman (1967)
perissodactyla	*Equus asinus*	150.000	40.0	3.60	1545	Maloiy (1970); Purohit (1974b)
	Equus caballus	300.000	37.0	3.40	1982	Sperber (1944); Rumbaugh et al. (1982)

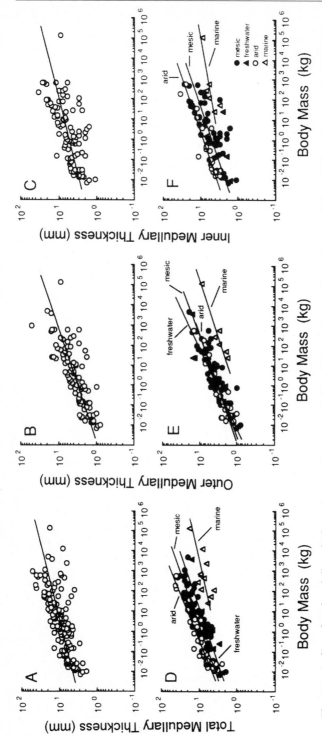

Fig. 6.11. Scaling of total medullary thickness (mm; A, D), outer medullary thickness (mm; B, E) and inner medullary thickness (mm; C, F) with body mass (kg) in all mammal species (*top*) and in mammal species differentiated by habitat (*bottom*). D–F It is apparent that, with the exception of marine mammals, most of the variability in medullary thickness among habitats is a consequence of differences in the scaling of the inner medullary thickness. The desert rodent *Psammomys obesus* (0.15 kg) has an exceptionally long thin ascending limb (11.5 mm), and this species appears as an outlier above the regression line for arid species in F. (Beuchat 1966)

gression for this relationship calculated for 78 mammalian species can be expressed as (Fig. 6.12A; Beuchat 1996):

$$U_{osm} = 524 \text{ RMT} + 154.$$

Most species in which RMT exceeds 7 are from arid habitats, but in these species, U_{osm} appears to be independent of RMT (Fig. 6.12B).

Greenwald (1989) examined the length of the loop of Henle and the metabolic intensity of mammal species. He pointed out that the volume of the kidney (H × D × W) is proportional to kidney mass and that the mass of a kidney in grams equals 3.66 times the animal's body mass in kilograms by the exponent 0.85. Therefore, by substituting in Sperber's equation:

$$RMT = L/(3.66 \ m_b^{\ 0.85})^{0.33},$$

and by collection:

$$RMT = (L \times m_b^{\ -0.283})/1.54.$$

Thus, the relative medullary thickness is directly proportional to the maximum length of the loop of Henle (L) times the mass-specific metabolic rate, and both are proportional to the maximum urine concentration of the mammal.

The exponent of -0.28 is similar to that of -0.27, which is often used to relate mass-specific metabolic rate to body mass. Urine concentration is dependent upon reabsorption of solutes by the ascending limb. This process is driven

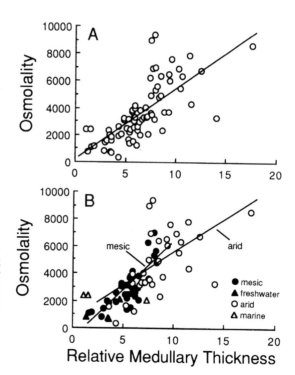

Fig. 6.12. The relationship between urine osmolality (mmol/kg H_2O) and relative medullary thickness (unitless) in all mammal species (A) and in mammal species differentiated by habitat (B). (Beuchat 1996)

by the active reabsorption of NaCl from the thick ascending limb and is therefore related to the mass-specific metabolic rate of the thick ascending limb tissue. If the metabolic rate of various tissues is similar to that of an entire animal,
it could explain why the urine concentrating ability is related to the product of
medullary thickness and metabolic intensity.

Yet RMT scales as $m_b^{-0.08}$ (Calder and Braun 1983) or as $m_b^{-0.11}$ (Beuchat
1996) and not as $L \times m_b^{-0.28}$. Greenwald (1989) theorized that the absolute
medullary thickness should scale as $m_b^{0.20}$, since $m_b^{0.20} \times m_b^{-0.28} = m_b^{-0.08}$.
Indeed, medullary thickness (L) on body mass scaled to a value close to this
exponent in two studies: as $m_b^{0.18}$ reported by Greenwald (1989) and as $m_b^{0.14}$
reported by Beuchat (1996).

Renal Pelvis. The size of the renal pelvis varies greatly among mammals. It is non-
existent in the beaver, an animal which does not possess an inner medulla and
does not have the ability to concentrate its urine greatly. The fat sand rat has
an extremely extensive renal pelvis in which the surface area is almost as large
as the outer surface of the entire kidney, and can concentrate its urine to a great
extent. However, gerbils and heteromyids show only a modest development of
the renal pelvis and yet have the ability to excrete more concentrated urine than
the fat sand rat.

What then is the role of the renal pelvis? Schmidt-Nielsen (1979, 1988) suggested that it may function to dilute urine after large intakes of water. The pelvic
urinary space reaches as far as the epithelium of both the inner and outer stripes
of the outer medulla. Urine can come into contact with the renal pelvis and exchange water and solutes with the tubules and capillaries. Deep pelvic refluxes
of urine that leave the collecting duct at the tip of the papilla are present only
when urine flows and urea concentration increases rapidly. Urea can then move
from the inner medullary tissue into the urine, lowering the urea concentration in the medullary tissue area. Shallow pelvic refluxes occur during falling
urine flow and increasing urea concentration. These refluxes bathe only the tip
of the papilla with urine. This urine has a higher urea concentration than the
papillary tissue and in this way increases the urea concentration in the tissue.

Habitat, Diet and Urine Concentration. There is a correlation between the length of the
loop of Henle and the aridity of the environment inhabited by an animal. For
example, the kidney of *Notomys alexis* is characterized by a highly elongated
medullary papilla, which reflects an extremely long loop of Henle of the juxtamedullary nephrons (Hewitt 1981). In contrast, mammals which live in
aquatic habitats have a very thin medulla and short loops of Henle. The length
of the loops of Henle in species from mesic habitats falls between these two
(Beuchat 1996). In addition, the volume of blood entering the glomeruli in juxtamedullary nephrons is greater than the volume entering superficial nephrons
in desert species, but not so in mesic species (Altschuler et al. 1979).

In many small desert mammals, the inner medulla, which is organized into
a single papilla, may extend well beyond the renal capsule into the ureters. This

was the case in seven rodent species of a total of 34 species examined by Sperber (1944). All seven species inhabit arid areas. An extended papilla has been described in some of the smaller heteromyids, such as the genera *Perognathus* and *Chaetodipus* (Altschuler et al. 1979), as well as in Australian hopping mice (MacMillen and Lee 1969), shrews (Lindstedt 1980a) and bats (Geluso 1978), suggesting that small desert mammals have the ability to concentrate their urine highly (French 1993).

The highest urine concentrations among mammals have been reported mainly in small desert rodents. Granivorous heteromyids are represented by species with the ability to excrete highly concentrated urine (French 1993). High concentrations have also been reported for Gerbillinae and Dipodidae from North Africa and Asia (Schmidt-Nielsen 1964), Muridae from Australia (MacMillen and Lee 1967) and Cricetidae from South Africa (Buffenstein et al. 1985). All these can survive entirely on a diet of seeds, but consume green vegetation and insects when they are available.

Insectivores obtain more preformed water from their food than do granivores, but are forced to excrete more nitrogenous waste from their high protein diets. In general, they do not concentrate their urine to the same extent as small mammals consuming only a dry diet. Arid zone insectivores have the ability to concentrate their urine to a greater extent than mesic species (Table 6.7; Fielden et al. 1990a). This is due, in a large part, to differences in the morphology of the kidney. This is clearly evident when comparing closely related species that inhabit areas of different aridity. The single papilla of the arid zone mole *Eremitalpa granti namibensis* is elongate and extends well into the ureter, whereas that of the mesic mole *Amblysomus hottentotus* stops at about the pelvis of the kidney (Fig. 6.13; Fielden et al. 1990a). A similar difference in morphol-

Table 6.7. Mean osmotic concentration of the urine of water-restricted insectivorous mammals maintained in the laboratory. (Fielden et al. 1990a)

Species	Urine osmolarity (mosmol/kg)	Reference
Arid zone		
Pipistrellus hesperus	4340	Geluso (1978)
Hemiechinus auritus	4010	Yaakobi and Shkolnik (1974)
Antrozous pallidus	3980	Geluso (1975)
Eremitalpa granti namibensis	3820	Fielden (1990a)
Paraechinus aethiopicus	3634	Yaakobi and Shkolnik (1974)
Macrotis lagotis	3566	Hulbert and Dawson (1974)
Onychomys torridus	3180	Schmidt-Nielsen and Haines (1964)
Sminthopsis crassicaudata	2322	Morton (1980)
Mesic zone		
Erinaceus europaeus	3062	Yaakobi and Shkolnik (1974)
Isoodon macrourus	2942	Hulbert and Dawson (1974)
Myotis volans	2910	Geluso (1978)
Planigale maculata	2317	Morton (1980)
Blarina brevicauda	1820	Deavers and Hudson (1979)

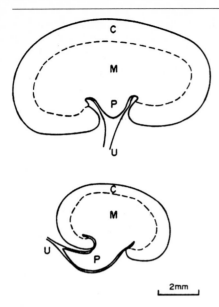

Fig. 6.13. Cross-section of the kidney showing the cortex (*C*), medulla (*M*), form of the papilla (*P*) and ureter (*U*) in the mesic *Amblysomus hottentotus* (*top*) and xeric *Eremitalpa granti namibensis* (*bottom*). (Fielden et al. 1990a)

ogy has been noted between the xeric hedgehog *Paraechinus aethiopicus* and the mesic species *Erinaceus europaeus* (Yaakobi and Shkolnik 1974).

Urine concentration in insectivorous bats also shows a relationship to habitat. The concentrating ability of Nearctic bats, as indicated by the inner medulla to cortex (IM/C) ratio, is significantly correlated with total annual precipitation (Fig. 6.14) and with the potential evapotranspiration to precipitation ratio (E/P;

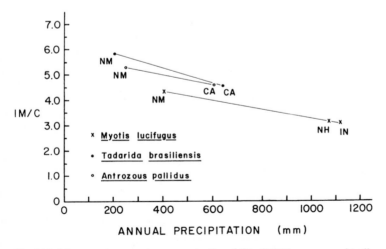

Fig. 6.14. Mean maximum urine concentrating ability (IM/C) on geographically separated populations of insectivorous bats as a function of the annual precipitation received by their habitat. *NH* New Hampshire; *IN* Indiana; *CA* California; *NM* New Mexico. (Bassett 1982)

Fig. 6.15) of the bats' habitats (Bassett 1982, 1986). However, habitat aridity explained only 25% of the variation among species. It was therefore concluded that factors other than aridity are also important in the urine-concentrating ability of insectivorous bats.

Study of the the kidney structure of dasyurids, which are insectivorous or carnivorous marsupials, showed that the relative medullay thickness ranged widely between 3.7 and 11.5, and that species from arid areas tended to have higher values than species from semi-arid, mesic and tropical areas (Brooker and Withers 1994). For example, arid-dwelling species such as *Ningaui ridei* and *Pseudantechinus macdonnellensis* had the highest relative medullary thickness, while the alpine *Antechinus swainsonii* had the lowest value. The inner medulla of the former two species extends well beyond the confines of the kidney, whereas that of the latter does not. In addition, values of the medulla to cortex ratio, inner medulla to cortex ratio and percent medullary thickness were greater in xeric than non-xeric species, but there was no difference related to habitat in percent medullary area and relative medullary area. Discriminant analysis revealed that the arid zone species were clearly separated from the rest, but that there was poor discrimination among the species from semi-arid, sub-humid and humid habitats.

Renal concentrating indices were correlated with body mass, and with climatic factors such as average daily maximum air temperature and average annual rainfall of a marsupial's habitat. However, renal indices were not related to phylogenetic affiliations, "suggesting that renal morphology is a phylogenetically plastic characteristic" (Brooker and Withers 1994).

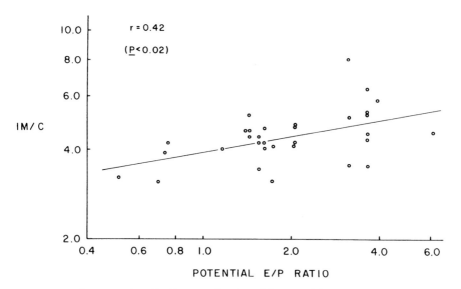

Fig. 6.15. Renal anatomy (IM/C) of bats as a function of the potential evapotranspiration to precipitation ratio (E/P) of the habitat in which they were captured. (Bassett 1986)

Herbivores usually have a high water intake coupled with a relatively low protein intake and therefore are not often required to concentrate their urine to a great extent. In a comparative study of kidney morphology of six species of otomyine rodents with differing habitats in southern African, xerophilic species were found to possess traits that indicated a higher concentrating ability than that of the mesophilic species (Pillay et al. 1994). The ratio of inner medulla to cortex increased from a mean of 1.17 in three mesophilic species to a mean of 3.33 in three xerophilic species. Secondary renal pyramids were present in the xerophilic species, but not so in the mesophilic species, and the percentage of long looped nephrons was higher in the xerophilic than in the mesophilic species. No mesic-xeric trend was noted in the density of glomeruli in either transverse or sagittal sections. However, the renal characteristics of xeric species do not appear to be optimally expressed for urine concentration. For example, the inner medulla to cortex ratio of 3.33 (range 3.01–3.56) for these otomyine rodents is considerably less than the 5.45 for *Psammomys obesus* and 6.60 for *Notomys alexis*.

An exception to the general rule that herbivores do not concentrate their urine highly appears to be the fat sand rat. This rodent has a high inner medulla to cortex ratio, a large renal plexus, and complex arterial and venous vasa recta. It has the ability to concentrate its urine to 6340 mOsm/kg H_2O, which is approximately 17 times the concentration of its plasma (Schmidt Nielsen 1964). This herbivore feeds only on chenopods such as *Atriplex halimus*, a plant that has a high preformed water content but also a very high salt content (Degen 1988). The fat sand rat, however, does not have the ability to concentrate its urea greatly (Schmidt-Nielsen and O'Dell 1961): Na^+, Cl^- and K^+ comprise about 80% of the total osmolytes of its urine (Kam and Degen 1988).

6.2.4
Nitrogen Excretion

Urea is the main end product of nitrogen metabolism in mammals and, as in most animals, is excreted almost entirely through the kidneys. It is highly soluble in water and permeable to membranes. It requires water for its excretion and generally comprises the largest part of the urine osmolytes. Urea is particularly important to ruminants where it can be used as a source of non-protein nitrogen for microbial protein synthesis, and therefore, its recycling from the kidneys to the gastrointestinal tract (the rumen) may be essential in the nitrogen balance of the animal (Harmeyer and Mertens 1980; Kennedy and Milligan 1980).

The amount of urea reabsorbed and recycled is dependent on the nitrogen or crude protein content of the diet. In the Sinai goat, a ruminant well known for its ability to thrive under harsh desert conditions, 40% of urea was recycled when the animal consumed a high-protein diet (alfalfa hay; 18.8% crude protein), 64% was recycled on a medium-protein diet (Rhodes grass; 9.6% crude protein) and 69% was recycled on a low-protein diet (wheat straw; 3.7% crude protein; Table 6.8). Four days of water deprivation had no effect on the proportion of urea that was recycled (Choshniak et al. 1988).

Table 6.8. Urea kinetics (Mean $\pm$ SD; mg/kg$^{0.75}$ per day) in Bedouin goats maintained on three different diets and given water either once every 4 days or once daily. (Choshniak et al. 1988)

Diet	Days between drinking bouts	Urea pool	Urea entry	Urine urea excretion	Urea recycling	Entry rate (%)
Alfalfa hay	1	290 ± 8	1018 ± 128	642 ± 92	376 ± 56	37.0 ± 3
	4	254 ± 58	1458 ± 205	870 ± 214	588 ± 202	40.3 ± 5
Rhodes grass	1	120 ± 13	302 ± 16	109 ± 13	194 ± 19	64.2 ± 4
	4	61 ± 14	264 ± 45	96 ± 28	168 ± 54	64.0 ± 6
Wheat straw	1	128 ± 14	279 ± 11	86 ± 15	193 ± 22	69.0 ± 6
	4	84 ± 41	241 ± 47	74 ± 19	167 ± 22	69.3 ± 9

Less information on nitrogen kinetics and urea recycling is available for small monogastric mammals. However, a study on rodents suggested that, when consuming a poor protein diet, desert species utilize the recycling of urea to a greater extent than do non-desert species (Yahav and Chosniak 1989). When the fat jird (*Meriones crassus*), a desert rodent, and the Levant vole (*Microtus guentheri*), a mesic rodent, were offered high-quality forage (alfalfa hay; 21% crude protein) plus agar water, the fat jird recycled 53% of its urea, the Levant vole recycled 89% of its urea, and both species were in positive nitrogen balance. However, when a poorer quality forage (*panicum* grass; 17% crude protein) plus agar water was offered, urea recycling increased to 67% in the fat jird but decreased to 63% in the Levant vole (Table 6.9). The fat jird was still able to maintain a positive nitrogen balance, whereas the Levant mole could not. True nitrogen digestibility did not change in the Levant mole on the low-quality diet but increased significantly in the fat jird (Table 6.10). The inability of

Table 6.9. Urine urea concentration, urea entry rate and urea recycling in fat jirds (*Meriones crassus*) and Levant voles (*Microtus guentheri*) fed alfalfa hay and *Panicum* grass, supplemented with agar water. Agar glucose was also offered to voles fed *Panicum* grass. Mean values ($\pm$ S.D; n = 6) marked by different letters are significantly different ($P < 0.05$) from each other. (Yahav and Choshniak 1989)

Animal	Urine urea concentration (mmol/l)	Urea entry rate (mmol/day)	Urea recycling (%)
	Alfalfa hay + agar water		
Fat jird	182.3 ± 29.4a	2.72 ± 0.60a	52.7 ± 5.4a
Levant vole	65.5 ± 28.6b	2.48 ± 1.05a	89.4 ± 11.6b
	Panicum grass + agar water		
Fat jird	149.4 ± 53.4a	2.07 ± 0.82a	67.3 ± 7.0c
Levant vole	235.7 ± 31.6c	4.56 ± 0.59b	63.2 ± 6.5c
	Panicum grass + agar glucose		
Levant vole	86.9 ± 7.8b	5.17 ± 3.65a,b	86.9 ± 7.8b

Table 6.10. Apparent and true nitrogen digestibilities in fat jirds (*Meriones crassus*) and Levant voles (*Microtus guentheri*) fed alfalfa hay and *Panicum* grass. Mean values ($\pm$ S.D; n = 6) marked by different letters are significantly ($P < 0.05$) from each other. (Yahav and Choshniak 1989)

Animal	Apparent nitrogen digestibility	True nitrogen digestibility
	Alfalfa hay	
Fat jird	57.9 + 4.8a	61.0 + 5.5a
Levant vole	54.9 + 3.4a	58.9 + 3.4a
	Panicum grass	
Fat jird	70.4 + 8.6b	74.2 + 7.0b
Levant vole	53.1 + 8.4a	59.6 + 6.9a

the Levent mole to recycle nitrogenous wastes better was attributed to its low energy intake since the transfer of urea to the digestive tract is also dependent upon an adequate supply of energy. When the agar water was replaced by agar glucose, the Levant vole recycled 90% of its urea.

The effect of dietary crude protein and water consumption on urea kinetics has also been determined in the rock hyrax (*Procavia habessina*), a herbivorous mammal of 2–4 kg that is widely distributed in the semi-arid and arid rocky regions of eastern and northern Africa and the eastern Mediterranean (Hume et al. 1980). Its nitrogen balance was positive on a diet containing 14.6% crude protein, but negative on a diet of 8.2% crude protein. The proportion of urea recycled to the gastrointestinal tract was 63 and 60% of the urea entry rate when the animals were fed diets with crude protein contents of 14.6 and 8.2%, respectively, but decreased to 40% on a low crude protein diet of 5.3%. This decrease was presumably a consequence of low energy intake. Urea recycling increased to 70% when the water intake was restricted to 60% of ad libitum intake. The utilization of urea in the gut was 59% and 53% of the urea degraded on the 14.6% and 8.2% crude protein diets, respectively. Utilization increased substantially to 71% on the low 5.3% crude protein diet and to 98% with water restriction (Table 6.11).

Allantoin Excretion. Mammals (except for man, apes and the dalmation dog) excrete their purine catabolites as allantoin. Allantoin is formed in the liver by the conversion of uric acid through the enzyme uricase. The water solubility range of allantoin is 6.3 mmol/l, and most mammals fall well within this range. In the rat, allantoin is filtered freely and is neither reabsorbed nor secreted along the nephron (Greger et al. 1975).

Crystalline allantoin could account for nitrogen excretion without the loss of water. In the shift to allantoin excretion from urea excretion, three moles of water are saved for each mole of urea. It is surprising then that the excretion of crystalline allantoin has been reported only in South African Cricetidae (Buffenstein et al. 1985) and is not more widespread among small desert mammals. Crystals were apparently reported in the urine of *Dipodomys merriami*,

Table 6.11. Urea kinetic parameters measured in the rock hyrax. Values are means $\pm$ SD from three animals. (Hume et al. 1980)

	Period			
Crude protein in diet	1 14.6%	2 8.2%	3 8.2%[d]	5 5.3%
Plasma urea (mmol N l^{-1})	9.8 ± 1.0^a	8.0 ± 1.6^a	9.6 ± 2.4^a	12.0 ± 1.2^b
Urea turnover time (h)	6.0 ± 0.5^a	6.4 ± 0.9^a	10.0 ± 1.7^b	9.7 ± 2.0^b
Urea pool (mmol N)	10.2 ± 2.4^{ax}	7.4 ± 0.8^{ax}	14.8 ± 4.0^{by}	13.6 ± 2.4^{by}
Urea entry rate (mmol N 24 h^{-1})	40.2 ± 7.2^a	27.2 ± 3.6^b	35.2 ± 4.2^b	33.8 ± 1.0^b
Urinary urea excretion rate (mmol N 24 h^{-1})	16.0 ± 2.8^{ax}	11.4 ± 1.6^{bx}	10.4 ± 2.8^{bx}	20.0 ± 3.8^{cy}
Urea degradation rate: (mmol N 24 h^{-1})	24.4 ± 4.8^a	15.8 ± 3.4^b	24.8 ± 2.4^a	14.0 ± 3.4^b
(% of nitrogen intake)	45.7 ± 7.3	56.8 ± 19.7	116.3 ± 35.7	120.8 ± 16.3
Urea recycled (%): (1)[e]	63.0 ± 3.1^a	57.7 ± 7.3^a	70.9 ± 5.2^a	41.1 ± 10.0^b
(2)	63.5 ± 5.9^{ax}	61.4 ± 4.5^{ax}	69.4 ± 4.2^{bx}	38.7 ± 5.6^{cy}
Urea-N turnover time (h)	8.0 ± 0.5^a	9.0 ± 0.8^a	9.7 ± 0.9^{ab}	11.0 ± 2.0^b
Urea-N reabsorbed (mmol 24 h^{-1})	10.0 ± 2.1^a	7.4 ± 1.7^a	0.6 ± 0.6^b	4.0 ± 1.7^b
Urea-N utilized (mmol 24 h^{-1})	14.4 ± 2.9^{ax}	8.4 ± 2.2^{bx}	24.3 ± 2.8^{cy}	9.9 ± 2.0^{abx}
% of degraded urea utilized	59.0 ± 3.1^{ax}	53.0 ± 5.6^{ax}	97.6 ± 2.8^{cz}	71.2 ± 7.4^{by}

[a, b, c]Means on the same line bearing different superscripts differ significantly ($P < 0.05$).
[d]Water intake restricted to 60% of ad libitum. In all other periods water was available ad libitum.
[e]Degradation rates as a % of entry rate (1), and % of injected ^{14}C-urea not recovered in urine (2).
[x, y, z]Means on the same line bearing different superscripts differ significantly ($P < 0.01$).

especially when water-stressed and on a high-protein diet (Carpenter 1966; quoted by French 1993). However, this phenonomen has not been reported in this species since Carpenter's (1966) study, nor has it been observed in other heteromyids (French 1993).

All South African desert cricetid species studied by Buffenstein et al. (1985) produced an off-white crystalline precipitate, allantoin, in their urine when deprived of water, whereas none of the desert murids did. The crystalline allantoin accounted for 29% of the urine mass and for about 47% of the total urine nitrogen (Table 6.12). In these cricetids, urea accounted for only 28–35% of the measured excreted nitrogen, whereas it accounted for about 67% in the murids (Table 6.13). By shifting nitrogen excretion from urea to allantoin, water-deprived cricetids are able to save more than 20% of their urine water (Table 6.13). Not only were the cricetids able to save water by allantoin excretion, they were also able to concentrate their urine to a greater extent than the murids could. In both groups, most of the urine osmotic pressure was due to urea, and both had similar uric acid outputs. Buffenstein et al. (1985) concluded that in South Africa the cricetids were better adapted to arid conditions than the murids. This

Table 6.12. Allantoin excretion in Namib desert rodents deprived of water for 4 weeks. (Buffenstein et al. 1985)

Species	Allantoin[a]		Crystalline allantoin			
	n	mM	n	mmol day^{-1}	mg day^{-1}	Urine mass (%)
Cricetidae						
Desmodillus auricularis	8	612 ± 126	9	0.35 ± 0.03	56.0 ± 5.2	29.63 ± 2.8
Gerbillurus paeba	8	788 ± 164	13	0.29 ± 0.002	45.3 ± 3.1	28.49 ± 2.1
Tatera leucogaster	11	870 ± 194	16	1 ± 0.43	175.0 ± 7.0	28.94 ± 9.3
Muridae						
Aethomys namaquensis	9	615 ± 202	28	0.00 ± 0.00	0.0 ± 0.0	0.00 ± 0.0

[a]Corrected values, urea deducted.

is in agreement with the general distribution of these rodents as the murids are restricted to the more mesic regions of the Namib Desert.

The synthesis and excretion of allantoin require the removal of twice as many carbon atoms as for urea (1C/1N for allantoin as opposed to 1C/2N for urea). As a result, the excretion of allantoin is more costly in energy terms than that of urea, since carbon loss is linked energetically. It would seem then that the water status, as well as the energy status of cricetid, is important in determining the route of nitrogenous waste excretion.

6.2.5
Unorthodox Method of Eliminating Electrolytes

Several unrelated desert rodents from different continents can thrive while consuming only chenopods, plants not only high in preformed water content, but also high in nitrogen and electrolyte content, which would thus require much urinary water for excretion. Such rodents include the North American heteromyid *Dipodomys merriami* which feeds on *Atriplex confertifolia* (Kenagy 1972, 1973b), the North African gerbillid *Psammomys obesus* which feeds on *A. halimus* (Degen 1988; Kam and Degen 1988) and the South American octodontid *Tympanoctomys barrerae* which feeds on *Heterostachys ritteriana* (Torres-Mura et al. 1989).

Kenagy (1972, 1973b) described a behavioural mechanism in which *D. merriami* scrapes off the cuticular and epicuticular layers of the leaf with its teeth before consuming them. The highest concentrations of electrolytes are present in these layers, and therefore *D. merriami* is able to lessen its electrolyte intake before consuming the plant. Similar leaf-scraping behaviour has been reported for *P. obesus* (Degen 1988). Both these species posses chisel-chaped lower incisors that are well adapted to scraping *Atriplex* leaves, and both species climb to the uppermost twigs to harvest leaves (Kenagy 1972, 1973b; pers. comm.). This leaf-scraping behaviour was not noted for *T. barrere*, and apparently, this rodent consumes the chenopod *in toto* (Torres-Mura et al. 1989).

Table 6.13. Daily nitrogen excretion in urea and allantoin and estimated water savings by excreting crystalline allantoin instead of urea. (Buffenstein et al. 1985)

	Aethomys namaquensis		*Desmodillus auricularis*		*Gerbillurus paeba*		*Tatera leucogaster*	
	$\bar{x}$	(%)	$\bar{x}$	(%)	$\bar{x}$	(%)	$\bar{x}$	(%)
Mass (g)	47.9	–	59.9	–	29.9	–	53.7	–
Urine vol. (10^{-1} ml day^{-1})	4.0	–	1.5	–	1.4	–	5.0	–
Total urea (10^{-3} mmol day^{-1})	979.6	79.97	550.1	55.45	443.6	52.50	1,796.0	53.74
Total uric acid (10^{-4} mmol day^{-1})	1.3	0.01	0.8	0.01	0.3	0.04	2.3	0.01
Total allantoin in solution (10^{-3} mmol day^{-1})	245.2	20.02	92.0	9.27	111.3	13.17	435.5	13.03
Total allantoin solids (10^{-3} mmol day^{-1})	0.0	0.0	350.0	35.28	290.0	34.32	1,110.0	33.22
Total nitrogenous wastes measured (10^{-3} mmol day^{-1})	1,226.1	–	992.9	–	845.2	–	3,343.8	–
Nitrogen in urea (10^{-3} mg day^{-1})	27.4	66.67	15.4	28.07	12.4	35.81	50.2	36.76
Nitrogen in dissolved allantoin (mg day^{-1})	13.7	33.33	19.6	35.74	6.2	17.90	24.4	17.86
Nitrogen in crystalline allantoin (mg day^{-1})	0.0	0.0	19.8	36.18	16.0	46.29	61.95	45.36
Measured urinary nitrogen (mg day^{-1})	41.1	–	54.8	–	34.6	–	136.55	–
Water saving from crystalline fraction (10^{-3} ml)	0.0	0.0	38.0	25.33	31.0	21.95	119.0	23.8

Psammomys obesus does not scrape all the leaves that it consumes, and it uses different patterns of scraping leaves. Sometimes it scrapes part of the leaf, consumes the scraped part, scrapes some more, consumes that, and so on. At other times, it scrapes large segments of a leaf at one time and then consumes it (Fig. 6.16; Degen 1988). The amount of leaf scraped off by *P. obesus* is related to the leaf's water content. Negligible amounts of leaf are scraped from leaves containing over 80% water; about 15% by mass is scraped off leaves containing 70% water (Degen 1988). Scraped leaf mass can reach close to 50% in leaves containing 50% water (unpublished data). In this way, *P. obesus* has some control over its electrolyte intake. *D. microps* also controls its electrolyte intake for it shows a seasonal shift in scraping leaves (Kenagy 1973b). The ash content of the scrapings of leaves left by *P. obesus* is approximately 117% greater and the gross energy content approximately 35% less than for whole halimus leaves.

Air temperature also affects the amount of *Atriplex* leaf scraped off by *P. obesus* (Kam and Degen 1992). When *P. obesus* were offered leaves with a water content of 77.5%, the rodents maintained at 15 °C scraped off 4.9% of the leaf dry matter and concentrated their urine to 2711 mOsm/kg, whereas others maintained at 34 °C scraped off 14.6% and concentrated their urine to 4476 mOsm/kg (Table 6.14). Thus, both the water content of the *Atriplex* leaves and the air temperature influence the amount of leaf and ash content removed before consumption (Table 6.15). Using the values in Table 6.15, a multiple regression of air temperature (T_a; °C) and water content of the leaves (WC; g H_2O g^{-1} fresh matter) on ash removal (ash_r; percent removed of percent in leaf) took the form:

$$ash_r = 2.251 + 0.0092\ T_a - 2.981\ WC,$$

Fig. 6.16. A non-scraped *Atriplex halimus* leaf and examples of leaves that were scraped, eaten, scraped, eaten, etc., and leaves that were almost scraped *in toto* before being eaten by *Psammomys obesus*. (Degen 1988)

Table 6.14. Effect of scraping off the outer layer of leaves on energy content of saltbush (*Atriplex halimus*) offered to fat sand rats (*Psammomys obesus*) that were maintained at different air temperatures, on water content of their faeces and on the concentration of their urine. Values (means $\pm$ SD) within rows with different superscripts vary ($P < 0.05$). (Kam and Degen 1992)

Air temperature (°C):	15	21	34
Number of animals	8[c]	9	9
Gross energy offered (kJ g^{-1})	14.2 $\pm$ 2.0	14.2 $\pm$ 1.6	14.2 $\pm$ 1.6
Dry matter (DM) scraped (g day^{-1})	0.75 $\pm$ 0.92[a]	1.02 $\pm$ 1.02[a]	2.08 $\pm$ 0.72[b]
(%)[d]	4.92 $\pm$ 6.64[a]	6.76 $\pm$ 5.38[a]	14.57 $\pm$ 3.68[b]
Scrapings energy (kJ g^{-1})	10.29 $\pm$ 0.61	9.36 $\pm$ 0.85	9.62 $\pm$ 0.88
Scrapings ash (% DM)	37.9 $\pm$ 3.2	40.1 $\pm$ 5.2	41.5 $\pm$ 3.8
Gross energy consumed (kJ g^{-1})	14.3 $\pm$ 0.9	14.6 $\pm$ 0.9	15.0 $\pm$ 0.9
Faecal water (% fresh matter)	66.4 $\pm$ 2.6	65.4 $\pm$ 3.4	67.5 $\pm$ 5.4
Urine concentration (mOsmol J kg^{-1})	2711 $\pm$ 560[a]	2942 $\pm$ 957[a]	4476 $\pm$ 1647[b]

[c] One fat sand rat died of unknown reasons during the study.
[d]% = [Dry matter scraped/(dry matter scraped + dry matter intake)] $\times$ 100.

Table 6.15. Effect of air temperature (T_a) and water content of *Atriplex halimus* leaves on ash removal due to fat sand rats (*Psammomys obesus*) scraping off the outer layer of the leaf. Included in the table are urine concentrations of the fat sand rats

T_a(°C)	Water content	Leaf scrapings[a](%)	Ash removed[b](%)	Urine concentration (mOsm kg^{-1})
34	0.775	14.6	25.4	4476
25	0.664	39.1	52.0	4500
25	0.744	14.4	20.0	2978
25	0.745	13.9	30.7	3700
25	0.776	6.4	14.6	3066
25	0.836	0.8	1.6	2241
21	0.775	6.8	11.8	2942
15	0.775	4.9	8.6	2711

[a] Leaf scraping was calculated as a percent of total dry matter handled: scraping mass/(dry matter intake + scraping mass).
[b] Ash removed was calculated as a percent of total ash handled: ash content in the scrapings times scrapings mass/ash content in the leaves times (dry matter intake + scraping mass).

where n = 8; SE of the intercept, T_a and WC were 0.258, 0.0029 and 0.321, respectively; $r^2 = 0.98$ and F < 0.05. Therefore, the percent of ash removed from the total ash increases by 0.9% for every 1 °C rise and decreases by 3.0% with an increase of 1% water content.

The effects of air temperature and leaf water content on urine osmolality (U_{osm}; mOsmol/kg) take the form:

$$U_{osm} = 10001 + 89.0\, T_a - 11620\, WC,$$

where n = 8; SE of the intercept, T_a and WC were 2560, 29.3 and 3188, respectively; $r^2 = 0.82$ and F < 0.05. Therefore, with an increase of 1 °C, urine osmolality increases by 89 mOsmol kg^{-1}, and with an increase of 1% leaf water content, urine osmolality decreases by 116 mOsmol kg^{-1}.

6.2.6
Hormonal Control of Urine Concentration

The concentration of urine in mammals is controlled by the antidiuretic hormone (ADH), also known as vasopressin. An increase in the osmolality of the extracellular fluid in the area of the hypothalamic receptors affects the release of ADH from the posterior lobe of the hypophysis (pituitary gland). This hormone then acts on the kidney to absorb more water and increase urine osmolality. As a consequence, the urine volume is reduced, and body fluids are conserved. The action of ADH is twofold: (1) it increases the permeability of the collecting duct to water and other certain non-electrolytes; and (2) it stimulates sodium transport. These two actions increase the osmotic pressure in the surrounding tissue and enhance the osmotic movement of water from the dilute urine entering the collecting ducts into the interstitium (Handler and Orloff 1973).

The normal plasma concentrations of ADH of desert rodents are much higher than those of non-desert rodents. Such was the case for *Acomys cahirinus* and *A. russatus* (Castel et al. 1974), *Gerbillus gerbillus, Jaculus jaculus* (El Husseini and Haggag 1974) and *J. orientalis* (Baddouri et al. 1984) where plasma levels are over 250 pg ml^{-1}. In comparison, the upper level for dehydrated rats is under 100 pg ml^{-1}. Moreover, the elevated levels in desert species can be maintained for extended periods, whereas extended water deprivation in rats causes ADH activity to approach zero (Randle and Haines 1976). However, water loading of *J. orientalis* results in a reduced plasma ADH concentration, a subsequent increase in urinary output and a decrease in urine osmolarity (Table 6.16). Nevertheless, even at this point, plasma ADH still remains higher than the upper limits reported for non-desert mammals (Baddouri et al. 1984).

The high ADH plasma levels in desert rodents suggest that they may have a role to play in conserving body water by maximizing urine concentration and minimizing urine output. What triggers the release of ADH in these rodents? Weaver et al. (1994) noted that plasma volume and osmolality remain constant during water deprivation, while plasma sodium actually declines in the xeric spinifex hopping mouse (*Notomys alexis*; Table 6.17), but that there are decreases in renal perfusion pressure. These authors concluded that vasopressin "would not appear to be the overriding fluid conserving adaptation because the normal stimuli for vasopressin release (volume and osmolality) are not obviously perturbed". They suggested that the moderate declines in renal perfusion pressure and probable reductions in glomerular filtration rate could "be an important factor facilitating fluid retention and urine concentration during water deprivation" and "seem, therefore, to be more important for the remarkable xeric-adaptation of *Notomys* in fluid balance and blood pressure maintenance than altered levels of the conventional fluid-retaining and pressor hormones".

Table 6.16. Effect of either a dry diet or chronic water intake on plasma antidiuretic hormone (ADH) and urine concentrations in the jerboa, *Jaculus orientalis*. (Baddouri et al. 1984)

Diet	Animal	V (μl h^{-1})	O_{osm} (mosmol liter^{-1})	P_{ADH} (pg ml^{-1})
Dry	1	–	–	471
	2	7.5	4044	668
	3	5.3	4641	236
	4	8.8	4635	248
	5	5.4	1948	238
	Mean	6.7[a]	3817	372
	$\pm$ SEM	0.8	638	86
Water enriched	6	479	393	31.6
	7	292	680	86.0
	8	329	649	13.5
	9	245	567	99.0
	10	158	945	151.0
	Mean	301**	647**	76.2*
	$\pm$ SEM	53	90	24.6

Note. For 4 to 7 weeks animals were given either a dry diet only, or the same diet plus fresh vegetables *ad libitum*. At the end of this pretreatment, urine was collected under oil for 24 h in metabolic cages. Animals were then sacrificed and blood was collected for plasma ADH measurement. *$P < 0.02$ and **$P < 0.01$ are compared with the corresponding mean value of the group fed a dry diet.
[a]This value is probably underestimated due to incomplete urine recovery in this group of jerboas.

Table 6.17. Mean ($\pm$ SE) body mass (BW), haematocrit (Hct%) and plasma renin-angiotensin system (RAS; PRC, plasma renin concentration, PRA, plasma renin activity) values during prolonged water deprivation in *Notomys alexis*. Values in parentheses are n (Weaver et al. 1994)

	Days without water				
	0	7	14	21	28
BW (g)	29.3 $\pm$ 1.0 (30)	24.1 $\pm$ 0.8** (30)	25.0 $\pm$ 1.2* (8)	26.3 $\pm$ 1.7 (6)	26.9 $\pm$ 1.3 (8)
Hct%	54.2 $\pm$ 0.6 (27)	59.4 $\pm$ 1.2* (9)	51.1 $\pm$ 0.9 (8)	51.9 $\pm$ 0.8 (6)	53.2 $\pm$ 1.0 (8)
PRC (mlU ml^{-1})	0.54 $\pm$ 0.125 (11)	2.2 $\pm$ 1.05** (9)	0.52 $\pm$ 0.086 (5)	0.74 $\pm$ 0.243 (6)	0.60 $\pm$ 0.013 (3)
Aogen (nmol l^{-1})	463 $\pm$ 37 (11)	676 $\pm$ 54** (9)	688 $\pm$ 155 (8)	379 $\pm$ 82 (8)	263 $\pm$ 30 (5)
PRA (nmolAI l^{-1} h^{-1})	11.7 $\pm$ 2.51 (10)	14.6 $\pm$ 3.22 (9)		14.0 $\pm$ 5.69 (6)	

Significance from day 0: *$P < 0.05$; ** $P < 0.01$.

Perhaps the high ADH levels may indicate that these xeric rodents are always near maximum urine concentration. It is only when the body has a surplus of water, which is not common in desert rodents, that they reduce their ADH concentrations and increase urine output. Reduced ADH levels have been measured in overhydrated desert rodents (Baddouri et al. 1984). Therefore, this system functions somewhat in reverse to what we are accustomed to in non-desert mammals, when ADH levels increase with water deprivation.

The renin-angiotensin-aldosterone system in mammals participates in the regulation of body sodium and potassium levels, of body fluid volumes, and of arterial blood pressure. Renin is synthesized and secreted by the juxta-glomerular cells of the kidney. It acts enzymatically on angiotensinogen, a plasma globulin, to release angiotensin which is hydrolysed by converting enzymes into angiotensin II. Angiotensin II and aldosterone are the two effector hormones in the system (Laragh and Sealey 1973). Angiotensin II acts in three principal ways: (1) as a potent constrictor of arterioles; (2) on the kidney to retain sodium; and (3) on the adrenal cortex to secrete aldosterone. Aldosterone acts primarily on the renal tubules to increase the reabsorption of sodium with chloride and to enhance the elimination of potassium ions. Furthermore, intravenous and intracranial injections of angiotensin II have elicited drinking in a number of mesic mammals; intracranial injections have elicited drinking in gerbils (see discussion by Wright and Harding 1980).

Weaver et al. (1994) reported that the maintenance of fluid volume and blood pressure during total water deprivation in *Notomys* was not dependent upon enhanced activities of the renin-angiotensin and ADH systems (Table 6.17). Wright and Harding (1980) also questioned the importance of renin and angiotensin in maintaining normal body water balance in dehydrated desert rodents. Their conclusions were based on studies of rats and two desert rodent species: the gerbil *Meriones unguiculatus* and the kangaroo rat *Dipodomys spectabilis*.

The concentration of plasma angiotensin II in normally hydrated rats was much lower than in the two desert species (Fig. 6.17). Following 4 days of food deprivation, the rats lost 18.8% body mass and 40.5% plasma volume. *M. unguiculatus* lost 18.8% body mass and 23.3% plasma, and *D. spectabilis* lost 15.1% body mass (plasma volume change was not measured; Table 6.18). Plasma osmolalities and sodium concentrations were slightly reduced and haematocrit slightly elevated in all three species. Water intake was reduced by 46.8% in the rats but was increased by 380 and 199% respectively, in the two desert species. Increased water intake, polydipsia, during starvation is common in xeric species of gerbils, kangaroo rats and hamsters, while reduced water intake during starvation is common in mesic rats and house mice. In spite of the different responses in water intake among the species and the loss in plasma volume, there was no difference in plasma angiotensin II concentrations between the controls and starved rodents within each species. There was also no difference in cerobrospinal fluid angiotensin II concentrations between control and starved rats and between control and starved *M. unguiculatus*. Therefore, the increased water intake in the desert species was not induced by the renin-angiotensin system.

Following 4 days of water deprivation in these three species, rats lost 16.9% body mass and 34.2% plasma volume, *M. unguiculatus* lost 14.9% body mass and 38.8% plasma, and *D. spectabilis* lost 12.6% body mass (plasma volume

Table 6.18. Plasma and cerebrospinal fluid measurements (mean ±SD) taken from *Rattus norvegicus*, *Meriones unguiculatus* and *Dipodomys spectabilis* when fed ad libitum, food-deprived or water-deprived. (Wright and Harding 1980)

Species and treatment condition	n	Plasma			Plasma volume[a] (ml 100 g⁻¹ BW)	Loss (%)	n	CSF AII (pg ml⁻¹)
		Osmolality (mOsm kg⁻¹)	Sodium (mEqa l⁻¹)	Haematocrit (% cells)				
R. norvegicus								
Ad libitum	8	302.0 ± 1.6	141.4 ± 1.5	45.3 ± 1.3	3.33	–	6	3.2 ± 1.0
Food-deprived	8	299.3 ± 2.4	143.9 ± 3.0	54.5 ± 1.3	1.98	40.5	6	3.0 ± 1.5
Water-deprived	8	311.1 ± 2.5	147.5 ± 3.6	53.1 ± 1.8	2.19	34.2	6	5.9 ± 1.9
M. unguiculatus								
Ad libitum	16	304.7 ± 3.8	147.6 ± 2.9	46.4 ± 1.2	4.17	–	6	4.7 ± 2.0
Food-deprived	16	296.5 ± 3.9	140.9 ± 1.8	51.3 ± 2.7	3.20	23.3	6	5.3 ± 1.5
Water-deprived	16	325.4 ± 6.4	153.0 ± 6.7	54.6 ± 2.1	2.55	38.8	6	2.8 ± 0.7
D. spectabilis								
Ad libitum	6	301.0 ± 1.3	148.5 ± 1.2	45.6 ± 1.4	–	–	–	–
Food-deprived	6	299.1 ± 2.2	146.1 ± 3.3	49.2 ± 1.5	–	–	–	–
Water-deprived	6	311.9 ± 1.9	151.8 ± 2.9	49.3 ± 2.3	–	–	–	–

[a] Derived from the regression equations PV = 9.94–0.146 (Hct) for rats (Wright et al. 1976) and PV = 13.31–0.197 (Hct) for gerbils (Wright and Donlon 1979). A similar equation is not available for kangaroo rats.

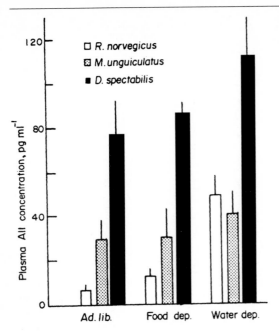

Fig. 6.17. Mean (±SD) concentrations of plasma angiotensin II measured for *Rattus norvegicus* (n = 8), *Meriones unguiculatus* (n = 16) and *Dipodomys spectabilis* (n = 6) at the completion of 4 days of food ad libitum, food deprivation or water deprivation. Ambient temperature was 21–23 °C, and RH was 40–50%. (Wright and Harding 1980)

change was not measured; Table 6.18). Plasma osmolalities, sodium concentrations and haematocrit were elevated in all three species. Plasma angiotensin II concentrations increased by approximately 500% in the rat, by a small but not significant amount in *M. unguiculatus* and by a relatively small but significant amount of 44% in *D. spectabilis*. Cerobrospinal fluid angiotensin II concentrations increased in rats but declined in *M. unguiculatus*. The different responses in angiotensin II levels during water deprivation led Wright and Harding (1980) to conclude that "xerophilous rodent species may be responding to water shortages with only moderate elevations in A2 (angiotensin II) which are physiologically relevant, or these species may be employing a body water conservation strategy that is very different from that observed in mesically adapted species".

6.2.7
Faecal Water Loss

The amount of water lost via the faeces is a function of both the water content of the faeces and the amount of faeces excreted. Granivorous desert rodents, particularly heteromyids, are known for their low faecal water loss as they produce small amounts of relatively dry faeces. The apparent dry matter digestibility of seeds is close to 90% in rodents, and therefore only 10% of dry matter is produced as faeces (Grodzinski and Wunder 1975; Degen and Kam 1991). However, the faecal water content as a fraction of fresh matter varies greatly among species, and there is much overlap among diets (Tables 6.19).

Table 6.19. Faecal water content in small eutherian mammals and in large desert eutherian mammals

Order	Species	Body mass (g)	Faecal (%)	water content (mg H$_2$O g^{-1}DM)	
Rodentia	Gerbillus henleyi	10	55.4	1242	Unpubl. data
	Reithrodontomys megalotis	12	44.0	786	Reaka and Armitage (1976)
	Reithrodontomys megalotis	13	53.0	1128	Reaka and Armitage (1976)
	Perognathus parvus	16	36.0	563	Withers (1982)
	Peromyscus maniculatus	19	44.0	786	Withers (1982)
	Gerbillurus tytonis	24	37.2	592	Downs and Perrin (1990)
	Gerbillurus paeba	24	28.8	404	Downs and Perrin (1990)
	Gerbillurus vallinus	27	32.4	479	Downs and Perrin (1990)
	Notomys alexis	29	48.8	953	MacMillen and Lee (1969)
	Notomys cervinus	34	51.8	1075	MacMillen and Lee (1969)
	Gerbillurus setzeri	35	37.0	587	Downs and Perrin (1990)
	Dipodomys merriami	35	45.0	818	Schmidt-Nielsen and Schmidt-Nielsen (1951)
	Desmodillus auriculatis	39	48.9	957	Christian (1979)
	Acomys russatus	44	48.0	923	Kam and Degen (1993b)
	Acomys cahirinus	44	52.0	1083	Kam and Degen (1993b)
	Liomys irroratus	45	37.0	587	Christian et al. (1978)
	Aethomys namaquensis	46	53.0	1128	Buffenstein (1985)
	Rhabdomys pumilio	47	43.6	773	Christian (1979a)
	Liomys pictus	50	35.0	538	Christian et al. (1978)
	Meriones crassus	75	42.3	733	Unpubl. data
	Ammospermophilus leucurus	86	43.0	754	Karasov (1983)
	Psammomys obesus	145	65.0	1857	Kam and Degen (1992)
	Rattus norvegicus	250	69.0	2226	Schmidt-Nielsen and Schmidt-Nielsen (1951)
Lagomorpha	Lepus californicus	1800	38.0	613	Nagy et al. (1976)
	Oryctolagus cuniculus	2000	48.0	923	Cooke (1982)
Perissodactyla	Equus asinus (Somali donkey)	150 000	60.9	1558	Maloiy (1970)
Artiodactyla	Gazella dorcas	16 000	51.0	1041	Ghobrial (1970)
	Capra hircus (Turkana goat)	18 000	51.5	1062	Maloiy and Taylor (1971)
	Capra hircus (Sinai goat)	20 000	40.0	667	Shkolnik et al. (1972)
	Capra hircus (E. African goat)	22 000	46.8	880	Schoen (1968)
	Ovis aries (Somali sheep)	23 000	48.2	931	Maloiy and Taylor (1971)
	Ovis aries (Awassi sheep)	45 000	45.0	818	Degen (1977)
	Ovis aries (Merino sheep)	50 000	44.3	795	Degen (1977)
	Aepycerus melampus	65 000	53.3	1141	Maloiy and Hopcraft (1971)
	Alcelaphus buscelaphus	100 000	51.9	1079	Maloiy and Hopcraft (1971)
	Taurotragus oryx	150 000	61.5	1597	Taylor and Lyman (1967)
	Bos taurus	230 000	75.1	3016	Taylor and Lyman (1967)
	Camelus dromedarius	450 000	43.6	773	Maloiy (1972); Schmidt-Nielsen (1964)

The water content of the faeces of kangaroo rats (*Dipodomys*) on a dry diet is only 45% of fresh matter (834 mg H_2O g^{-1} faecal dry matter). In comparison, the white rat on the same diet excreted faeces with a water content of 69% (2246 mg H_2O g^{-1} dry matter; Schmidt-Nielsen 1964). Dehydrating heteromyids can further reduce the water content of their faeces. For example, when water-stressed, *Perognathus parvus* reduced its faecal water content from 40% (667 mg H_2O g^{-1} dry matter) to 36% (562.5 mg H_2O g^{-1} dry matter) of the fresh matter (Withers 1982), while *Liomys pictus* and *L. irroratus* were able to produce faeces of 35 and 37% water content (540 and 575 mg H_2O g^{-1} dry matter), respectively (Christian et al. 1978). These latter values appear to be the lowest reported for any mammal. This is of interest, as these heteromyids do not penetrate deserts of the same aridity as other heteromyids.

Notomys alexis and *N. cervinus*, two murids that can survive on certain dry diets and that have the ability to concentrate their urine greatly, produce faeces with a water content of 48.8–51.8% when deprived of drinking water (MacMillen and Lee 1969). Gerbilline rodents are also known for their granivory and their ability to conserve water. The faecal water content of *Gerbillus henleyi* was 55.4 (SD = 12.0; n = 14) and of *Meriones crassus* 42.3; ($\pm$7.15; n = 8) when offered only millet seeds. When offered green vegetation in addition to the seeds, faecal water increased to 65.8% ($\pm$10.2%; n = 6) and 55.5% ($\pm$2.40%; n = 17), respectively (unpubl. data).

Faecal water content as a percentage of fresh matter is higher on a high-fibre diet that on one of seeds. Furthermore, the dry matter digestibility of vegetation is much lower than that of seeds – amounting to approximately 67% of dry matter intake. Consequently, far more faeces are produced than on a seed diet. The rock rat (*Aethomys namaquensis*) and the pigmy gerbil (*Gerbillurus paeba*) inhabit the Namib Desert. When offered millet seeds and water ad libitum, they produced faeces containing 53–54% water content; however, when offered bran instead of seeds, the water content of the faeces rose to 58–61%. When deprived of water on these diets, the faecal water content was reduced to 45–46% for millet and 50–51% for bran (Buffenstein 1985). *Psammomys obesus* produced faeces with 65–68% water content when offered *Atriplex halimus* leaves with a water content of 77.5%. Air temperatures ranging between 15 and 34 °C and reduction in the water content of the leaves had no effect on faecal water content in this species (Kam and Degen 1992, pers. observ.). However, the black-tailed jackrabbit (*Lepus californicus*) can reduce its faecal water to 38% when consuming dry pasture (Nagy et al. 1976; Nagy 1994a).

The water content of the faeces of *Acomys cahirinus* appears to be similar for either a meat or a seed diet. When offered snails plus millet seeds, the faecal water content was 56% of fresh matter, and when offered snails only, it was 52–58%. When seeds plus water or snails plus water were offered, the faecal water contents were 68 and 61%, respectively (Kam and Degen 1991, 1993b). A lower faecal water content was found for *A. russatus*, a spiny mouse that coexists with *A. cahirinus* but can penetrate harsher deserts. This species produced faeces with 48% water content when offered snails and seeds (Kam and

Degen 1993b). The omnivorous antelope ground squirrel (*Ammospermophilus leucurus*) can reduce its faecal water content to 43% of fresh matter when water is restricted (Karasov 1983).

The Colon and Water Absorption. Forman and Phillips (1993) reported that the proximal colon of mammals is important in the absorption of water from faecal material, and particularly so in desert species. It also absorbs numerous other molecules and ions such as amino acids, urea, ammonia, sugars, volatile fatty acids, sodium, potassium, chloride and bicarbonate. The morphology of the lower colon of several heteromyids was found to differ from that of other rodents, and it was suggested that the changes may be related to their enhanced ability to reabsorb water (Forman and Phillips 1988, 1993). When viewed in cross-section, the mucosa distal to the caecum is asymmetrical. This colonic asymmetry is more pronounced in species inhabiting more arid areas. The mesenteric mucosa shows a reduction in mucus-producing goblet cells and an increase in thick absorptive-type columnar cells. This results in decreased mucus production and, consequently, in a reduced loss of mucus through the faeces, and in an increased absorption area. In addition, "broad glandular pits, particularly evident in *Dipodomys merriami* and *D. spectabilis*, could serve as 'sinks' for the collection and absorption of water, or as means of increasing absorptive surface area by way of the epithelial projections which would lie between them" (Forman and Phillips 1993). The vasculature on the mesenteric side is prominent, indicating its role in transporting water from the colon.

Murray et al. (1995) described the distal and proximal colons of *Notomys alexis* and *Pseudomys hermannsburgensis*. Their study revealed that the distal colon of these two desert species "is relatively greater than that of many North American forest rodents, and is suggestive of a greater ability to absorb water in the desert species which would facilitate water conservation". Their study also revealed that the mucosal surface of the proximal colon is organized into oblique folds, which greatly increases the surface area. These folds separate the fluid and small particles from the large particles and enhance retrograde transport to and retention of the fluid and fine particles in the caecum. It is possible that the increased surface area increases water absorption, which would agree with the reports of Forman and Phillips (1988, 1993). However, as net water absorption occurs in the distal rather than the proximal colon, Murray et al. (1995) postulated that the main function of the folds and retrograde transport is to improve digestion by increasing the retention time of the small particles.

6.3
Total Body Water Volume and Its Distribution

Total body water volume of an animal is composed of intracellular and extracellular fluid volumes. Extracellular fluid can be divided into plasma water and interstitial water, the latter including gastrointestinal fluid. In order to study body water balance and fluid compartment regulation, it is essential to measure total body water and its distribution. A fluid space is usually measured by the dilution technique. This consists of administering a known amount of an appropriate marker that penetrates the total space in question. The concentration of the space marker declines at a constant exponential rate. By taking serial samples of fluid after marker administration, dilution at $t = 0$ can be extrapolated from the intercept of the regression line on the y-axis. Single samples at a time that approximates to equilibration of the marker within the space being measured could also be used. The space in question is then calculated by dividing the total amount of marker administered by the concentration at time zero.

For total body water volume estimation, tritiated or deuteriated water are commonally used. These isotopes can be administered either intravenously, intramuscularly, intraperitoneally or orally. For small mammals, administration of the isotope is generally done intramuscularly or intraperitoneally, and single blood samples are collected after equilibration of the isotope with body fluids. For extracellular fluid volume, ^{14}C-labelled inulin and sodium thiocynate (SCN^-) have been used for small mammals. These are introduced into the plasma and remain in the extracellular fluid because of their large molecular size. Evan's blue or T-1824 is usually used to measure plasma volume. T-1824 binds onto albumin within the plasma fluid and, in addition to estimating the plasma volume space, can also be used to estimate the outflow of albumin from the plasma space. Serial samples and extrapolation are usually done for extracellular and plasma volume space estimations, although single samples taken after 20 and 10 min, respectively, have been used. The intracellular fluid volume can then be calculated from the difference between total body water space and extracellular fluid volume space, and interstitial fluid space by the difference between extracellular fluid volume and plasma volume.

Total body water as a proportion of body mass varies inversely with body lipid, being relatively high in animals with little body lipid content and low in animals with much body lipid. As a result, young, growing animals usually have a relatively large total body water volume which decreases as the animal matures and its lipid content increases. In mature animals, total body water can be used as an indicator of body lipid content and as an index of body condition. This was demonstrated in two sheep breeds in the Negev Desert whose total body water increased in the summer when pasture was poor and the sheep were mobilizing body lipid for energy, and decreased in the winter when pasture was lush and the sheep were adding to their energy reserves (Degen 1977). Tritiated water has been used to estimate the total body water volume and body composition (lipid, protein and ash contents) as well as total energy yield of the body (Benjamin et al. 1993). However, this technique has been employed mainly with large animals. In all mammals, including small-sized ones, total body water volume (TBWV) has been used to estimate the fat free-free mass (FFM) by the relationship (Prentice et al. 1952):

FFM (fraction of body mass) = TBWV (fraction of body mass)/0.73.

Animals that have adapted to desert conditions show large variability in total body water volume. Camels (Macfarlane 1964) and sheep (Hansard 1964; Anand et al. 1966) have a relatively low total body water volume, whereas Sinai goats (Shkolnik et al. 1972), elands (Macfarlane and Howard 1972) and hartebeests (Maloiy and Hopcraft 1971) have large total body water volumes. Nonetheless, all these mammals thrive under harsh, dry conditions. It would seem that animals with large amounts of body solids, especially those with localized fat deposits, maintain an energy reserve and have adapted to food shortages.

Unlike the numerous studies available on total body water volume and its seasonal changes in large mammals, few studies are available for small mammals. Schmidt-Neilsen (1964) reported a total body water volume of about 66% body mass for laboratory-maintained kangaroo rats *Dipodomys merriami*. This volume, as a percentage of body mass, remained constant for 7 weeks during which time the kangaroo rats were kept on a dry diet of barley with no drinking water. A total body water pool of about 60% body mass was reported for three species of captive rodents (body mass = 21–298 g) and rabbit (Richmond et al. 1962). Holleman and Dieterich (1973) reported a tritiated water space of 61–71% body mass for 11 species of rodents ranging in body mass from 18.5 to 753 g. A twelfth species, the vesper mouse (*Calomys ducilla*; body mass = 26.6 g), had a tritiated water space of only 44.6% body mass.

Energy requirements and feed availability for free-living animals usually vary from season to season. This can affect their total body water volume and body composition. However, results of measurements of seasonal changes in total body water in small desert mammals have not been equivocal. Tritiated water space in free-living fat sand rats ranged between 70 and 75% body mass in adults, and between 73 and 79% body mass in juveniles, and did not differ significantly within groups between seasons (Degen et al. 1990). Similarly, tritiated water spaces in two gerbillinae, *Gerbillus allenbyi* and *G. pyamidum*, did not change between the winter and summer. The water contents of these two ranged from 71.2 to 73.7% body mass, and also did not differ between the species (Degen et al 1992). Furthermore, the total body water volume of free-living *Dipodomys merriami* altered relatively little throughout the year and ranged between 65.1 and 69.4% body mass (Nagy and Gruchasz 1994). It can be reasoned that the small size of most rodents does not afford them the same option of storing large amounts of energy and significantly changing their total body water volumes.

Seasonal changes in the total body water volume of small desert mammals have nonetheless been reported. Grenot and Buscarlet (1988) measured tritiated water space during the dry and rainy seasons in five sympatric rodent species of the Chihuahuan Desert: the pocket mouse (*Perognathus penicillatus*), the kangaroo rats *Dipodomys nelsoni* and *D. merriami*, the spotted ground squirrel (*Spermophilus spilosoma*) and the packrat (*Neotoma albigula*). The tritiated water space of these species ranged between 64.8 and 67.6% in the dry season and increased to between 76.6 and 82.9% in the rainy season. In Blanford's fox (*Vulpes cana*; body mass = 1 kg), the summer tritiated water space

Table 6.20. Plasma volume (PV) changes in different species of mammals under rapid dehydrating conditions (Horowitz and Borut 1994)[a]

Species	PV conserver/non-conserver	Reference
A. cahirinus[b]	PV conserver	Horowitz and Borut (1970)
M. crassus[b]	PV non-conserver	Horowitz and Borut (1970)
P. obesus[b]	PV conserver	Horowitz et al. (1978)
R. norvegicus[b]	PV non-conserver	Horowitz and Borut (1970)
		Horowitz and Samueloff (1979)
Camelus dromedarius[d]	PV conserver	Schmidt-Nielsen (1964)
Lama guanacoe	PV non-conserver	Rosenmann and Morrison (1963)
Equus asinus[e]	PV conserver	Yousef et al. (1970)
		Kasirer-Israeli et al. (1994)
Somali donkey[f]	PV non-conserver	Maloiy and Boarer (1971)
Papio hamadryas[f]	PV conserver	Zurovsky et al. (1984)
Dog	PV non-conserver	Meir and Shkolnik (1984)
slughi[g]		
pointer[g]		
canaan[g]		
mongrel[o]		Horowitz and Nadel (1984)
Capra hircus[e]	PV non-conserver	Shkolnik et al. (1980) Shkolnik
		and Choshniak (1987)
Merino sheep	PV non-conserver	Macfarlane et al. (1959)
Capra ibex nubiana[g]	PV non-conserver	Shkolnik et al. (1980)
Gazelle dorcas[f]	PV conserver	Peled (1989)
Gazelle gazelle[f]	PV non-conserver	Peled (1989)
Man[h]	PV non-conserver	Adolph (1947); Senay and
		Christensen (1964)

[a]Species exhibiting a ratio of plasma water loss: total body water loss (PV:TBW loss) < 1 are defined as PV conservers, and > 1, as PV non-conservers.
Dehydrating conditions. [b]Thermal dehydration: 37 °C, 20–30% RH. [c]Thermal dehydration: 40 °C, 20–30% RH. [d]8 days without water, summer desert. [e]Walk in summer desert without drinking, following food and water deprivation. [f]Water deprivation, high temperature outdoors. [g]Climatic chamber, 30–32 °C, dry food. [h]Hot climatic chamber, exercise.
In the above list: *A. cahirinus, P. obesus, C. dromedarius, E. asinus, P. hamadryas,* and *G. dorcas* have a long evolutionary history in harsh arid, hot climates.

of 73.3% body mass was higher than the winter space of 66.1% (Geffen et al. 1992a). The lower volume of total body water during the rainy season in the former studies and during the winter in the latter indicated higher body lipid contents and, presumably, better body condition during these periods.

6.4
Body Fluid Regulation

Most studies on fluid compartment regulation in mammals during thermal heat stress and dehydration have been carried out on large animals. Horowitz and Samueloff (1979) divided mammals into two main groups with respect to dehydration: (1) plasma volume conservers and (2) plasma volume non-conservers. Mammals within these groups are not related on either the generic or family level (Table 6.20; Horowitz and Borut 1994). Mogharabi and Haines (1973) divided dehydrated mammals into three categories along the same lines: (1) those in which the plasma space loses proportionately less water than either the intracellular or interstitial space; (2) those whose plasma space loses proportionately more water than either the intracellular or interstitial space; and (3) those in which all water fluid spaces lose proportionately similar amounts.

The circulatory system is important in thermoregulation during thermal stress and dehydration. Blood is the major means by which heat is conducted from the core of the body to the periphery where it is dissipated. Plasma water is lost during sweating, saliva spreading and respiration, and may be severely depleted in hot desert environments. The plasma volume lost is replenished from the interstitial fluid. Thus, the relationship between plasma fluid loss and extracellular fluid loss indicates the rate of plasma fluid replacement.

The ability of mammals to conserve plasma volume during dehydration is important in the maintenance of circulatory function and body temperature control. It is considered to be an adaptation to desert conditions which allows mammals to tolerate heat stress. The inability of a mammal to maintain plasma volume results in more viscous blood, failure in thermoregulation, an explosive heat rise and eventual death.

Mammals are able to maintain their plasma volume up to a certain level of dehydration. Beyond this point, regulatory mechanisms fail, and the plasma volume loss is actually enhanced. In general, desert mammals are able to tolerate a higher degree of dehydration and are able to maintain their plasma volume for a longer period of time (Horowitz and Samueloff 1979). There are numerous exceptions to this rule in the literature. Some of the differences are due to the degree of acclimatization to heat before the study was conducted, and some are due to the method of dehydrating the animal.

Perhaps the best-known example of a plasma conserver is the one-humped dromedary (*Camelus dromedarius*; Schmidt-Nielsen et al. 1956; Schmidt-Nielsen 1964). Following 8 days of water deprivation, the camel lost 17% of its body mass, mainly due to a reduction in total body water volume. Of the total body water loss, there was a 24% reduction in intracellular water volume (tritiated water space – SCN^- space), a 38% reduction in interstitial water volume (SCN^- space – T-1824 space), but only a 10% reduction in plasma water volume (T-1824 space). Thus, the camel was able to maintain a near normal plasma volume at the expense of other body fluid volumes. This ability to conserve plasma volume has also been demonstrated in the burro (Yousef et al. 1970), the kangaroo (Denny and Dawson 1975) and the Marwari sheep (Taneja 1973), all of which are desert animals.

However, the ability to maintain plasma volume at the expense of other body fluids has not been demonstrated in all desert mammals (Table 6.21). Australian merinos dehydrated for 5 days lost 25% of their body mass and 45% of both their T-1824 space and SCN^- space (Macfarlane et al. 1956). Furthermore, the same species or breeds of animals show different responses to dehydration. For

Table 6.21. Total body water loss and its compartments in eutherian mammals and plasma (p) ume (ECFV) loss

Species		Condition	m_b	% loss m_b	% loss TBW
Acomys cahirinus	Spiny mouse	WD 10	40	12.8	11.2
Meriones unguiculatus	Fat jird	FD 1	70	14.9	
Meriones unguiculatus	Fat jird	WD 1	70	14.9	
Meriones crassus	Fat jird	WD 10	92	18.8	3.3
Ammospermophilus leucurus	Antelope ground squirrel	WD 9	100	10.0	
Ammospermophilus leucurus	Antelope ground squirrel	WD 9	100	20.0	
Ammospermophilus leucurus	Antelope ground squirrel	WD 9	100	30.0	
Rattus norvegicus	Rat	WD 10	120	19.0	21.0
Rattus norvegicus	Rat	WD LD 10	140	15.5	18.5
Rattus norvegicus	Rat	WD HD 10	140	20.5	25.5
Rattus norvegicus	Rat	WD ACC 10	141	21.0	22.2
Spermophilus beldingi beldingi	Belding ground squirrel	WD 9	250	10.0	
Spermophilus beldingi beldingi	Belding ground squirrel	WD 9	250	20.0	
Spermophilus beldingi beldingi	Belding ground squirrel	WD 9	250	30.0	
Rattus norvegicus	Rat	FD 1	475	18.8	
Rattus norvegicus	Rat	WD 1	475	16.9	
Citellus tridecemlineatus	Squirrel	WD 1	160	23.0	26.0
Psammomys obesus	Fat sand rat	WD LD 10	230	10.1	14.3
Psammomys obesus	Fat sand rat	WD HD 10	230	13.8	21.5
Psammomys obesus	Fat sand rat	WD ACC 10	230	12.9	16.0
Rattus norvegicus	Rat	WD 1	290	23.0	26.0
Lepus californicus	Jackrabbit	WR 1	2 300	10.6	10.7
Ovis ovis	Marwari sheep	WD 8	38 000	18.0	31.0
Ovis ovis	Awassi	WR 7	40 000	21.0	25.8
Ovis ovis	GMM	WR 7	47 000	27.5	35.4
Equus asinus	Somali donkey	WD 3	155 000	15.5	21.0
Equus asinus	Burro	WD 4	195 000	18.0	25.0
Bos taurus	Desert cattle	WD 6	438 000	15.0	16.0
Camelus dromedarius	Camel	WD 2	521 000	17.0	
Camelus dromedarius	Camel	WD 5	619 000	17.0	23.0

WD = water deprived: WR = water restricted; FD = food deprived; LD = low dehydration; HD = high dehydration; ACC = acclimatized; 1, at room temperature; 2, 3–8 days in summer desert; 3, 22–40 °C; 4, 1–1.5 days in summer desert than walk; 5, 7–10 days in summer desert; 6, 3–4 days in summer desert; 7, in summer desert; 8, 3–5 days in summer desert; 9, 21 °C–45% RH; 10, 37 °C–18 RH; TBW, total body weight loss as a fraction of body mass (m_b) loss, total body water (TBW) loss and extracellular fluid vol-

% loss ECFV	% loss plasma	P/m_b	P/TBW	P/ECFV	Reference
0.0	4.9	0.4	0.4		Horowitz and Borut (1970)
	23.3	1.3			Wright and Harding (1980)
	38.8	2.6			Wright and Harding (1980)
18.0	11.0	0.6	3.3	0.5	Horowitz and Borut (1970)
	8.0	0.8			Hartman and Morton (1973)
	4.0	0.2			Hartman and Morton (1973)
	12.0	0.4			Hartman and Morton (1973)
32.0	34.0	1.8	1.6	1.1	Horowitz and Borut (1970)
22.0	15.0	1.0	0.8	0.7	Horowitz and Samueloff (1979)
34.0	45.0	2.2	1.8	1.3	Horowitz and Samueloff (1979)
20.3	NC				Horowitz and Samueloff (1979)
	8	0.8			Hartman and Morton (1973)
	24	1.2			Hartman and Morton (1973)
	22	0.7			Hartman and Morton (1973)
	40.5	2.2			Wright and Harding (1990)
	34.2	2.0			Wright and Harding (1990)
	36.0	1.6	1.4		Mogharabi and Haines (1973)
32.4	NC				Horowitz et al. (1978)
53.0	37.0	2.7	1.7	0.7	Horowitz et al. (1978)
NC	NC				Horowitz et al. (1978)
	39.0	1.7	1.5		Mogharabi and Haines (1973)
10.7	10.7	1.0	1.0	1.0	Reese and Haines (1978)
33.0	44.0	2.4	1.4	1.3	Purohit et al. (1972)
15.6	15.7	0.7	0.6	1.0	Degen (1977)
20.3	24.5	0.9	0.7	1.2	Degen (1977)
29.0	35.4	2.3	1.7	1.2	Maloiy and Boarer (1971)
14.0	7.0	0.4	0.3	0.5	Yousef et al (1970)
21.0	25.0	1.7	1.6	1.2	Siebert and Macfarlane (1975)
	9.0	0.5			Schmidt-Nielsen and Schmidt-Nielsen (1952)
30.0	17.0	1.0	0.7	0.6	Siebert and Macfarlane (1972)

example, Marwari sheep dehydrated for 3 days lost 18% body mass, 45% of their T-1824 space and 33% of their SCN⁻ space. Dehydrated donkeys also lost 21% of their total body water but 31% of their plasma volume (Maloiy and Boarer 1971). The Marwari sheep and donkeys seemed to lose plasma volume preferentially, which is contrary to the results reported by Taneja (1973) and Yousef et al. 1970), respectively. Perhaps these differences can be explained by the various methods used in dehydrating the animals, since Macfarlane et al. (1961) reported that plasma loss is much smaller during slow than during rapid dehydration.

In comparison with the number of studies on losses of body fluid spaces during dehydration in large animals, relatively few have been made on small mammals. This is due mainly to the difficulty in making the necessary measurements. As reported in large animals, some small mammals tend to conserve plasma volume during dehydration better than others. The counterpart to the camel among small mammals with regard to conserving plasma volume appears to be the desert spiny mouse *A. cahirinus*. This mouse has the ability to maintain its plasma volume relatively unchanged during severe dehydration, which causes about a 11.5% body mass loss (Fig. 6.18), representing the upper limit that *A. cahirinus* can tolerate and still survive (Horowitz and Borut 1994). However, unlike the camel in which intracellular fluid comprise less than 30%

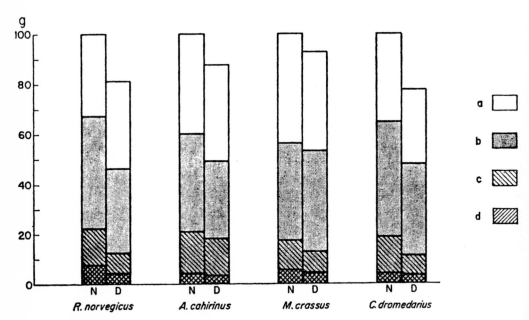

Fig. 6.18. Body fluid compartments in non-dehydrated and dehydrated *Rattus norvegicus, Acomys cahirinus, Meriones crassus* and *Camelus dromedarius* (from Schmidt-Nielsen 1964), calculated as g/100 g initial body mass of non-dehydrated animal. *a* Body mass; *b* total body water volume; *c* extracellular fluid volume; *d* plasma volume; *N* non-dehydrated; *D* dehydrated. (Horowitz and Borut 1970)

of total body water loss, 77% of the total water loss in *A. cahirinus* was from the intracellular fluid.

In general, small desert-adapted mammals conserve plasma volume to a greater degree during body water loss than non-desert species. Hartman and Morton (1973) examined plasma volume losses in dehydrated desert-dwelling antelope ground squirrels (*Ammospermophilus leucurus leucurus*) and in non-desert, high altitude-dwelling Belding ground squirrels (*Spermophilus beldingi beldingi*). Water was denied, and measurements were made at 0, 10, 20 and 30% body mass losses. At 10% body mass loss, both species lost 8–9% plasma volume but at 20 and 30% body mass losses, the desert species lost proportionately much less. Plasma conservation was noted at 10% body mass loss in the desert species and at 20% body mass loss in the non-desert species (Fig. 6.19). It was concluded that the desert species was much better able to conserve plasma volume and began to conserve plasma volume earlier than did the non-desert species.

The black-tailed jackrabbit (*Lepus californicus*), which inhabits hot, arid areas of the American Southwest, also minimizes plasma water loss during dehydration. It is considered to be a plasma conserver (Reese and Haines 1978). Following chronic water restriction, achieved by maintaining rabbits at an air

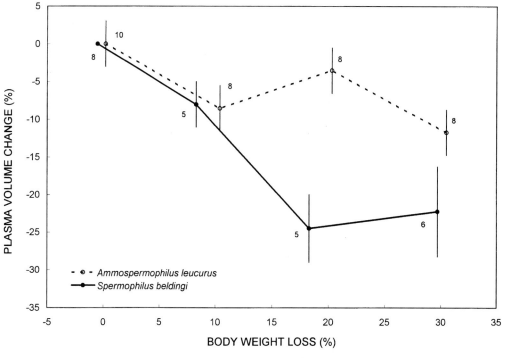

Fig. 6.19. Mean (± 2 SE; *numbers* is sample size) plasma volume changes in ground squirrels following body mass loss due to dehydration. (Hartman and Morton 1973)

temperature of 20–25 °C and gradually reducing their drinking water until the body mass had been reduced by 10.6%, total body water reduced by only 2.5%. Both extracellular and plasma fluid volumes declined in the same proportion to body mass loss. With acute dehydration, during which all drinking water was denied until 10.7% body mass had been lost, rabbits reduced the amount of water in all compartments by the same proportion. With acute thermal dehydration, in which water was denied and the air temperature kept at 35–40 °C, rabbits could tolerate only a 5.5% body mass loss. The plasma volume was reduced by the same proportion.

Preferential plasma volume conservation is not demonstrated by all small desert mammals. Mogharabi and Haines (1973) dehydrated 13 lined ground squirrels (*Citellus tridecemlineatus*), whose distribution includes arid regions, and the laboratory rat (*Rattus norvegicus*) by completely removing drinking water. Dehydration was stopped when the animals had lost 20–25% of their body mass. This took about 5 days in the rats and 23 days in the squirrels. Proportional body fluid losses, however, were similar in both species. Both species lost 26% of their total body water volume; the squirrels lost 36% of their plasma volume and the rats, 39%. Thus, xeric squirrels can resist dehydration much longer than rats can, but the body fluid shifts in the two species were similar. Neither species conserved plasma volume, but both lost proportionately more plasma water than total body water.

Changes in body fluid compartments following acute thermal dehydration were measured in two desert-dwelling rodent species, the spiny mouse *Acomys cahirinus* and the gerbil *Meriones crassus*, and in the laboratory rat (Horowitz and Borut 1970). Animals were maintained at 37 °C without water "until they suffered the maximal weight loss from which they were able to recover when offered water or fresh vegetables." The rat lost body mass at a faster rate than did the two desert species, but was able to survive a proportionately much greater loss in body mass: 14–22% vs 11–14% for *A. cahirinus* and only 5–9% for *M. crassus*. All the body mass loss in the rat and *A. cahirinus* was accounted for by a reduction in total body water, whereas only 50% of the loss in *M. crassus* was body water. The rat suffered significant reductions in plasma and extracellular fluids: by 34 and 32%, respectively. In *M. crassus*, the extracellular fluid volume was significantly reduced by 18% and the plasma volume by 11%, but this latter was not statistically significant. None of these fluid spaces was reduced in *A. cahirinus* (Fig. 6.18).

The conservation of plasma volume in rodents is also dependent upon the degree of dehydration and whether the animal had been acclimated to heat stress. Non-acclimated fat sand rats (*Psammomys obesus*) conserved plasma volume during acute dehydration as long as body mass loss did not exceed 10–11%. In fact, plasma volume, as a proportion of body mass, increased compared with control animals. This degree of body mass loss took 16–30 h and was considered as low dehydration (LD). Total body water volume and extracellular fluid volume were reduced by 14.3 and 32.4%, respectively. Plasma conservation was therefore at the expense of interstitial fluid. However, at 14% body mass loss, *P. obesus* could not maintain its plasma volume, and this space was reduced by 37%. Such a loss of body mass took 25–44 h and was considered to be high dehydration (HD). Concomitantly, total body water and extracellular fluid volumes were reduced by 21 and 53%, respectively. The plasma volume loss was therefore proportionately less than the extracellular fluid volumes loss, but was greater than the losses of body mass and total body fluid (Fig. 6.20).

The importance of the degree of acclimation in relation to thermal stress previous to dehydration has been well illustrated in the fat sand rat (Horowitz and Samueloff 1979). In fat sand rats acclimated to heat stress, dehydration to 12–13% loss in body mass, which required 25–48 h, led to a 16% reduction in total body water volume; but plasma and extracellular fluid volumes were maintained (Fig. 6.21). Furthermore, the total body water volume of dehydrated acclimated fat sand rats was higher than that of dehydrated LD and HD animals (70.6 vs 67.9% body mass), but the total body lipid content was lower (12.26 vs. 15.66%).

6.4.1
Colloid Osmotic Pressure

The ability of small desert mammals to conserve plasma volume is a consequence of an increase in colloid osmotic pressure (COP) due to an increase in albumin mass. This can be accomplished by: (1) a decrease in the permeability of the blood capillaries to serum albumin; and (2) enhanced albumin syn-

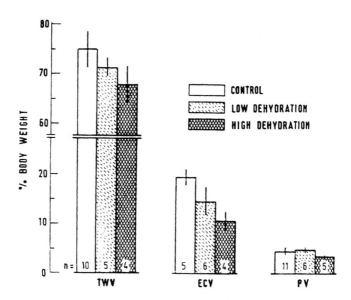

Fig. 6.20. Body water compartment volumes ($\pm$SD) in non-acclimated *Psammomys obesus* before and after low and high dehydration at 37 $\pm$ 1 °C *PV* plasma volume; *ECV* extracellular volume; *TWV* total body water volume. (Horowitz et al. 1978)

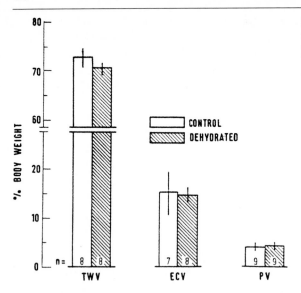

Fig. 6.21. Body water compartment volumes ($\pm$SD) in acclimated *Psammomys obesus* before and after dehydration at 37 °C (Horowitz et al. 1978)

thesis. The former can be accomplished by changes in the morphology of the capillaries to minimize the outflow of albumin. In *Acomys cahirinus*, a murid known for its ability to conserve plasma volume (Horowitz and Borut 1994), electron microscopy revealed that the capillary endothelium closed into tight junctions. It was suggested that "during dehydration more junctions close into tight junctions and reduce the permeability of the vascular system to small proteins" (Borut et al. 1972). An increase in albumin mass can also be accomplished by the redistribution of cardiac output to capillaries which are less permeable to albumin outflow. For example, fenestrated capillaries in the gastrointestinal tract, and particularly the liver vascular bed, provide a permeable vascular area for the outflow of large protein molecules. As the capacity of the splanchic area equals about 20% of the total blood volume, it is likely that the leakage of blood proteins from this area accounts for the highest proportion of protein efflux from the circulation (Horowitz et al. 1985). In contrast, capillary beds in the skin are less permeable to plasma proteins and, following acclimation to heat, cutaneous capillaries further reduce their permeability to large molecules (Senay 1972).

Both a decrease in albumin outflow and enhanced albumin synthesis are used in the maintenance of albumin mass to conserve plasma volume, but large differences exist among species. For example, in the spiny mouse, the maintenance of plasma protein mass and colloid osmotic pressure to conserve plasma volume during dehydration is a consequence of a 45% decrease in protein outflow from the circulatory system and a 21% increase in plasma albumin synthesis (Horowitz and Borut 1970, 1975). In the fat sand rat, albumin outflow from the capillary bed almost ceased when plasma was conserved during low dehydration, but albumin synthesis did not increase (Horowitz and Adler 1983).

During high dehydration, when the plasma volume was not maintained but declined drastically, albumin outflow increased above that of controls. This bimodal pattern of changes in plasma volume and albumin outflow with the degree of dehydration was also observed in the rat. The fat sand rat, however, was able to conserve plasma volume during the first phase for a long period, whereas the rat could only maintain plasma volume for a short time. However, unlike in the desert spiny mouse and fat sand rat, protein synthesis in the rat contributes largely to the maintenance of albumin mass for water retention (Horowitz and Adler 1983). In discussing the two methods of maintaining albumin mass, Horowitz and Adler (1983) concluded "that retention of albumin and plasma volume brought about by changes in capillary permeability seems to be more effective, more rapid and is possibly less energy consuming than increasing albumin production. This suggests that the mechanism of decreased capillary permeability has advantages as an adaptation to desert conditions."

The change in permeability of the capillaries to plasma albumin is indicated by the rate of disappearance of T-1824 from the plasma volume, since T-1824 couples with the blood albumin. ^{14}C-dextran, which has a molecular weight similar to that of plasma albumin, can also be used. With thermal stress, albumin efflux increased slightly in *A. cahirinus* but decreased significantly during thermal dehydration. In the rat, the flux increased with both thermal stress and thermal dehydration (Fig. 6.22). Differences in albumin efflux were found in two desert gerbilline rodents: in *M. crassus*, the rate of disappearance of T-1824 increased significantly during dehydration, indicating a leakage of plasma proteins from the circulatory system, wheras the rate decreased in *P. obesus*, indicating a lower loss of plasma proteins. *M. crassus* is considered to be a non-conserver of plasma, whereas *P. obesus* is a plasma conserver.

The rate of disappearance of T-1824 at approximately 15%/h was the same in hydrated and dehydrated mesic *Spermophilus beldingi beldingi*. This was not the case in the xeric *Ammospermophilus leucurus leucurus*. The rate of loss of T-1824 from plasma was 18%/h when receiving water ad libitum and when dehydrated to a reduction of 10% in body mass, but declined to 7–9%/h when water deprivation resulted in 20–30% body mass loss (Hartman and Morton 1973).

The incorporation of ^{3}HL-leucine into the plasma albumin is used to measure albumin synthesis. Change in the specific activity of plasma ^{3}H increased greatly in dehydrated rats, slightly in dehydrated *A. cahirinus*, and decreased in dehydrated *P. obesus* (Fig. 6.23; Horowitz and Samueloff 1987).

6.5
Cardiac Output and Its Distribution During Dehydration

Changes in cardiac output during dehydration vary widely among mammals. Some species show no change, others show a decrease, and still others show an increase. These changes do not appear to be related to the habitat (Horowitz and Samueloff 1988).

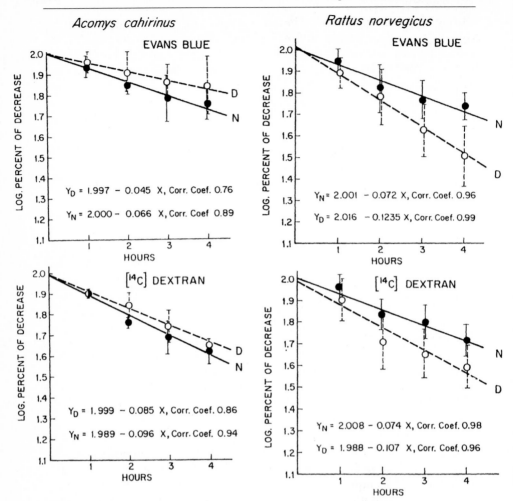

Fig. 6.22. Decrease in concentration of ^{14}C-dextran (*below*) and Evan's blue (*above*) in plasma of normal (●) and dehydrated (o) *Acomys cahirinus* ($n = 6$) and *Rattus norvegicus* ($n = 6$). Data given as the logarithmic value ($\pm$ SD) of the percentage decrease in concentration of the tracers. Dehydrating conditions: 37 °C and 15–20% RH. (Borut et al. 1972)

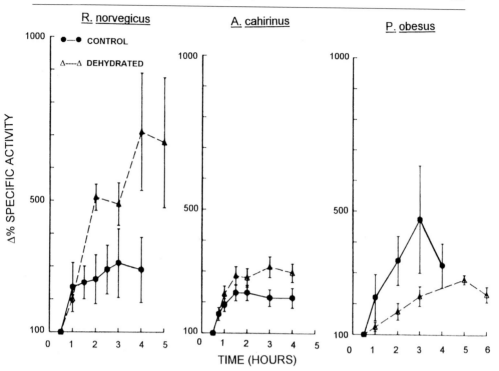

Fig. 6.23. Plasma albumin synthesis in control and thermally dehydrated (low dehydration) *Rattus norvegicus, Acomys cahirinus* and *Psammomys obesus*. Albumin synthesis is expressed as the increase (%) in ³HL-leucine incorporated into plasma albumin. (Horowitz and Samueloff 1987)

In the conscious rat, cardiac output showed a slight decline with low dehydration, and a large decline with high dehydration, whereas in the conscious *P. obesus*, cardiac output increased by 30–40% with both low and high dehydration. At low dehydration of 7–9% body mass, rats showed peripheral vasodilation concomitantly with splanchnic vasoconstriction and a reduction of blood flow to the splanchnic organs. A large portion of the blood was directed to the arteriovenous anastomoses (AVA) and shunted from the muscles and splanchnic organs. Fat sand rats also showed peripheral vasodilation, and a smaller percentage of the cardiac output reached the splanchnic organs. However, because of the increased cardiac output in *P. obesus* and a decrease in the volume of blood to the muscles, there was actually a slight increase in the volume of blood perfusing the splanchnic area. With high dehydration and a failure to maintain plasma volume and body temperature, both *P. obesus* and rats exhibited vasodilation in the splanchnic areas and decreased blood flow to AVA.

Maintenance of splanchnic perfusion and peripheral vasodilation during thermal dehydration has also been reported in *Acomys cahirinus* (Horowitz and Samueloff 1987). Horowitz and Samueloff (1988) suggested that "internal blood perfusion together with peripheral thermoregulatory vasodilation might have an advantage over peripheral vasodilation at the expense of splanchnic blood flow" in that a decrease in blood flow to the splanchnic organs can lead to an impairment in heat dissipation from this area.

6.6
Water Flux

The water flux of an animal, also referred to as water turnover, equals the total volume of water passing through the animal. This can be differentiated into water influx, or the water gained by the animal, and water efflux, or the water lost by the animal. Water influx includes preformed and metabolic water from food intake, drinking or free water intake, and the water produced from the catabolism of body tissue. Water efflux includes water leaving the body through respiratory and cutaneous evaporation, urine, faeces, saliva and milk (Fig. 6.5). In free-living animals, tritiated or deuteriated water is often used to estimate both water influx and efflux.

As found in many ecophysiological measurements in mammals, water turnover scales with body mass. Mass specifically, the water flux of smaller mammals is greater than that of larger mammals. Macfarlane and Howard (1972) reported that water flux scales to body mass in mammals to an exponent of 0.82, and this exponent is frequently used when comparing water fluxes among mammals. Furthermore, they found that the water turnover of arid zone mammals, expressed in relation to body mass to the exponent 0.82, was lower than that of non-arid zone mammals. Exponents of 0.80 for 7 species of captive eutherians (Richmond et al. 1962) and of 0.78 for 12 species of captive rodents (Holleman and Dieterich 1973) have also been reported.

Nicol (1978) calculated an allometric regression of water flux (WF; ml/day) on body mass (m_b; kg) using measurements on 41 captive marsupial and eutherian mammal species and found the same exponent as that reported by Macfarlane and Howard (1972). In this regression analysis, six arid and seven non-arid zone marsupials and 11 arid and 28 non-arid zone eutherian mammals were included. The relationship had an $r = 0.98$ and took the form:

$$WF = 102.2 \ kg^{0.82}.$$

When water flux was expressed per body mass in kg to the power of 0.82, there was no difference between marsupials and eutherians, but arid zone mammals had significantly lower water flux than non-arid zone mammals (76.7 vs 136.7 ml $kg^{-0.82}d^{-1}$).

Nagy and Peterson (1988) analyzed the data from 562 water flux measurements, on 99 captive eutherian species, which covered all the points mentioned in the previous studies. The data included measurements on animals differing widely in their maintenance and in their physiological states. The slope of the relationship differed from that of the previous studies, and took the form:

$$WF = 0.159 \ m_b{}^{0.946}.$$

This equation indicates the water flux relates linearly to body mass (g). That is, mass-specific water flux is about the same for mammals of different sizes.

The slope calculated by Nagy and Peterson (1988) for captive marsupials differed from that of eutherian mammals, but was similar to that for marsupials and eutherians generated by Nicol (1978). The equation took the form:

$$WF = 0.547 \, m_b^{0.771}.$$

6.6.1
Water Flux of Free-Living Mammals

Nagy and Peterson (1988) generated an allometric relationship between water flux (ml day^{-1}) and body mass (g) for free-living eutherian mammals using 43 species (115 points). The regression equation took the form:

$$WF = 0.326 \, m_b^{0.818}.$$

The water flux of granivores did not differ from that of carnivores, and both had a rate approximately one-third that of herbivores. Herbivores were expected to have the highest water flux due to the high preformed water content of vegetation. However, it was rather surprising that granivores and carnivores had similar rates of water flux considering that seeds have relatively little preformed water while meat contains relatively large quantities. Omnivores had a water flux that was intermediate between those of granivores-carnivores and herbivores.

The regression slope for desert eutherians differed from that for non-desert eutherians: 0.954 vs 0.734, respectively. This indicates that small desert mammals have a lower water flux than small non-desert mammals of the same body mass when compared to larger animals. As pointed out by Nagy and Peterson (1988), the small desert mammals are mainly granivorous and omnivorous rodents that are well known for their low water flux.

An allometric relationship just for rodents was generated by Morris and Bradshaw (1981) using 12 species and 26 points. The equation took the form:

$$WF = 0.21 \, m_b^{0.90}.$$

An allometric relationship between water flux and body mass just for small mammals (Table 6.22) took the form (Fig. 6.24; unpublished data):

$$WF = 0.251 \, m_b^{0.848},$$
$$(n = 43; SE_a = 0.058; SE_b = 0.068; SE_{y.x} = 0.301; r^2_{adj} = 0.78; F = 153.3; P < 0.01).$$

Only one value was taken per species, and where possible, it was chosen during the summer (or dry season) at a time when the animal was maintaining its body mass.

Desert species (0.21 ± 0.10 ml g$^{-0.85}$ day^{-1}; n = 35) have lower water influx rates when expressed per g$^{0.85}$ than non-desert species (0.70 ± 0.58 ml g$^{-0.85}$ day^{-1}; n = 8), but no difference was found between rodents (0.28 ± 0.25 ml g$^{-0.85}$ day^{-1}; n = 34) and non-rodents (0.34 ± 0.11 ml^{-1} g$^{-0.85}$ day^{-1}; n = 9) or among dietary groups (omnivores: 0.25 ± 0.10 ml g$^{-0.85}$ day^{-1}; n = 22; granivores:

Table 6.22. Total water influx (*TWI*), body mass, diet and habitat of small free-living eutherian mammals. Measurements were taken during the dry season where possible and were done using labelled water

Species	Body mass (g)	TWI (ml day^{-1})	Diet[a]	Habitat[b]	TWI pred[c] (ml day^{-1})	TWI/ TWI pred	TWI (ml g$^{-0.848}$ day^{-1})	Reference
Rodentia								
1. *Gerbillus henleyi*	9.7	1.1	G	D	1.71	0.64	0.16	Degen et al. (unpubl.)
2. *Mus musculus*	13.9	3.25	O	ND	2.36	1.38	0.35	Morris and Bradshaw (1981)
3. *Clethrionomys rutilus*	16	5	H	ND	2.67	1.87	0.48	Holleman et al. (1982)
4. *Perognathus formosus*	18	1.31	G	D	2.97	0.44	0.11	Mullen (1970)
5. *Perognathus penicillatus*	18	1.31	G	D	2.97	0.44	0.11	Grubbs (1980)
6. *Petromyscus collinus*	19	0.8	O	D	3.12	0.26	0.07	Withers et al. (1980)
7. *Pseudomys albocinereus*	19.3	3.47	O	D	3.16	1.10	0.28	Morris and Bradshaw (1981)
8. *Peromyscus truei*	20.2	3.22	O	D	3.29	0.98	0.25	Bradford (1924)
9. *Perognathus fallax*	21.3	2.61	G	D	3.45	0.76	0.20	MacMillen and Christopher (1975)
10. *Gerbillus allenbyi*	22.8	2.78	G	D	3.66	0.76	0.20	Degen et al. (1992)
11. *Gerbillurus paeba*	23.4	1.91	O	D	3.75	0.51	0.13	Downs and Perrin (1990)
12. *Peromyscus maniculatus*	25	5.28	O	D	3.98	1.33	0.34	Grubbs (1980)
13. *Gerbillurus tytonis*	25.1	2.47	O	D	3.99	0.62	0.16	Downs and Perrin (1990)
14. *Gerbillus pyramidum*	31.8	3.24	G	D	4.92	0.66	0.17	Degen et al. (1992)
15. *Gerbillurus setzeri*	33	2.55	O	D	5.09	0.50	0.13	Downs and Perrin (1990)
16. *Dipodomys merriami*	34.8	2.3	G	D	5.33	0.43	0.11	Nagy and Gruchacz (1994)
17. *Acomys cahirinus*	38.3	5.06	O	D	5.81	0.87	0.23	Degen et al. (1986)
18. *Sekeetamys calurus*	41.2	5.89	O	D	6.19	0.95	0.25	Degen et al. (1986)
19. *Acomys russatus*	45	5.65	O	D	6.70	0.84	0.22	Degen et al. (1986)

Species								Reference
20. *Aethomys namaquensis*	46	2.2	O	D	6.83	0.32	0.09	Withers et al. (1980)
21. *Notomys mitchelli*	46.9	8.01	O	D	6.95	1.15	0.31	Morris and Bradshaw (1981)
22. *Dipodomys microps*	55	6.2	O	D	8.01	0.77	0.21	Mullen (1971b)
23. *Praomys natalensis*	57.3	12.33	O	ND	8.30	1.48	0.40	Green and Rowe-Rowe (1987)
24. *Rattus fuscipes*	63.5	16.9	O	ND	9.10	1.86	0.50	Morris and Bradshaw (1981)
25. *Meriones crassus*	77.6	3.81	G	D	10.87	0.35	0.10	Degen et al. (unpubl.)
26. *Meriones libycus*	85	11	O	D	11.78	0.93	0.25	Bradshaw et al. (1976)
27. *Ammospermophilus leucurus*	86	13	O	D	11.91	1.09	0.30	Karasov (1981)
28. *Arvicola terrestris*	88.1	68.88	H	ND	12.17	5.66	1.54	Grenot et al. (1984)
29. *Dipodomys nelsoni*	92.6	7.2	G	D	12.72	0.57	0.15	Grenot and Serrano (1979)
30. *Thomomys bottae*	103.8	26.39	H	ND	14.07	1.88	0.51	Gettinger (1984)
31. *Meriones shawii*	108.4	17.5	O	D	14.63	1.20	0.33	Bradshaw et al. (1976)
32. *Petromys typicus*	130	4.4	O	D	17.19	0.26	0.07	Withers et al. (1980)
33. *Neotoma albigula*	144	19.9	O	D	18.82	1.06	0.29	Grenot and Serrano (1979)
34. *Psammomys obesus*	202	36.36	H	D	25.42	1.43	0.40	Degen et al. (1991)
Lagomorpha								
35. *Oryctolagus cuniculus*	1500	83	H	D	150.79	0.55	0.17	Richards (1979)
36. *Lepus californicus*	1800	82.4	H	D	177.29	0.46	0.14	Nagy et al. (1976)
Chiroptera								
37. *Anoura caudifer*	11.5	13.4	N	ND	1.99	6.72	1.69	Helversen and Reyer (1984)
38. *Macrotus californicus*	13	2.4	C	D	2.22	1.08	0.27	Bell et al. (1986)
Insectivora								
39. *Eremitalpa granti*	19.8	1.73	C	D	3.23	0.54	0.14	Fielden et al. (1990a)

Table 6.22 (contd.)

Species	Body mass (g)	TWI (ml day^{-1})	Diet[a]	Habitat[b]	TWI pred[c] (ml day^{-1})	TWI/ TWI pred	TWI (ml g$^{-0.848}$ day^{-1})	Reference
Edentata								
40. *Bradypus variegatus*	4335	153	H	ND	386.94	0.40	0.13	Nagy and Montgomery (1980)
Carnivora								
41. *Vulpes cana*	1015	106.67	C	D	106.60	1.00	0.30	Geffen et al. (1992)
42. *Vulpes macrotis*	1850	137	O	D	181.66	0.75	0.23	Golightly and Ohmart (1984)
43. *Vulpes velox*	1990	289	C	D	193.81	1.49	0.46	Miller et al. (unpubl.)

[a]Diet: C, carnivore, G, granivore, H, herbivore, O, omnivore, N, nectarivore
[b]Habitat: D, desert, ND, non-desert
[c]TWI pred, TWI predicted $= 0.228 m_b^{0.848}$.

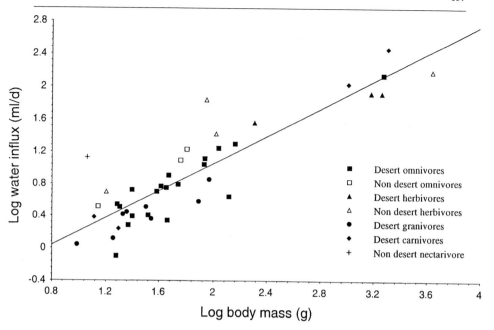

Fig. 6.24. Allometric regression of water influx on body mass in small eutherian mammals

0.15 ± 0.03 ml^{-1} g$^{-0.85}$; n = 9; herbivores: 0.48 ± 0.50 ml g$^{-0.85}$ day^{-1}; n = 7; carnivores: 0.23 ± 0.14 ml g$^{-0.85}$day^{-1}; n = 4 (Fig. 6.25). Of interest is the small standard deviation found among desert species, indicating that they are rather similar in their relatively low water influx rates, and the large standard deviation among non-desert species, indicating a large range of values. Similarly, herbivores had a large standard deviation compared with mammals of other dietary habits. This indicates that the water content of vegetation is much more varied than that of other diets.

A scattergram of the residuals of water flux on body mass depicting desert and non-desert species with different diets segregated the desert from the non-desert species (Fig. 6.26). Interestingly, the three Namib Desert rodent species measured by Withers et al. (1980), before the advective fog, had the lowest relative water influx rate of all species. In the present analysis, these three rodent species were classified as omnivores, as there appear to be no granivores in the South African deserts (Kerley and Whitford 1994), and most mammals appear to be omnivores (Kerley 1989, 1992b; Kerley et al. 1990). However, Downs and Perrin (1990) classified *Petromyscus collinus* and *Aethomys namaquensis* as granivores. Furthermore, the six omnivores from the South African deserts have the lowest water influxes of all omnivores. It should be noted that, other than in these six measurements, there is basically no overlap between granivores and herbivores.

When considering rodents only, the slope of the regression is 1.07, and that for desert mammals is 0.94 (Table 6.23). This indicates that, mass-specifically,

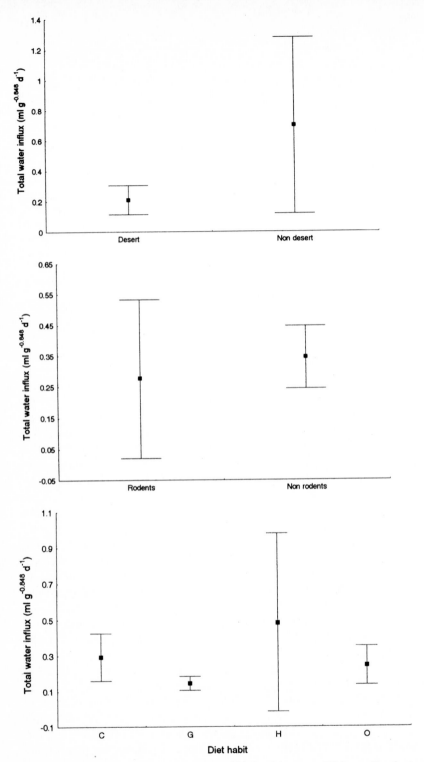

Fig. 6.25. Comparison of water influx in ml g$^{-0.848}$ day^{-1} (means ± SD) in small eutherian mammals in which measurements were done with labelled water: between desert and non-desert species ($P < 0.05$); between rodents and non-rodents; and among carnivores-(*C*), granivores-(*G*), herbivores-(*H*) and omnivores-(*O*)

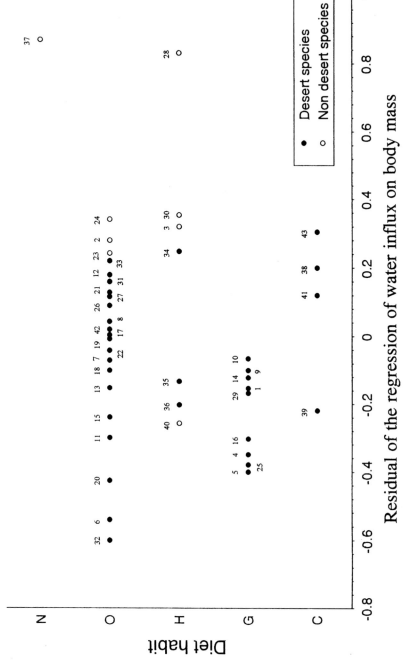

Fig. 6.26. Residuals of the regression of log water influx on log body mass for desert and non-desert eutherian mammalian species with different dietary habits. Numbers refer to species listed in Table 6.22

Table 6.23. Coefficients of the linear regression equations which take the form log y = a + blog (body mass; g) where y is water flux (ml day^{-1}) for non-reproducing small eutherian mammals. Measurements were taken using labelled water and, where possible, during summer (n, sample size; S_a, S_b, $S_{y \cdot x}$, SE of a, b and of estimate; r^2_{adj}, coefficient of determination; F, F-statistic for the linear term; P, significance level of regression)

y	n	a	b	S_a	S_b	$S_{y \cdot x}$	r^2_{adj}	F	P
All animals	43	−0.642	0.848	0.132	0.068	0.301	0.78	153.3	0.000
Desert animals	35	−0.884	0.936	0.104	0.054	0.207	0.90	297.7	0.000
Non-desert animals	8	0.214	0.576	0.275	0.139	0.304	0.70	17.1	0.006
Rodents	34	−2.347	1.071	0.553	0.146	0.641	0.62	53.9	0.000
Non-rodents	9	−0.460	0.700	0.711	0.112	0.782	0.83	39.2	0.000
Omnivores	22	−1.701	0.892	0.478	0.118	0.558	0.73	57.5	0.000
Granivores	9	−1.629	0.750	0.465	0.136	0.278	0.79	30.3	0.001
Herbivores	7	0.965	0.499	0.724	0.119	0.589	0.73	16.9	0.009
Carnivores	4	−2.035	0.992	0.496	0.090	0.407	0.98	121.2	0.008

small rodents have water influx rates similar to those of larger mammals, and similar to those of larger rodents. This is unlike most other relationships, where smaller mammals have a higher mass-specific water influx rate than larger mammals. This anomaly could be due to the measurements in the nine granivore desert rodents in the analysis, since these species are characterized by low body mass and low water requirements. Support for this reasoning is strengthened when analysing the nine rodents alone. The slope of the line for these nine species is 0.75, indicating that small granivorous rodents have a higher rate of mass-specific water influx than larger granivorous rodents.

Nagy and Peterson (1988) also related water flux to body mass in free-living marsupials. For 19 species and 57 points, the equation took the form:

$$WF = 2.488 \, m_b^{0.602}.$$

The slope for 28 herbivorous marsupials was similar to that for 23 carnivorous marsupials, but the intercept for the carnivores was higher than that for the herbivores. Therefore, unlike that found for eutherians, water flux in carnivorous marsupials was higher than in herbivorous marsupials.

Seasonal Water Flux. Seasonal water influx changes within a species often reflect changes in the water content of the diets of small desert mammals. Water influx rates in the ash-grey mouse (*Pseudomys albocinereus*) were positively correlated with the moisture content of the vegetation and with rainfall (Morris and Bradshaw 1981). However, some of the change in seasonal water influx rate could perhaps be associated with differences in total food intake, as this also affects water intake.

Granivorous rodents often show a large increase in water influx when they shift their diet mainly of seeds to one that contains more green vegetation. Such was the case for *Gerbillus allenbyi* and *G. pyamidum* when water influx rates in winter were 1.63 and 1.67 times higher than their respective summer rates (Degen et al. 1992). A shift from a diet mainly of seeds in the dry season to one of green vegetation in the rainy season resulted in an increase in water influx in the two heteromyids, *Perognathus pencillatus* and *Dipodomys merriami* (Grenot and Buscarlet 1988). Nagy and Gruchacz (1994) reported that the water influx of *D. Merriami* was 93.9 and 83.9 ml kg^{-1} day^{-1} during August and November, respectively, when seed intake comprised 100% of the diet, and rose to 189.4 ml kg^{-1} day^{-1} in March when seed intake was 0%, green vegetation composed 90% and arthropods 10% of the diet.

Desert herbivores also show seasonal shifts in water influx, with the lowest rates when the pasture is dry and highest when the pasture is moist. This is well illustrated by the Australian desert rabbit (*Oryctolagus cuniculus*; Richards 1979) and the North American jackrabbit (*Lepus californicus*; Nagy et al. 1976) which increased their water influx rates by 363% and 862%, respectively, when the pasture was moist compared to when it was dry. A smaller increase was reported for the fat sand rat which had a higher water influx in the spring than in the winter and summer (0.43 vs 0.27 and 0.18 ml H_2O g^{-1} day^{-1}). The fat sand rat consumes relatively moist *Atriplex halimus* leaves all year round, but the moisture content can vary between 40 and 90% of the fresh matter and is highest in the spring (Degen et al. 1990).

Omnivorous small mammals select a diet that will satisfy water requirements, often adding insects to supplement water and protein. Dry conditions should have less effect on the water influx rate of these animals than on those consuming plant material only. The largest seasonal difference in water influx rates for the ground squirrel *Ammospermophilus leucurus* was between the spring and winter (Karasov 1983). The spring rate was 2.21 times higher than the winter rate (0.20 vs 0.91 ml g^{-1} day^{-1}).

Carnivorous and insectivorous desert mammals should show the smallest seasonal changes in water influx rate unless they either drink in certain seasons, add additional items to their diet and/or differ in food intake in different seasons. Examples of such occurrences are presented by Blanford's fox (*Vulpes cana*) and the bat *Macrotus californicus*. Blandford's fox eats mainly insects and small vertebrates, but also consumes some fruit. Energy intake was similar in this species in the summer and winter, but the summer water influx rate was 1.38 times higher. It was calculated that, on a dry matter basis, the diet of the fox included 31.5% fruits in the summer and 13.4% fruits in the winter (Geffen et al. 1992a). The water content of the fruit was higher than that of meat, and as a consequence, the summer water influx rate was higher than the winter rate. *Macrotus californicus* increased its water influx rate from, 0.13 ml $g^{-1}day^{-1}$ in the winter to 0.24 ml $g^{-1}day^{-1}$ in the summer (times 1.85), although the bat was consuming only insects in both seasons. It was calculated that the fresh food intake in the winter was about 2.08 times that in the summer, or about the same ratio of winter: summer water influx rate (Bell et al. 1986). Therefore, the water influx rate reflected total food intake.

The occurrence of advective fog can provide a source of water for desert animals living close to the sea (Downs and Perrin 1990). An increased water influx in desert rodents after fog compared with the influx before fog was reported to occur in the rock mouse *Petromyscus collinus* (1.4 vs 0.8 ml day^{-1}), the rock mouse *Aethomys namaquensis* (3.2 vs 2.2 ml day^{-1}) and the rock rat *Petromus typicus* (5.1 vs 4.4 ml day^{-1}) (Withers et al. 1980).

CHAPTER 7

Energy Requirements and Energy Flux

"The measurement of energy intake and utilization by animals has been a major area of study within ecological physiology. This area can assume such a prominence that it is sometimes regarded as the principal subject and accomplishment of the entire field. Through energetics, ecological physiology finds its strongest links with ecology and behavior. In these fields, energy exchange assumes a major role in many hypotheses (e.g., life history strategy and foraging theory). Most often, ecological physiology has provided the methodology and data to test such hypotheses empirically. It consistently asks how much processes or activities cost and proceeds to make the appropriate determinations. The field has been successful in energetic studies because of its insistence on quantification of energy exchange."

(Bennett 1987)

7.1
Digestibility and Energy Value of Food

The energy yield and nutritive value of food is a function of its composition and digestibility. The chemical energy content, or gross energy, of food (Table 7.1) is equivalent to the heat produced on combustion of its total organic matter. It is expressed in joules; 4.184 joules (1 calorie) is the amount of heat required to raise the temperature of 1g of water by 1 °C from 14.5 °C to 15.5 °C.

Bomb calorimetry is commonly used to measure the energy content of a sample. In this method, the sample is ignited in excess oxygen, and the amount of heat generated is measured and compared with that generated by the combustion of a sample of known energy content, usually benzoic acid (26 453 J g^{-1} DM). From Table 7.1, it can be seen that the minimum energy content of a 1-g sample of organic dry matter is about 17 kJ and the maximum energy content about 39 kJ.

The dry matter (DM) digestibility of food consumed by an animal is calculated from its DM intake and faecal DM output as:

DM digestibility = (DM intake − faecal DM output)/DM intake.

The gross energy intake (GEI) of an animal is the potential energy of the total organic matter intake (Fig. 7.1). Some of this potential energy is lost as faecal energy (FE); the resulting energy is the apparent digestible energy (DE) and given as:

DE (kJ) = GE(kJ) − FE(kJ).

Furthermore, some of the GEI is lost as urine energy (UE) and as the energy in combustible gases (CGE). This is mainly due to methane, and could be considered to be zero in monogastric animals. The resulting energy is the apparent

Table 7.1. Gross energy content, metabolic water produced, O_2 consumption, and CO_2 production for different dietary components

	Carbohydrates	Lipids	Protein[a]
Gross energy (kJ g^{-1})	17.4	39.6	23.8
Metabolic water (ml g^{-1} digested)	0.56	1.07	0.40
O_2 uptake ($1\,g^{-1}$)	0.84	1.96	1.19
kJ $1^{-1}O_2$	20.9	19.7	18.8
ml $CO_2\,ml^{-1}\,O_2$	1.00	0.71	0.81
kJ $1^{-1}\,CO_2$	20.8	27.7	23.1
Metabolic H_2O ($\mu l\,ml^{-1}\,CO_2$)	0.662	0.754	0.509

[a]With urea as end product.

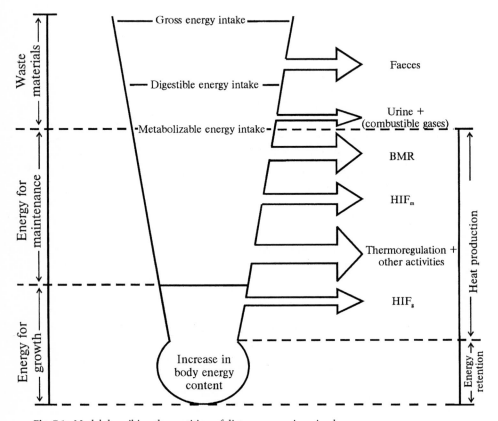

Fig. 7.1. Model describing the partition of dietary energy in animals

metabolizable energy (ME) and can be calculated as:

$$ME = GE - (FE + UE + CGE) \text{ or}$$
$$ME = DE - (UE + CGE).$$

ME can be corrected for the zero nitrogen retention (NR) of the animal, and this N-corrected ME (ME_n) can be estimated as:

$$ME_n = ME - a(6.25 \, NR)$$

where $NR = g \, d^{-1}$ and a is a constant close to 5.63 kJ g^{-1} (Emmans 1994).

MEI is the energy available to an animal for maintenance and production (growth, milk production, etc.). If the MEI of an animal equals maintenance energy requirements, then there is no change in the energy content of the animal. This means that there is zero energy retention (ER), and all the chemical energy intake is lost as heat production (HP). If the MEI is below maintenance requirements, then the animal has to make up the difference by catabolism of tissue from its body energy. If MEI is above maintenance requirements, however, the animal can then add to its body reserves. These relationships may be somewhat more complicated when, for example, the intake of a lactating female is below requirements for both maintenance and milk production, but nonetheless, milk is produced by the mobilization of body energy (Kam and Degen 1993a). This, of course, can only continue for a limited period of time. Energy mobilized from tissue or retained as body energy is in the form either of lipid or of protein, which yields approximately 39.6 and 23.8 kJ g^{-1}, respectively (small amounts of carbohydrates are involved as well).

The relationship between MEI, ER and HP can be presented as:

$$MEI = HP \pm ER,$$

where ER is the combined energy of lipid retention (LR) and protein retention (PR). This equation can then be modified to differentiate between lipid and protein retention as:

$$MEI = HP \pm (39.6 \, LR + 23.8 \, PR).$$

In a caged animal maintained at thermoneutral temperatures and at maintenance energy intake (ME_m), HP is composed of the basal metabolic rate (BMR) plus, mainly, heat increment of feeding (HIF). The latter is the heat produced due to the intake and digestion of food. This relationship could be expressed as:

$$ME_m = BMR + HIF_m,$$

where HIF_m is heat increment of feeding for maintenance. The ratio of BMR to ME_m is the efficiency of utilization of energy for maintenance (k_m), and ($1 - k_m$) is the proportion of MEI lost as HIF_m. HIF_m is composed of the heat produced due to: (1) the production of faecal output (FO); (2) the excretion of urine N(UN); and (3) the production of combustible gases (GP). Assuming that heat production is linear with the production of each of these factors, then:

$$HIF_m = w_f \cdot FO + w_u \cdot UN + w_g \cdot GP,$$

where w_p, w_u and w_g are the heat costs associated with the production of faeces, excretion of urine and the production of gases, respectively. Values for these coefficients, at least for ruminants, have been found to be: $w_f = 3.80$ (kJ g^{-1} DM), $w_u = 29.2$ (kJ g^{-1} N) and $w_g = 0.616$ (kJ kJ^{-1} combustible gas; Emmans 1994).

At energy intake above maintenance, part of the energy is retained (ER) either as lipid or as protein, and the rest is released as heat production. In this case, the heat produced is due to the energy cost of maintenance plus the heat produced due to energy retention, or the heat increment of feeding for growth (HIF_g). This can be expressed as:

$$MEI = ER + BMR + HIF_m + HIF_g.$$

This equation combines lipid and protein in the energy retention. However, the costs are not the same for these two, and therefore this relationship can be expressed as:

$$MEI = w_l \cdot LR + w_p \cdot LP + BMR + HIF_m + HIF_l + HIF_p.$$

The efficiency of the utilization of energy for growth (k_g) can be calculated as:

$$HIF_g = [ER/(MEI - ME_m)],$$

and HIF_g can be expressed as $(1 - k_g)$. Values for the coefficients of w_l and w_p, at least for ruminants, were found to be 16.4 and 36.5 (kJ g^{-1}), respectively. The efficiency of utilization of energy intake for lipid retention (k_l) equals:

$$k_l = LR/(LR + HIF_l).$$

By substituting reported values, this equation reads as:

$$k_l = 39.6/(39.6 + 16.4), \text{ or } 0.706.$$

Similar calculation for protein yields a k_p of 0.395. It is therefore more costly, energetically, to retain a unit energy of protein than a unit energy of lipid.

The value of k_m is usually higher than that of k_g, indicating that an animal uses energy more efficiently for maintenance than for growth. The relationships between ER and MEI, both below and above maintenance needs, are generally assumed to be linear. This is convenient for calculations and is also close to the truth. In reality, however, they are both slightly curvilinear (Webster 1979; Emmans 1994).

In general, the gross energy content of most plant parts, such as seeds, leaves, stem, branches and roots, yields between 15 and 22 kJ/g DM. Samples with high ash content, such as chenopod leaves, have lower energy yields, while samples with high fat content, such as sunflower seeds, have higher energy yields. The energy content of meat, which is rich in protein, is higher than that of plant parts and usually ranges between 23 and 30 kJ/g DM. Again, values could be lower or higher if DM includes large amounts of ash or fat, respectively.

Digestible energy and metabolizable energy, as fractions of gross energy, vary greatly among dietary items (Fig. 7.2). For example, in small mammals the

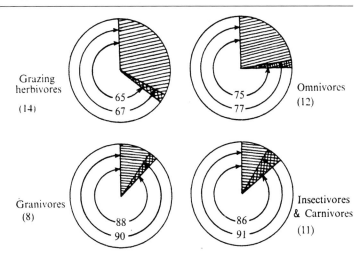

Fig. 7.2. Utilization of food energy by small mammals. *Top* and *bottom numbers* on each scheme are the apparent digestible and metabolizable energies as percentages of gross energy. The *striped areas* represent energy lost through faeces and the *cross-hatched areas*, energy lost through urine. (Grodzinski and Wunder 1975)

digestible energy as a proportion of gross energy averages 0.91 for meat (including insects), 0.90 for seeds and 0.67 for vegetation, whereas metabolizable energy averages 0.86, 0.88 and 0.65 for these dietary items, respectively (Grodzinski and Wunder 1975). In addition, digestibilities differ within each dietary item among animal species, depending upon such factors as body size and dietary experience.

Ingested metabolizable energy is oxidized during respiration, and this leads to the production of CO_2 and metabolic water. The energy produced by this catabolism is lost to the environment as heat (heat production, HP). When an animal is in thermal equilibrium – that is, no storage or loss of body heat, and the MEI equals energy lost as heat production – then the animal is in energy balance (Fig. 7.3). This is considered the maintenance energy requirements under the defined conditions of measurement.

The volumes of CO_2 produced and O_2 consumed, and therefore the ratio $CO_2:O_2$, is dependent upon the composition of the material being oxidized (Table 7.1). Knowledge of the composition of the energy intake and volumes of O_2 consumed and CO_2 produced (or any two of these variables) can be used to estimate the metabolic rate of an animal (HP). In addition, the uptake of 1 ml O_2 has been equated to the heat production of approximately 20.08 J, and only this measurement has been used to estimate heat production. If an animal is in both thermal and energy balance, this also equals MEI, and estimations of dry matter intake can be made.

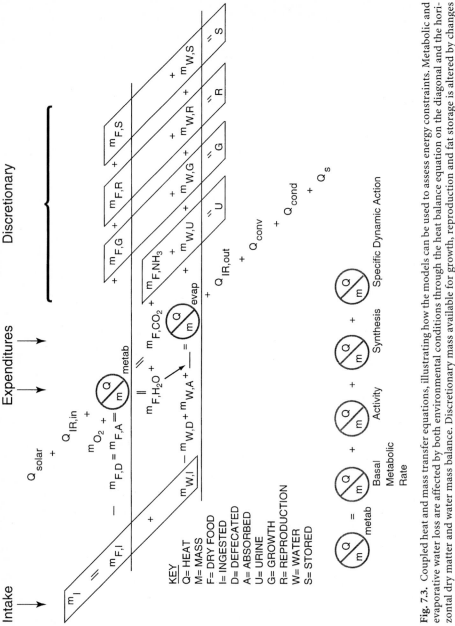

Fig. 7.3. Coupled heat and mass transfer equations, illustrating how the models can be used to assess energy constraints. Metabolic and evaporative water loss are affected by both environmental conditions through the heat balance equation on the diagonal and the horizontal dry matter and water mass balance. Discretionary mass available for growth, reproduction and fat storage is altered by changes in the environment and by ingested mass and depends on the mass balance of both the dry matter and water. (Porter et al. 1994)

7.1.1
Digestibility of Seeds

The gross energy of seeds varies considerably from species to species (between 15.6 and 26.4 kJ g^{-1} dry matter; Table 7.2). In general, however, seeds provide a highly digestible, concentrated energy source for animals. Dry matter digestibility and metabolizable energy each average about 0.90 of dry matter and gross energy intakes, respectively (Grodzinski and Wunder 1975).

Table 7.2. Gross energy (GE), mass and dry matter content (DMC) of seeds of different plant species

Plant species	GE (kJ g^{-1})	Mass (g seed^{-1})	DMC (%)	Reference
Acacia karoo	20.47	0.0520	90.30	Kerley and Erasmus (1991)
Tetragonia hirsuta	16.84	0.1120	88.90	Kerley and Erasmus (1991)
Tetragonia fruticosa	18.83	0.0590	88.40	Kerley and Erasmus (1991)
Eriocephalus umbellulatus	17.96	0.0150	91.00	Kerley and Erasmus (1991)
Cuspidium cernua	19.38	0.1440	91.50	Kerley and Erasmus (1991)
Emex australis	17.14	0.0430	90.00	Kerley and Erasmus (1991)
Gazania krebsianna	19.86	0.0050	89.90	Kerley and Erasmus (1991)
Dorotheanthus bellidiformis	19.24	0.0003	90.00	Kerley and Erasmus (1991)
Eragrostis curvula	19.39	0.0003	90.00	Kerley and Erasmus (1991)
Osteospermum sinautum	19.80	0.0160	91.10	Kerley and Erasmus (1991)
Hirpicium integrifoolium	17.28	0.0330	90.30	Kerley and Erasmus (1991)
Phalaris arundinacea	19.66			Shuman et al. (1988)
Helianthus maximiliani	23.43			Shuman et al. (1988)
Panicum miliaceum	18.83			Shuman et al. (1988)
Cirsium spp.	25.94			Shuman et al. (1988)
Phleum pratense	19.66			Shuman et al. (1988)
Triticum spp.	18.41			Shuman et al. (1988)
Desmanthus illinoensis	20.08			Shuman et al. (1988)
Cassia marilandica	18.83			Shuman et al. (1988)
Ampelamus albidus	26.36			Shuman et al. (1988)
Fagopyrum esculentum	19.25			Shuman et al. (1988)
Sorghum vulgare	19.25			Shuman et al. (1988)
Sporbolus cryptandrus	17.99			Saunders and Parrish (1987)
Sorghum vulgare	17.99			Saunders and Parrish (1987)
Sorghum spp.	18.41			Saunders and Parrish (1987)
Pennisetum glaucum	18.83			Saunders and Parrish (1987)
Amaranthus spp.	18.83			Saunders and Parrish (1987)
Amaranthus spp.	19.25			Saunders and Parrish (1987)
Phalaris canariensis	19.25			Saunders and Parrish (1987)
Lespedeza stipulacea	19.25			Saunders and Parrish (1987)
Panicum virgatum	19.66			Saunders and Parrish (1987)
Helianthus spp.	25.52			Saunders and Parrish (1987)
Polygonum arenastrum	15.59		91.10	Kam et al. (1987)
Reamuria mirtella	16.41		92.10	Kam et al. (1987)
Erodium ciconium	21.34		95.20	Kam et al. (1987)
Calendula aegyphaca	20.95		94.20	Kam et al. (1987)
Trigonella arabica	15.90		92.60	Kam et al. (1987)
Amaranthus spp.	16.71		92.50	Kam et al. (1987)
Medicago spp.	19.71		93.10	Kam et al. (1987)
Triticum spp.	18.60		92.90	Kam et al. (1987)

A large disparity may, however, exist between intact seeds and what is actually ingested (Jenkins 1988; Kerley and Erasmus 1991). For example, the husks of seeds are more fibrous and less digestible and yield less energy than do the kernels (Khokhlova et al. 1995). Husking of seeds has been reported among heteromyids (Randall 1993), and it appears that larger heteromyids husk seeds faster than smaller heteromyids, but the smaller heteromyids are more apt to husk seeds (Rosenzweig and Sterner 1970). In contrast, *Tatera leucogaster* consumes millet seeds whole, without husking them (Neal and Alibhai 1991). These different strategies, that is, non-husking and husking (total or partial), can be potentially important in determining the energy consumption and utilization of an animal. Nevertheless, this option has thus far been largely ignored in rodent studies.

The intake of husks and its effect on energy intake and digestibility were examined in two different sized, co-existing Gerbillinae, *Gerbillus henleyi* (adult body mass 8–12 g) and *Meriones crassus* (50–110 g), offered whole millet seeds (Khokhlova et al. 1995). *G. henleyi* and *M. crassus* females removed practically all the husks before consuming seeds, whereas *M. crassus* males removed only about 50.0%. The greater proportional consumption of husks by *M. crassus* males resulted in reduced dry matter and energy digestibilities when compared with those of females and of *G. henleyi* (Table 7.3).

7.1.2
Digestibility of Vegetation

Herbivores can utilize the fibrous portion of plants. They are represented in 11 of the 20 mammalian orders, five of which contain only herbivores (Stevens 1989). Fibrous portions, such as cellulose and hemicellulose, are not digested by enzymes secreted by mammals but are broken down by microbial fermentation, producing mainly short-chain fatty acids such as acetic, propionic and butyric. Fermentation of polysaccharides involves both bacteria and protozoans. Within each species it occurs principally in a single segment of the digestive tract, the fermentation chamber. It takes place in a modified part of the stomach in ruminants, camels, hippopotamuses, colobus and langur monkeys, sloths, sirenians and kangaroos. These mammals are known as foregut fermenters. Fermentation takes place in the posterior part of the gut in horses, rhinoceroses, elephants, howler monkeys, some arboreal marsupials, lagomorphs and rodents. These mammals are referred to as hindgut fermenters. In this latter group, the larger herbivores use the colon as the main fermentation chamber, while the smaller herbivores mainly use the caecum (Hume 1989). Foregut fermenters, principally ruminants, predominate in the 10 to 600 kg group size, while hindgut fermenters predominate among small and very large herbivores (Demment and Van Soest 1985).

Digestibility of forage is dependent upon its composition, including lignin and fibre content, and factors within the animal, such as fermentation rate and the mean retention time of digesta. Foregut fermenters are considered to be

Table 7.3. Mean body mass, husks removed, and digestibilities in *Meriones crassus* (M.c.) and *Gerbillus henleyi* (G.h.) when offered different levels of whole millet seeds. Values are mean ± SE. Difference between means are significant at $P < 0.05$; ns, not significant (Khokhlova et al. 1995)

	Meriones crassus		*Gerbillus henleyi*		Differences between			
					Species		Sexes	
	Male	Female	Male	Female	M.c.	G.h.	Male	Female
Sample size	8	7	7	5				
Body mass (g)	86.2 ± 4.8	57.0 ± 1.2	9.20 ± 0.34	8.85 ± 0.56	*	NS	*	*
Husks removed (%)	50.2 ± 17.6	99.4 ± 0.6	99.8 ± 0.2	98.2 ± 1.8	*	NS	*	NS
Dry-matter digestibility (%)	88.3 ± 1.4	94.7 ± 0.6	94.1 ± 0.5	94.5 ± 0.7	*	NS	*	NS
Faecal energy (kJ g^{-1} dry matter)	18.8 ± 0.3	20.4 ± 0.5	20.2 ± 0.2	19.8 ± 0.2	*	NS	*	NS
Digestible energy (% gross energy)	88.6 ± 1.3	94.4 ± 0.6	93.8 ± 0.5	94.4 ± 0.7	*	NS	*	NS

more efficient utilizers of cell wall constituents than hindgut fermenters. The passage rate of digesta is slower in foregut fermenters, and the cell walls are therefore more fully digested. In hindgut fermenters, however, digesta passes through more quickly, which allows a greater intake of feed.

Among foregut fermenters, cows generally have a higher dry matter digestibility than sheep and goats, and goats have a slightly, but significantly, higher digestibility than sheep (Tolkamp and Brouwer 1993). It appears that for poor roughage, the difference between goats and sheep increases. Furthermore, desert and tropical breeds of cattle and goats on poor quality diets show higher digestibilities than breeds from temperate zones.

In a comparison between the black Bedouin goat, a breed that is raised in extremely harsh deserts, and the Swiss Saanan goat, a temperate breed, dry matter digestibility was always higher in the Bedouin goats when both breeds were offered the same feed, but the difference between breeds was greatest when the fodder was of the poorest quality. For example, when good quality alfalfa hay was offered, dry matter digestibility was 71.6% for Bedouin goats and 66.8% for Saanan goats; however, when poor quality wheat straw was offered, these respective values were 53.5% for Bedouin goats and only 38.9% for Saanan goats (Silanikove 1986). Bedouin goats have an enhanced ability to digest lignin and structural carbohydrates compared with Saanan goats. This is related to their faster fermentation rate and to the slower passage rate of the digesta (Silanikove and Brosh 1989).

Studies on herbivory have centred mainly on large animals, in particular ruminants. Yet the highest dry matter digestibilities of fibrous diets have been reported in small herbivores, the voles. Voles are usually able to achieve dry matter digestibilities of over 70%, whereas among domestic ungulates they range between 50 and 60% (see Lee and Houston 1993a,b). When consuming a fibrous diet, the high digestibilities of voles are accomplished mainly through coprophagy and an increase in gut capacity.

Three species of British voles, the bank vole (*Clethrionomys glareolus*; body mass 25–35 g), the field vole (*Microtus agrestis*; body mass 30–40 g) and the water vole (*Arvicola terrestris*; body mass 200–400 g) were offered a diet of either seeds or plant leaf material (Lee and Houston 1993a,b). The total length and dry mass of the intestine, in particular the caecum, were larger when a leaf rather than a seed diet was provided. These responses were proportionately greatest in the smallest species (vole) and least in the largest, where the changes showed trends to be larger, but were not statistically significant. Interestingly, the mean retention time of the digesta was less with the leaf than with the seed diet, and here the proportional difference between diets was more pronounced for the largest vole species than for the smallest. Actual mean retention time of the digesta in the three vole species ranged between 5.6 and 9.2 h with the leaf diet and 11.9 and 30.2 h with the seed diet (Table 7.4; Lee and Houston 1993a). Retention time of the leaf diet in the caecum was only 1–2 h (Lee and Houston 1993b). However, the contents were continually recycled through coprophagy which contributed towards the high digestibilities.

Table 7.4. Estimate of the rate constant (k) and mean retention time (MRT) in hours from $\log_{10}$ concentration curves of marker in the faeces of the bank vole, field vole and water vole. (Lee and Houston 1993a)

Diet	n	Mean $t_{1/2}$	Mean k	MRT (1/k)
Bank vole				
Seed	5	8.25	0.087	11.9
Leaves	5	6.38	0.111	9.2
Field vole				
Seed	4	10.65	0.074	15.3
Leaves	4	3.9	0.219	5.6
Water vole				
Seed	4	21.0	0.035	30.2
Leaves	4	5.7	0.124	8.1

Digestibility trials comparing large mammals inhabiting desert and non-desert regions, in particular domestic animals, are readily available on a variety of forages (Silanikove 1986; 1992). Such is not the case for small mammals. However, in one study, the desert fat jird (*Meriones crassus*) and the mesic Levant vole (*Microtus guetheri*) were offered the same diets, and their digestibilities were compared (Choshniak and Yahav 1987; Yahav and Choshniak 1989). When both were offered alfalfa hay, a high quality roughage diet (25% fibre), the apparent energy digestibility was equal in the two species and amounted to about 69%. However, when tested on dry Rhodes grass, a low quality roughage diet (30% fibre), the apparent digestibility in *Me. crassus* remained at about 69%, while in *Mi. guentheri* it dropped to 49%. *Me. crassus* was able to maintain its body mass on both diets, whereas *Mi. guentheri* could do so on the alfalfa hay but not on the Rhodes grass.

The ability of *Me. crassus* to utilize the poor roughage diet better than *Mi. guentheri* was attributed to differences between the species in the retention time of digesta in the gastrointestinal tract, particularly in the caecum. The mean retention time of digesta along the entire tract and in the caecum increased about twofold in *Me. crassus* when consuming Rhodes grass compared with alfalfa, but it actually decreased slightly in *Mi. guentheri* (Table 7.5). The pattern of decreased mean retention time of digesta with a poorer diet in *Mi. guentheri* was similar to that of the three British voles. Moreover, the actual mean retention time of digesta by *Mi. guentheri* was 9.5 h, similar to that of the bank vole, but the retention time of digesta in its caecum (6.6 h) was considerably longer. The study on *Mi. guentheri*, therefore, suggests that these voles either do not practice coprophagy or else did not have access to their faeces in their metabolic cages.

7.2
Anti-Nutritional Compounds

Plants possess anti-nutritional secondary compounds, notably tannins, which protect them from grazing animals. Tannins are chemically diverse, naturally

Table 7.5. Mean retention time (MRT) of digesta along the tract (T) and in the caecum (C) of the fat jird (*Meriones crassus*) and the Levant vole (*Microtus guentheri*) tested on different diets. Values within columns that are marked by the same letter are significantly different from each other ($P < 0.01$). (Choshniak and Yahav 1987)

Diet	Animal	Body mass (g)	MRT(h)			
			T	C	T–C	C/T
Alfalfa hay	Fat jird	82.4 ± 5.2	$12.5 \pm 3.4a$	$10.4 \pm 2.1b$	2.4 ± 3.0	0.84
	Levant vole	38.8 ± 9.8	13.2 ± 3.4	9.6 ± 2.0	3.2 ± 2.4	0.72
Rhodes grass	Fat jird	72.3 ± 0.6	$23.2 \pm 2.5a$	$19.5 \pm 0.5b$	3.3 ± 2.0	0.84
	Levant vole	41.1 ± 1.5	$9.5 \pm 2.5a$	$6.6 \pm 1.6b$	2.5 ± 1.1	0.70

occurring, water-soluble polyphenolics widespread in dicotyledonous forbs, trees and shrubs, but usually absent in grasses. They are also present in many cereal grains, fruit, berries and legumes (Mehanso et al. 1987). Tannins are complex polymers divided into two main types, condensed and hydrolysable forms, which can be differentiated by their structure and reactivity toward hydrolytic agents. Both types are generally present in tree leaves and browse, and they may differ in their effect on digestive parameters and in their toxicity. Tannin concentrations are affected by the maturity of the plant and the position of the leaf within the plant. Furthermore, tannin levels fluctuate seasonally within plants and can change daily in response to alterations in temperature, light intensity and phenological status of the forage (Furstenburg and van Hoven 1994).

Condensed tannins (molecular weight, 1000–20 000), also referred to as proanthocyanidins, are polymers of flavon-3-ols or flavon-3,4-diols such as catechin. Heating in acid causes the oxidative depolymerization of condensed tannins to anthocyanidins and other polymers. Hydrolysable tannins (molecular weight, 500–3000), also referred to as tannic acids, contain a central core of glucose or a polyhydric alcohol esterified with gallic acid or its condensation product, ellagic acid.

Tannins have wide-ranging effects that reduce the consumption and digestibilities of forage and can be toxic to herbivores (Mehanso et al. 1987; Clausen et al. 1990; Makkar 1993). They form precipitates with proteins, resulting in the formation of indigestible tannin-protein complexes (Robbins et al. 1987a; Hagerman 1989; Makkar 1993). They also form complexes with carbohydrates, cellulose, hemicellulose and amino acids, thereby reducing their digestibilities. In addition, tannins combine with the protein of the outer cellular layer of the gut, hindering the permeability of nutrients; they combine with digestive enzymes, rendering them ineffective; and they inhibit microorganisms leading to a decrease in the production of gases and volatile fatty acids. Tannins can also damage the liver and kidneys (Makkar 1993).

The amount of tannins that herbivores can tolerate differs greatly from species to species (Robins et al. 1987b, 1991). For example, laboratory rats are more resistant to the negative effects of tannins than are hamsters (*Mesocricetus auratus*) and prairie voles (*Microtus ochrogaster*), while mule deer and goats are more resistant than cows and sheep. Two main methods have been described by which herbivores can neutralize tannins: (1) the secretion of tannin-binding proteins, mainly proline-proteins, in saliva (Mehanso et al. 1987) and (2) detoxification.

Proline has a high affinity for tannins and neutralizes their effects. Browsers generally have saliva which is richer in proline than that of grazers, and as a result, they ingest more tannins (Austin et al. 1989) while being less affected by them. In addition, their parotid glands are larger than those of grazers (Hofman 1989). Tannase has been identified in the rumen of goats; it hydrolyses various gallic acid esters, such as gallotannin and tannic acid, and thus detoxifies tannins (Begovic 1978). Furthermore, gut surfactants have been found that inhibit the precipitation of protein (Lindroth 1988). An anaerobic enterobacterium colonizes the caecal wall of koalas (*Phascolarctos cinereus*), a marsupial which feeds almost exclusively on the leaves of *Eucalyptus* spp., which are known to contain high concentrations of tannins (Osawa et al. 1993). The enterobacterium is capable of degrading tannin-protein complexes and of producing tannase (Osawa and Walsh 1993).

Consumption of tannins increases the amounts of faecal material and faecal nitrogen of endogenous origin, interferes with mineral absorption and reduces total body sodium. Rats consuming diets containing relatively high levels of tannins, in particular hydrolysable tannins, evidence a depressed food consumption and growth rate. Similarly, the growth rates of hamsters and mice are slowed when consuming tannins (Lindroth and Batzli 1984; Mehanso et al. 1987; Vallet et al. 1994). However, rats have rather high concentrations of proline proteins in their saliva. The relative size of the parotid glands and the quantity of proline-rich proteins in the glands increase linearly with the intake of condensed tannins (Jansman et al. 1994). Mexican woodrats (*Neotoma mexicana*) fed a diet containing 7% quebracho (condensed tannin) had a reduced food intake, lost body mass and suffered 50% mortality. However, when they were gradually acclimated to a diet of 5% quebracho, the masses of their parotid and submandibular glands and kidneys were similar to those of controls, although their livers were smaller (Voltura and Wunder 1994).

Not all studies have demonstrated a negative effect of tannins and indeed, *Microtus pennsylvanicus* regularly incorporates plant species containing tannins in its diet (Jean and Bergeron 1986; Bergeron and Jodoin 1987). Lindroth and Batzli (1984) showed that there were no digestive disturbances in prairie voles when their dietary protein was more than 12%, probably a result of a surplus of binding sites. However, the detoxification of tannins requires energy, and this increase in energy ranged between 13.6 and 22.6% for *M. pennsylvanicus* (Breton et al. 1989). On the other hand, the dry stomach mass of animals receiving a diet of 0.3% tannin was 54% higher than those receiving 6% phenols, and 100% higher than controls.

Many desert plants, especially perennials, remain green all year round but are not often eaten. Secondary compounds have been suggested as the reason for this. However, very little is known about the ability of desert wildlife, and small mammals in particular, to use plant material containing relatively high concentrations of anti-nutritional secondary compounds or whether they have developed a special mechanism to detoxify them. The livers of desert wood rats (*Neotoma lepida*) fed on a diet treated with 7% creosote resin were larger than those of control wood rats, suggesting that the liver may be involved in the detoxification process (Meyer and Karasov 1991). Kerley and Erasmus (1991) found no correlation between seed selection and polyphenol concentration (expressed as % tannic acid equivalents) in three Karoo Desert rodent species, *Gerbillurus paeba*, *Mastomys natalensis* and *Mus minutoides*, when given a choice of 11 seed species. However, as was also pointed out by the authors, concentrations of polyphenols were low. One seed species contained 2.41% polyphenols, and the others contained less than 1.0%; six had no detectable polyphenols at all. In the Negev Desert, three Gerbillinae, *Gerbillus henleyi*, *G. dasyurus* and *Meriones crassus*, tended to avoid plants high in phenolic compounds when offered a range of plant species (Kam, Khokhlova and Degen, unpubl. data).

7.2.1
Bezoars and the Ibex

Since ancient times man has known of objects that are occasionally formed in the abomasum of ruminants and ruminant-like animals. These bezoar stones, found mainly in the desert ibex in Persia, were attributed to have powerful healing powers: they could even stop bleeding. As a consequence, bezoars were in great demand among the nobility, and large numbers of ibexes were slaughtered for these valuable commodities. If we attempt to interpret such classical myths and writings and assume that genuine bezoars are composed primarily of ellagic acid, then a pattern quickly emerges. Ellagic acid is an exceptionally powerful blood coagulant, and it is this property that may have led to the legends attributed to bezoars. In fact, some Bedouins in the Negev Desert today cut up tanned leather into fine pieces and apply them to cuts to stop bleeding. How are benzoars formed in ibexes? This, too, has a plausible explanation. After annual plants had been eaten during the dry summer and when the pasture that remained was scarce and poor, these caprids would migrate to a hilly area and graze on the leaves and bark of trees, which are rich in tannins. These tannins precipitate proteins and other plant components, thus forming bezoars.

7.3
Digestive Tract and Absorption of Nutrients

Desert mammals are often forced to consume and survive on poor quality forage. They are credited with the ability to extract more digestible energy from such food than their non-desert counterparts. Physiological adaptations such as digesta passage rate have been related to these differences. Anatomical differences have also been described between desert and mesic species.

The morphology, ultrastructure and function of the jejunum were compared between desert and non-desert species, using the Mongolian gerbil (*Meriones unguiculatus*) and the fat sand rat (*Psammomys obesus*) as examples of desert species and the New Zealand white rabbit (*Oryctolagus cuniculus*) and the hooded Lister rat (*Rattus rattus*) as representatives of non-desert species (Buret et al. 1993). Gerbils and fat sand rats had longer microvilli and villi and larger brush border surface areas (Fig. 7.4) than rabbits and rats. As a result of this larger mucosal absorptive surface area, absorption of Na^+ and Cl^- was found to be greater in gerbils and fat sand rats than in rabbits and rats, and the absorption of 3-0-methyl-D-glucose was greater in gerbils than in non-desert species. Thus, Buret et al. (1993) concluded that desert-dwelling animals demonstrate enhanced absorption of electrolytes and nutrients in the small intestine and suggested "that intestinal mucosal adaptation allows desert dwelling animals to compensate for the limited availability of water and nutrients in an arid environment".

7.4
Basal Metabolic Rate

The basal metabolic rate (BMR) is the most common measure of energy expenditure in animals, and is usually determined by their rate of oxygen (O_2)

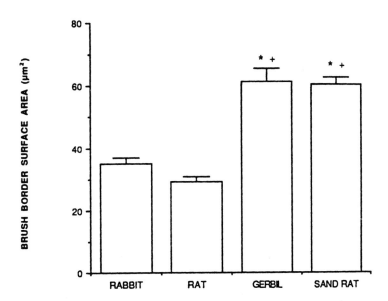

Fig. 7.4. Microvillus brush border surface area per μm^2 of cell surface in the jejunum of rabbits (n = 6), rats (n = 6), gerbils (n = 6) and sand rats (n = 5). Values are means ± SE. $P < 0.01$ compared with rabbits (*) and rats (+). Values between rabbits and rats or between gerbils and sand rats were not significantly different. (Buret et al. 1993)

consumption. In fact, the use of O_2 uptake to estimate BMR is so common that BMR is often expressed in terms of the rate of O_2 uptake. It is the rate measured under specified laboratory conditions in which the animal is: (1) mature and not active reproductively; (2) kept in a cell with black walls; (3) not allowed much movement and is lying down, resting and comfortable but awake; (4) after absorption of food; and (5) kept at an air temperature within its thermal neutral zone. Black walls are used so that radiation from them does not affect the measurement (Porter 1969). Examining the animals after food absorption eliminates heat produced by the specific dynamic action of the food (SDA) as well as the energy required in its handling, chewing and swallowing. The time required to achieve this post-absorptive state varies greatly from animal to animal and is affected mainly by their body size. It could be less than 2 h in small rodents and 4–6 days in large ruminants. The post-absorptive state may be difficult to achieve in some small mammals (Speakman et al. 1993a). For example, shrews become hyperactive when deprived of food (Hanski 1985), and some die after 2 h of food deprivation. Furthermore, some small mammals cannot maintain their body temperature when deprived of food: this produces abnormal physiological conditions and, in some cases, results in torpor. The post-absorption time is also affected by the passage rate of feed and the diet prior to measurements.

BMR represents a baseline of minimal energy expenditure for the body functions of an animal while awake, since almost zero energy expenditure is used in movement, thermoregulation and food absorption. However, energy expenditure rate measurements differ within species, even when all standard conditions are met, and are dependent upon such variables as the time of day, the season, food consumed before measurements were made and food availability. Some authors argue that BMR is the lowest of these rates; others claim that each state has its own BMR.

Measurements of BMR have been largely standardized among laboratories and are known for a large number of mammal species. For example, Hayssen and Lacy (1985) examined 293 species, Elgar and Harvey (1987) 265 species, McNab (1988) 321 species and Heusner (1991) 391 species in their respective comparative studies of BMR among mammals. This large data-base and the importance of energy expenditure in ecophysiological studies make BMR a very attractive measurement for intraspecific and interspecific comparisons, and indeed it has been used extensively.

In most interspecific studies, body mass spans six orders of magnitude – from several grams to thousands of kilograms – and the largest mammal is approximately one million times heavier than the smallest. Like most physiological responses, mass-specific BMR is much greater in small mammals than in large ones. As such, BMR for these mammals extends over four orders of magnitude, from less than one to several hundred watts per animal. Hayssen and Lacy (1985) found that the Edentata have the lowest relative BMR, followed by Marsupialia, Primata and Chiroptera, then Rodentia, Lagomorpha and Carnivora (but these do not differ significantly from Primata and Chiroptera), while Artiodactyla have the highest, but this is not significantly different from those of Lagomorpha and Carnivora.

The scaling of BMR to body mass, the proper exponent, and the reasons for residual variations have been discussed extensively. It is generally accepted that BMR is related to body mass raised to the power of 0.75. This value (or a figure close to this value) was first proposed by Brody and Proctor (1932) and Kleiber (1932) and was derived from the famous "mouse to elephant" curve (also the Kleiber curve) in which a log-log regression of BMR relative to body mass (m_b; kg) in 26 points yielded the following relationship:

$$BMR\ (W) = 3.39\ m_b^{0.75}.$$

The reasons for this exponent is, however, uncertain, but this has not prevented other authors from using the magical "predicted" value to compare values measured in ecophysiological studies. Such a comparison allows conclusions to be drawn as to whether the BMR is lower or higher than that predicted for an animal based on its body mass. In a number of cases it has been used to explain habitat distributions and adaptations of animals to their environment.

Heusner (1982) argued that the exponent of 0.75 as the power function is actually a statistical artifact. Sets of data from seven species showed that the intraspecific relationships lay on parallel lines, each with a slope of 0.7, and that the mass coefficients of the groups were significantly different. The mass coefficient increased threefold with the size of the animal and ranged from 1.91 in *Peromyscus* weighing 0.018 to 0.040 kg to 6.06 in cattle weighing 207 to 922 kg, when BMR was expressed in watts and body mass in kilograms. In a later analysis of data, Heusner (1991) found that in effect there were two allometric regression lines of BMR relative to body mass. The first included 363 mammals ranging in body mass from 2.5 to 407 000 g, resulting in an exponent of 0.68. The second consisted of 25 outliers of the first regression line, also resulting in an exponent of 0.68. When all points were used, the exponent found was 0.71. Heusner (1991) further pointed out that when only small mammals ($< 20\ 000$ g; $n = 362$) were used in the regression analysis, the exponent was 0.66, whereas when only large mammals ($> 20\ 000$ g; $n = 29$) were used, the exponent was 0.79. The difference in exponent due to size in his analysis was not an isolated occurrence: several other examples were cited, and it was also emphasized that small mammals compose some 85 to 95% of the mammals used in the studies.

Since the early studies of Brody and Kleiber, numerous allometric equations have been generated relating BMR to body mass. In all studies reported thus far, differences in body mass explained most of the variation in BMR, with r^2 values being generally above 0.90 and close to 1.0. Nonetheless, in spite of the high coefficient of determination, BMR differs by a factor of at least 2 to 3 at any given body mass and can vary by a factor of 10 (Weiner 1989; McNab 1992b). Concomitantly, body mass can differ by 20 to 30 times at any given BMR (Weiner 1989). Although early studies ignored this "noise", more attention is now being given to it in an effort to explain the residual variation.

7.4.1
Comparison of BMR Among Mammals

McNab (1988) found that, in general, the BMRs of eutherian mammals, especially medium-sized to large species, that feed on vertebrates, herbs and nuts are high, whereas those of species that feed on invertebrates, fruit and the leaves of woody plants are low. In addition, seed-eating desert rodents, especially small Heteromyidae, have low BMRs. In a later study, McNab (1992b) examined the effect of body mass and taxonomic affiliation (genus, family, order and infraclass) as well as ecophysiological variables such as dietary habits (seeds, vertebrates, invertebrates, fruit, leaves, grasses and forbs, nectar or blood), climate (desert, arctic, temperate, tropical, subterranean or aquatic), activity level (active or sedentary), torpor (spontaneously entering daily or seasonal torpor, or not doing so) and reproduction (laying eggs, giving birth to virtually embryonic young, producing altricial or precocial young). In all these analyses, body mass was important in explaining BMR. Climate proved to be a significant factor in a number of analyses. In Chiroptera, BMR was strongly affected by body mass, family and diet, but not by climate. In Rodentia, body mass, family, diet and climate were significant factors affecting BMR; but, climate was not when family was excluded. However, when family was excluded in the last analysis, whether the rodents enter torpor or not became a significant factor (Table 7.6). McNab (1992b) concluded that "this substitution (torpor for diet) reflects the clear correlation that exists in rodents between the occurence of torpor and familial affiliation, although why food habits lost their significance is obscure, unless it represents an interaction between food habits and torpor".

It has been argued that a lower than predicted BMR would be advantageous for animals inhabiting arid areas as it would reduce energy requirements, heat production and evaporative water loss (Borut and Shkolnik 1974). Indeed, many desert mammals have been reported to have a BMR lower than that predicted either by the Kleiber curve or by other established allometric relationships. Goyal and Ghosh (1983) reported that, among desert rodents, the maximum and average decrease of BMR from expected values were 37 and 16%, respectively, whereas in non-desert rodents, the maximum and average increase of BMR from expected values were 88% and 34%, respectively (Fig. 7.5). Similarly, McNab's (1966) data showed that the BMR of desert-dwelling rodents was, on average, 10% below that predicted from their body mass according to the Kleiber curve, whereas for non-desert rodents it was 10% above that predicted.

A lower than predicted BMR has been reported for desert mammals in all deserts of the world. For example, in India, four rodent species from the Rajasthan Desert had lower than expected BMR, ranging from a reduction of 13% for the 56 g bush rat (*Golunda ellioti gujerati*) to 37% for the 72 g Indian Desert gerbil (*Meriones hurrianae*; Goyal et al. 1981). In the South African deserts, rodents similarly showed reduced BMRs: in Namibia, the 21.7 g big-eared desert mouse (*Malcothrix typica*) had a BMR that was 42% lower than predicted for a mammal of its body mass (Knight and Skinner 1981); four om-

Table 7.6. Statistically significant factors associated with $\log_{10}$ basal metabolic rate (BMR) in various mammalian taxa. (McNab 1992b)

Taxon	n	$\log_{10}$m	Subgroup	Food	Climate	Activity	Torpor	Reproduction	r^2
Family									
Phyllostomidae	17	0.0001	Genus NS	0.0010	NS	*	*	*	0.954
Arvicolidae	24	0.0001	Genus NS	NS	0.0551	*	*	*	0.914
	24	0.0001	Genus NS	NS	NS	*	*	*	0.885
	24	–	Genus 0.0006	NS	0.0017	*	*	*	0.959
Order									
Xenarthra	15	0.0001	Family NS	NS	NS	NS	NS	*	0.954
Carnivora	30	0.0001	Family NS	0.0001	NS	0.0069	NS	*	0.986
	30	0.0001	Family NS	0.0001	0.0437	–	NS	*	0.989
Chiroptera	30	0.0001	Family 0.0490	–	NS	0.0396	NS	*	0.983
	40	0.0001	Family 0.0035	0.0004	NS	*	NS	*	0.982
	40	0.0001	–	0.0001	NS	*	NS	*	0.963
Rodentia	106	0.0001	Family 0.0001	0.0384	0.0001	*	NS	NS	0.980
	106	0.0001	–	NS	0.0001	*	0.0001	NS	0.964
Infraclass									
Metatheria	43	0.0001	Order NS Family 0.0427	NS	NS	NS	NS	*	0.993
	43	0.0001	Order NS	NS	NS	NS	NS	*	0.988
	43	0.0001	–	NS	NS	NS	NS	*	0.988
Eutheria	251	0.0001	Order 0.0001 Family 0.0001	0.0001	NS	NS	0.0001	NS	0.993
	251	0.0001	Order – Family 0.0001	0.0001	NS	NS	0.0001	NS	0.993
	251	0.0001	Order 0.0001	0.0001	0.0001	NS	0.0001	NS	0.989
	251	0.0001	–	0.0001	0.0001	0.0001	0.0001	NS	0.978

Table 7.6 (*contd.*)

Taxon	n	$\log_{10}m$	Subgroup	Food	Climate	Activity	Torpor	Reproduction	r^2
Subclass									
Theria	294	0.0001	Order 0.0001 Family 0.0001	0.0006	0.0353	NS	0.0135	NS	0.993
	294	0.0001	Order– Family 0.0001	0.0006	0.0353	NS	0.0135	NS	0.993
	294	0.0001	–	0.0001	0.0001	0.0001	0.0006	0.0023	0.977
Class									
Mammalia	297	0.0001	Order 0.0001 Family 0.0001	0.0001	NS	NS	0.0001	NS	0.993
	297	0.0001	Order– Family 0.0001	0.0001	NS	NS	0.0001	NS	0.993
	297	0.0001	–	0.0001	0.0001	0.0001	0.0003	0.0001	0.977

The covariant analysis for each taxonomic group continued by sequentially withdrawing factors that were statistically insignificant one at a time to obtain the maximal number of statistically significant factors. Those factors that remained significant have their level of significance given; those factors that were included in the analysis but did not show a significant effect on $\log_{10}$ basal rate are indicated by NS; those factors that were not included because they were inappropriate for the taxon (e.g. no arvicolids enter torpor or are sedentary) are indicated by *; and those factors that were arbitrarily eliminated from an analysis to explore other interactions are indicated by –. The number of species and the r^2 of each model are indicated. Note that r^2 corresponds to the model after all insignificant factors were discarded.

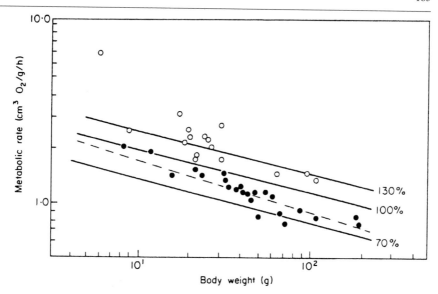

Fig. 7.5. Relationship of basal metabolic rate (BMR) to body mass in burrowing desert and non-desert rodent species. *Solid lines labelled* 100, 130 and 70% represent the Kleiber standard curve, and 130 and 70% of that value, respectively. *Broken line* is from study of Goyal and Ghosh (1983). For the solid line labelled 100%, BMR = 3.4 $g^{-0.25}$ and for the broken line, BMR = 3.334 $g^{-0.295}$; o = non-desert rodents; ● = desert rodents. (Goyal and Ghosh 1983)

nivorous *Gerbillurus* species ranging in body mass from 30 to 46 g had BMRs that were 25–39% lower than predicted (Downs and Perrin 1990); and in the Kalahari Desert, the 130 g *Thallomys paedulcus* and 65 g *Aethomys namaquensis* had rates that were 46 and 36% lower than those expected, respectively. In South America, the BMR of two herbivorous rodents of the Monte Desert were measured: that of *Octomys mimax*, weighing 119 g, was 6% lower than predicted, and in *Tympanoctomys barrerae*, a 91 g individual had a BMR 19% lower and a 52 g individual had a BMR 2% lower than predicted (Bozinovic and Contreras 1990). The springhare (*Pedetes capensis*) of East African deserts, weighing 2.2–2.4 kg, has a BMR 25–35% lower than predicted (Muller et al. 1979), while the fennec fox (*Fennecus zerda*), a small carnivorous canid weighing 1.1–1.2 kg which inhabits the sandy deserts of the Sahara, Arabia and Sinai, has a BMR 25–39% lower than expected (Noll-Banholzer 1979a; Maloiy et al. 1982).

Differences in BMR among animals, when compared allometrically have also been related to differences in distribution. In small mammals, a number of studies have shown that species or subspecies from arid environments have a lower BMR than their counterparts from a mesic or less arid environment. For example, heteromyids from desert areas have lower BMRs than species from "intermediate" (less arid) and coastal areas (least arid; Hinds and MacMillen 1983, 1985). A decrease in BMR with an increase in aridity has also been found among the subspecies of *Peromyscus* (McNab and Morrison 1963).

In a comparison of hedgehogs (Insectivora) from different habitats, it was found that the Ethiopian hedgehog (*Paraechinus aethiopicus*; $m_b = 453$ g) had a lower BMR than the long-eared hedgehog (*Hemiechinus auritus aegyptius*); ($m_b = 397$ g) which in turn had a lower BMR than the European hedgehog (*Erinaceus europaeus*; $m_b = 749$ g). BMR in these hedgehogs was associated with the level of aridity of their habitats. The lowest BMRs were recorded in hedgehogs from the extreme deserts and steppes of the Middle East, East Africa and Sahara, middle values in hedgehogs from semi-arid areas on the verge of the desert in North Africa and the Middle East, and highest ones in hedgehogs from temperate areas in northern Europe to the Mediterranean Sea (Shkolnik and Schmidt-Nielsen 1976). The golden spiny mouse (*Acomys russatus*; $m_b = 50$ g), a diurnal murid, and the common spiny mouse (*A. cahirinus*; $m_b = 41$ g), a nocturnal murid, are sympatric in many rocky areas in deserts. However, whereas *A. russatus* inhabits only extreme deserts, mainly in eastern Egypt, including the Sinai and Negev Deserts and the Arava and Judean Deserts in Israel, *A. cahirinus* is widely distributed from eastern and northern Africa to Iran, but does not penetrate the extreme deserts inhabited by its congener. Both murids had BMRs lower than predicted for a mammal of their body mass: for *A. russatus* it was 34.5% lower but for *A. cahirinus*, only 13.5% lower (Shkolnik and Borut 1969).

In general, small mammals from hot deserts have a BMR lower than expected, a relatively narrow thermal neutral zone, a high lower critical temperature and a steep thermoregulatory curve. Small mammals from cold deserts are exposed to environmental conditions that differ substantially from those of hot deserts and, in consequence, tend to show different metabolic responses. For example, let us examine the BMRs of five small mammal species from the arid steppes of eastern Mongolia, an area where seasonal air temperatures have been measured to span -45 to $+36$ °C (Weiner and Gorecki 1981). The area is characterized by hot arid summers, sparce and poor vegetation, little precipitation and long, cold winters in which permafrost can occur. These mammals thus face typical hot desert conditions during the summer as well as arctic conditions during the winter.

The responses of these five small mammal species were diverse, leaning to one or other of the two climatic extremes. Two rodent species (Brandt's vole, *Microtus brandti*, and the Asiatic mountain vole, *Alticola argentatus*) as well as one lagomorph (the Dahurian pika, *Ochotona daurica*) had BMR values higher than those predicted from their body masses, while two rodents (the Djungarian hamster, *Phodopus sungorus*, and the Mongolian gerbil, *Meriones unguiculatus*) had BMR values close to predicted values (Table 7.7). In addition to high BMRs, *A argentatus* and *O. daurica* showed relatively low lower critical temperatures (T_{lc}), broad thermoneutral zones and low thermal coefficients-typical adaptations to cold climates. *A. argentatus* is an alpine species, and the eastern Mongolian steppes are probably its most southern and lowest limits of occurrence. Weiner and Gorecki (1981) suggested that the responses measured

Table 7.7. A comparison of the measured and predicted (according to Hart 1971 values of basal metabolic rates of Mongolian small mammals. Observed values are means $\pm$ SE (n). (Weiner and Gorecki 1981)

Species	Body Weight (g)	Basal metabolic rate (ml O_2 $g^{-1}h^{-1}$)	
		Observed	Predicted
Microtus brandti	40.3	1.91 ± 0.24 (3)	1.41
Alticola argentatus	37.7	3.21 ± 0.094 (9)	1.44
Phalopus sangorus	33.2	1.60 (2)	1.48
Meriones unguiculatus	66.9	1.15 ± 0.13 (5)	1.22
Ochotona daurica	127.7	1.95 ± 0.14 (7)	1.03

for *O. daurica* are typical for Lagomorpha and may not be adaptive to cold desert.

The BMR of *Me. unguiculatus* was slightly lower than that predicted, its thermoregulatory line was steep, and its thermal neutral zone was narrow. These latter two characteristics are typical features of hot desert adaptations. The BMR of *P. sungorus* was only slightly above that predicted, the thermal coefficient was close to that predicted, and the T_{lc} was relatively high. The responses of *P. sungorus* lie between those of the "hot desert" adaptive responses of *M. unguiculatus* and the "arctic" adaptive responses of *A. argentatus*.

7.4.2
Seasonal Changes in BMR

Seasonal measurements of BMR have been made on a large number of small mammals, in particular rodent species. In general, BMR is higher in winter- than in summer-acclimatized animals. A number of North American leporids have demonstrated this shift (Fig. 7.6), including desert cottontails (*Sylvilagus audubonii*), black-tailed jackrabbits (*Lepus californicus*), antelope jackrabbits (*Lepus alleni*) and white-tailed jackrabbits (*Lepus townsendi*; Hinds 1973, 1977, Rogowitz 1990). The increase in BMR in the winter is accompanied by a decrease in the T_{lc} and an increase in summit metabolism.

The two main factors that trigger thermoregulatory responses to seasonal shifts are photoperiod and air temperature. Levent voles (*Microtus guentheri*) acclimatized to long scotophase (16L:8D) had a resting metabolic rate that was 19% higher than that of individuals acclimatized to long photophase (16D:8L Banin, et al. 1994). This is consistent with the notion of a lower BMR in the summer. However, the T_{lc} did not differ between the two groups, and the thermal neutral zone was narrower in the long scotophase group (30–34 °C vs 30–36 °C). The long photophase voles were able to maintain their body temperature at higher air temperatures: in fact, long scotophase voles could not survive at an air temperature of 36 °C. At an air temperature of 6 °C, however, both groups

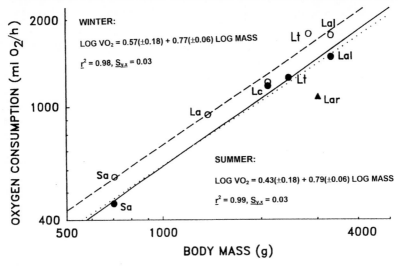

Fig. 7.6. Standard rates of oxygen consumption in North American leporids. Data are means for summer-(●) or winter-(○) acclimatized animals. Least-squares regression lines for standard metabolic rate in winter (*dashed line*) and summer (*solid line*) are shown. Data are for *Sylvilagus audubonii* (*Sa*; Hinds 1973), *Lepus americanus* (*La*; Irving et al. 1957), *L. californicus* (*Lc*; Hinds 1977), *L. townsendii* (*Lt*; Rogowitz 1990), *L. alleni* (*Lal*; Hinds 1977), and *L. arcticus* (*Lar*; ▲; not included in summer equation; Wang et al. 1973). *Dotted line* indicates Kleiber estimate of BMR (ml O_2/h = 3.49 $g^{0.75}$, assuming 19.67 kJ l^{-1} O_2). (Rogowitz 1990)

were able to maintain a stable body temperature, but the long photophase voles had a body temperature of 37 °C while the long photoperiod voles were hypothermic and had a body temeperature of 35.8 °C.

7.4.3
Maximum Metabolic Rate

The meaning and measurement of the maximum metabolic rate (MMR) of an animal have presented considerable confusion (see discussion by Weiner 1989). It is usually determined by measuring the rate of maximum oxygen consumption (V_{O_2}) over relatively short periods and is often referred to as maximum aerobic capacity (Langman et al. 1981). Measurements are carried out while forcing the animals to expend energy through either (1) intense muscular activity (exercise induced) or (2) thermoregulation (thermogenically induced). Exercise-induced maximum V_{O_2} is usually measured as the animal runs on a treadmill (Seeherman et al. 1981) and is also referred to as maximum metabolism for muscular effort. In this method, animals are trained to run at increasingly faster speeds until maximum V_{O_2} is attained. Thermogenically induced maximum V_{O_2} is measured while exposing the animals to low air temperatures or by maintaining them in a medium of high thermoconductivity such as helium-oxygen (Rosenmann and Morrison 1974). It is also referred to as maximum thermogenic capacity (Wickler 1980). This method is best suited to small animals as it is necessary to subject them to an environment in which aerobic metabolism cannot maintain their body temperature. There is some debate as to whether these two methods give compatible results. Chappell (1984) found that maximum V_{O_2} uptake measured during exercise was similar to values measured during cold exposure in deer mice (*Peromyscus maniculatus*). However, Hinds and Rice-Warner (1992) reported that, at least in some rodents, the exercise-induced MMR was higher than the cold-induced MMR.

The MMR achieved by an animal is dependent upon the length of time of measurement and is often presented as a rate above BMR or resting metabolic rate. Peak power outputs are achieved with brief bursts of activity. This activity is fueled by anaerobic respiration and is also referred to as peak rate of aerobic and anaerobic metabolism (Taylor 1982). It can only be maintained for a very limited time, often for only seconds, due to the buildup of toxic lactic acid.

Aerobic respiration can support energy expenditure for longer periods, but not indefinitely, since the animals are not in energy balance and are depleting their energy reserves. Peterson et al. (1990) suggest that the ceiling metabolic rate an animal can sustain indefinitely (sustained metabolic rate) is up to approximately 7 times the resting metabolic rate, and that values for field metabolic rates of animals generally are between 1.5 and 5.0 times those of the resting metabolic rates.

Independent of the method of measurement, the MMR is higher in larger than in smaller animals, but is not so mass-specifically. In studies where low air temperature or a helium-oxygen medium was used to measure the MMR (in watts) of mammals (27 species; 56 points) ranging in body mass (m_b; g) from 3.3 to 1100 g, the generated allometric regression took the form (Weiner 1989):

$$\text{MMR (thermogenic induced)} = 0.168 \, m_b^{0.670}.$$

The relationship for rodents (39 species) ranging in body mass between 7 and 1100 g (Hinds and Rice-Warner 1992) took the form:

$$\text{MMR (thermogenic induced)} = 0.153 \, m_b^{0.674}.$$

When just rodents from different habitats in South America and ranging in body mass from 7 to 195 g were used, (24 species; 25 points), the allometric relationship (Fig. 7.7) took the form (Bozinovic and Rosemann 1989):

$$\text{MMR (thermogenic induced)} = 0.158 \, m_b^{0.662}.$$

According to this last relationship rodents from cold climates have relatively high MMR, whereas tropical species have relatively lower rates. The temperate and desert rodents lie in an intermediate position.

The relationship between exercise-induced MMR and body mass in mammals (45 species; 67 points) ranging in body mass from 7.2 to 667 000 g took the form (Weiner 1989):

$$\text{MMR (exercise induced)} = 0.101 \, m_b^{0.857}.$$

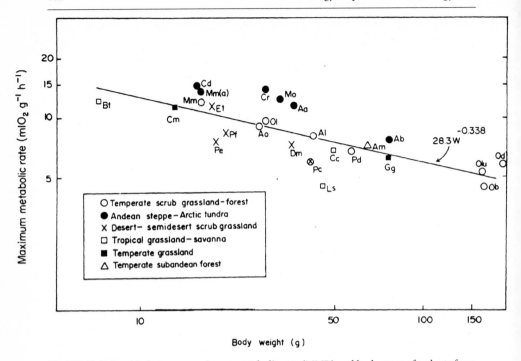

Fig. 7.7. Relationship between maximum metabolic rate (MMR) and body mass of rodents from different environments. *Line* is fitted by least-squares regression and represented by MMR = 28.3 $g^{-0.338}$. (Bozinovic and Rosenmann 1989)
Octodonitidae: *Octodon bridgesi* (Ob); *Octodon degus* (Od); *Octodon lunatus* (Olu).
Heteromyidae: *Dipodomys merriami* (Dm); *Liomys salvini* (Ls); *Perognathus fallax* (Pf).
Cricetidae: *Abrothrix andinus* (Aa); *Abrothrix longipilis* (Al); *Abrothrix olivaceus* (Ao); *Auliscomys boliviensis* (Ab); *Auliscomys micropus* (Am); *Baiomys taylori* (Bt); *Calomys callosus* (Cc); *Calomys ducilla* (Cd); *Calomys musculinus* (Cm); *Clethrionomys rutilus* (Cr); *Eligmodontia typus* (Et); *Graomys griseoflavus* (Gg); *Oryzomys longicaudatus* (Ol); *Peromyscus californicus* (Pc); *Peromyscus eremicus* (Pe); *Phyllotis darwini* (Pd); *Microtus oeconomus* (Mo).
Muridae: *Mus musculus* (Mm(a) wild, Andean geographic race); *Mus musculus* (Mm wild)

and that for 18 rodent species took the form:

MMR (exercise induced) $= 0.122 \ m_b^{0.773}$.

Maximum metabolism has been related to minimum metabolism (Hinds and MacMillen 1984). This is based on the aerobic capacity model, which states that selection for enhanced aerobic metabolism, required to support high levels of sustained activity, also results in high levels of metabolism at rest (Bennett and Ruben 1979). Since desert rodents usually have a lower minimum metabolism than their non-desert counterparts, this would suggest that desert rodents have a relatively low maximum metabolism. However, at least among herteromyids – rodents known for inhabiting deserts – this is not the case. Indeed, these heteromyids have reduced BMRs (Hinds and MacMillen 1985; Hinds and Rice-

Fig. 7.8. Relationship of body mass to minimum and maximum metabolism in 47 species of rodents including 10 species of heteromyids (*circles*). Maximum metabolism induced by either cold (*open symbols*) or exercise (*filled symbols*). Other symbols: *rectangles,* cricetids; *diamonds,* murids; *triangles,* other rodents. Equations for metabolism (in ml O_2 min^{-1}): minimum = 0.080 $g^{0.687}$; maximum cold-induced = 0.456 $g^{0.674}$; maximum exercise-induced = 0.336 $g^{0.773}$. (Hinds and Rice-Warner 1992)

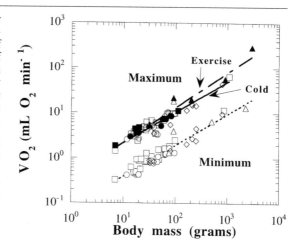

Warner 1992), but their cold-induced MMR is about 6.5 times as great and not lower than that of other rodents (Fig. 7.8; Hinds and Rice-Warner 1992).

7.5
Average Daily Metabolic Rate

Average daily metabolic rate (ADMR) is the metabolizable energy intake (MEI) required by a caged animal to maintain a constant body energy content. In this measurement, the animal is maintained under laboratory conditions so that the energy requirements refer to those needed for maintenance in the specified environmental conditions of the study. ADMR includes the BMR, the heat increment of feeding for maintenance, some minimal locomotory costs and possibly some thermoregulatory costs. Although this measurement could serve a very useful purpose in understanding and evaluating the energy needs and efficiency of utilization of energy, it has been largely ignored in studies of energy expenditure in wild animals. Part of the reason for this lies in the difficulty in measuring it accurately: no standard methods have been established. To date, three main methods have been used which can be referred to as: (1) linear regression analysis,(2) total heat production and(3) energy balance and body mass change.

7.5.1
Linear Regression to Estimate ADMR

The ADMR of a species can be estimated by offering different levels of metabolizable energy to a number of animals and measuring changes in their body energy content (energy retention, ER). Linear regression analysis is then applied to the points at which ER is negative, that is, the metabolizable energy intake (MEI) is below maintainance energy requirements, and to the points at which ER is positive, that is, MEI is above maintenance energy requirements, on MEI. In both cases, the regression equation takes the form (Fig. 7.9):

$$\text{ER (kJ g}^{-1}\text{ day}^{-1}) = a + b \text{ MEI (kJ g}^{-1}\text{ day}^{-1}).$$

The regression lines for the points above and below maintenance are somewhat different. The slope of the latter is steeper than that of the former (Fig. 7.9). In each case, however, ADMR or metabolizable energy required for maintenance (ME_m) is the point at which ER equals zero. For the points below maintenance, a is the intercept on the y-axis at zero MEI, and in its absolute form can be considered the fasting metabolic rate. Some researchers consider this point to approximate the standard metabolic rate (SMR) or BMR. The slope b is a measure of the efficiency of utilization of energy for maintenance (k_m), and ($1 - k_m$) is the proportional cost of the heat increment of feeding for maintenance (HIF_m).

Therefore, ADMR can be expressed as:

$$\text{ADMR} = \text{BMR} + \text{HIF}_m.$$

A good estimate of k_m is:

$$k_m = \text{BMR/ADMR}.$$

For the points above maintenance energy intake, the slope b is a measure of the efficiency of utilization of energy for growth (k_g), while ($1 - k_g$) is the propor-

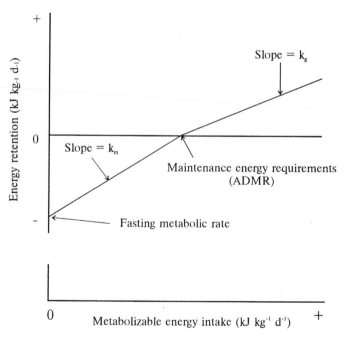

Fig. 7.9. Relationship between energy retention and metabolizable energy intake in captive homeotherms. The slopes k_m and k_g indicate the efficiency of utilization of energy for maintenance and growth, respectively ($k_m > k_g$)

tional cost of the heat increment of feeding for growth (HIF_g). The intercept of the regression line for points below maintenance is lower than that for points above maintenance. In addition, k_g can be determined as:

$$k_g = [ER/(MEI - ADMR)].$$

The difficulty in the estimation of ER change for each level of MEI has limited the use of regression analysis in determining ADMR. The change in ER can be estimated by the comparative slaughter technique. In this, groups of animals on different levels of feeding are slaughtered, and their energy content determined after feeding trials. The energy contents of these animals are then compared with a control group of animals slaughtered prior to the feeding trials. The drawbacks of this method are that: (1) many animals have to be used in order to obtain reliable results; and (2) the same animals cannot be monitored in a series of treatments.

Measurments of ER can also be calculated from the difference between MEI and HP. HP can be determined by indirect calorimetry through the continual measure of O_2 consumption (and/or CO_2 production). On a long-term basis, such measurements are often very inconvenient, while extrapolating HP from spot measurements can lead to erroneous results. HP can also be determined from CO_2 production using doubly labelled water (see Sect. 7.6.2). This is more convenient but also much more expensive.

As a result of these difficulties in determining ER change, ADMR is commonly calculated from the regression of body mass change on MEI (Robbins 1983). This method can provide accurate results but assumes that body mass change is a reliable indicator of body energy change. While this is usually the case, there are exceptions, such as during compensatory growth. In a study on two species of gerbils, ADMR was measured simultaneously by ER change related to MEI and by body mass change related to MEI (Degen and Kam 1991). In that study, HP was determined using doubly labelled water. ADMR determined from body mass change related to MEI was 7.9% higher than that of ER related to MEI for one gerbil species and 2.4% lower for the other gerbil species, that is both methods gave similar estimates of ADMR (Table 7.8).

7.5.2
Total Daily Heat Production to Estimate ADMR

ADMR has been estimated by determining the total heat production of a feeding animal over at least 1 day of measurement. Heat production is usually estimated from the rate of O_2 consumption of an animal and is equated with ADMR (French et al. 1976). Body mass change is usually not taken into account. However, if the MEI of an animal is below maintenance energy requirements, body tissue is then mobilized, and the total heat production will be near ADMR, but will under estimate it. If the MEI is above maintenance requirements on the other hand, the heat increment of feeding for growth or fattening will be in-

Table 7.8. Effect of body mass change ($\triangle m_b$) or energy retention (ER) on metabolizable energy intake (MEI) using regression equation in *Gerbillus pyramidum* and *G. allenbyi* and their calculated average daily metabolic rates (ADMR). (Degen and Kam 1991)

Dependent variable	Independent variable	Constnt (a)	Slope (b)	S.E estimate	r^2	ADMR (kJ.Kg$^{-0.75}$·day^{-1})
		Gerbillus pyramidum				
$\triangle m_b$	MEI	-3.59	0.0081	1.13	0.71	444.0
ER	MEI	-253.5	0.62	56.2	0.85	410.1
		Gerbillus allenbyi				
$\triangle m_b$	MEI	-3.01	0.0080	0.84	0.74	378.1
ER	MEI	-216.4	0.54	62.6	0.70	397.3

The regression equations are: $\triangle m_b$ (%·day^{-1}) $= a + b$.MEI (kJ·Kg$^{-0.75}$·day^{-1}) and ER (kJ·kg$^{-0.75}$·day^{-1}) $= a + b$·MEI (kJ.kg$^{-0.75}$·day^{-1})

cluded in the estimation of the rate of heat production, and as a consequence, ADMR will be overestimated.

7.5.3
Energy Balance and Body Mass Change to Estimate ADMR

ADMR has also been determined by offering an estimated near maintenance ration to animals and measuring body mass change. Zero change in body mass indicates that the energy intake of the animals equals maintenance energy requirements. Where change in body mass occurs, ADMR is determined by adding energy to or subtracting energy from MEI when the animals lose or gain body mass, respectively (Withers 1982; Buffenstein 1985; Downs and Perrin 1990). The difficulty in these calculations lies in the determination of the energy content of the body mass gained or lost and the efficiency in conversion. Buffenstein (1985) made her calculations assuming that all body loss or gain was fat and also assumed that 1 g of mobilized fat yielded 23.36 kJ. Slight errors in these assumptions could lead to large errors in ADMR calculations if the animals show substantial changes in body mass or if the loss or gain in body mass includes much body water.

7.5.4
Comparison in ADMR Among Small Mammals

Allometric relationships between ADMR and body mass have been calculated for rodents and shrews by French et al. (1976). The equation for rodents took the form:

$$\text{ADMR (kJ day}^{-1}) = 8.56 \, m_b \, (\text{g})^{0.54}.$$

There were 47 points in the regression, taken from 36 species with a range of body masses from 7.5 g in a pocket mouse (*Perognathus flavus*) to 370 g in a hamster (*Cricetus cricetus*).

For shrews, the equation took the form:

$$\text{ADMR (kJ day}^{-1}) = 14.28 \, m_b \, (g)^{0.43}.$$

In this regression, 15 points from 8 species were used, and body masses ranged from 2.9 g in the lesser shrew (*Sorex minutus*) to 21.2 g in a short-tailed shrew (*Blarina brevicauda*).

Data for the above regression analyses were collected from measurements of ADMR made at an air temperature of 20 °C because this temperature was considered to be close to that experienced under natural conditions by the animals. Nests were available in some studies, and food consumption was such that body mass was maintained near constant. Thus, these measurements include BMR, heat increment of feeding for maintenance, part of the activity cost and, in some cases, thermoregulatory costs. This latter expenditure could differ greatly among species. The T_{lc} of homeotherms are reduced by feeding (Kleiber 1975) and by the insulating effects of nests. Therefore, 20 °C may well be within the thermoneutral zone of some small mammals, although it is still below the thermoneutral zone for most of them.

A number of other studies have followed the suggestion of French et al. (1976) to use 20 °C in estimating ADMR (Table 7.9). Adding these measurements to those listed by French et al. (1976) and using only one point per species, the allometric regression equation for rodents takes the form:

$$\text{ADMR (kJ day}^{-1}) = 8.79 \, m_b \, (g)^{0.521},$$
$$(n = 53; \, SE_{y \cdot x} = 0.058; \, SE_a = 0.031; \, SE_b = 0.020; \, r^2_{adj} = 0.93; \, F = 701; \, P < 0.01).$$

Where more than one ADMR measurement is available for a species, a mean of the values was taken.

The conditions under which ADMR is measured should be standardized with regard to particular environmental variables. This would allow proper comparisons among species from different studies, as is common for BMR measurements. And, as for BMR, measurements should be made at air temperatures that would not require the animal to expend energy for thermoregulation. In some cases where the BMR curve of the animal has been established, the approximate T_{lc} can be calculated. As a general rule, however, an air temperature of about 25 °C should not require small mammals to expend energy for thermoregulation.

Some ADMR measurements were made on rodents maintained at 25 °C (Table 7.9). All were classified as desert species, and the regression for these measurements takes the form:

$$\text{ADMR (kJ d}^{-1}) = 7.10 \, m_b \, (g)^{0.552},$$
$$(n = 13; \, SE_{y \cdot x} = 0.127; \, SE_a = 0205; \, SE_b = 0.127; \, r^2_{adj} = 0.59; \, F = 18.3; \, P < 0.01).$$

The regression equation generated for all rodents measured at either 20 or 25

Table 7.9. Average daily metabolic rate (ADMR), basal metabolic rate (BMR), diet and habitat of

Species	T^a (°C)	Diet[b]	Habitat[c]	Body mass (ADMR; g)	ADMR (kJ day^{-1})
Insectivora					
1. *Blarina brevicauda*	20	I	ND	20.77	56.43
2. *Neomys anomalus*	20	I	ND	13.60	43.10
3. *N. fodiens*	20	I	ND	15.80	42.60
4. *Sorex hoyi*	20	I	ND	3.50	26.50
5. *S. arcticus*	20	I	ND	5.40	26.05
6. *S. araneus*	20	I	ND	10.46	44.30
7. *S. cinereus*	20	I	ND	3.70	26.86
8. *S. cinereus*	10	I	ND	3.80	31.91
9. *S. cinereus*	15	I	ND	4.10	32.23
10. *S. minutus*	20	I	ND	4.32	24.75
Rodentia					
11. *Ammospermophilus leucurus*	20	O	D	76.60	75.20
12. *Spermophilus tereticaudcus*	20	O	D	125.00	117.57
13. *Tamiasciurus hudsonicus*	10	G	ND	245.40	245.01
14. *T. hudsonicus*	15	G	ND	252.30	212.96
15. *T. hudsonicus*	20	G	ND	252.00	173.81
16. *Glaucomys sabrinus*	15	O	ND	174.40	147.21
17. *G. sabrinus*	20	O	ND	174.00	138.48
18. *G. volans*	20	G	ND	72.50	89.17
19. *Perognathus amplus*	20	G	D	13.00	33.36
20. *P. flavus*	20	G	D	7.50	25.00
21. *P. longimembris*	20	G	D	8.20	27.17
22. *P. parvus*	20	G	D	16.30	34.56
23. *Chaetodipus baileyi*	20	O	D	31.00	53.38
24. *C. formosus*	20	O	D	18.00	34.73
25. *C. intermedius*	20	O	D	13.00	30.29
26. *Reithrodontomys megalotis*	20	O	D	9.00	27.43
27. *Peromyscus leucopus*	20	O	ND	22.00	48.81
28. *P. maniculatus*	20	G	D	19.40	38.2
29. *Oligoryzomys destructor*	20	G	ND	27.80	36.80
30. *Akodon lanosus*	20	H	ND	25.20	40.70
31. *A. longipilis*	20	H	ND	49.50	57.20
32. *A. olivaceus*	20	H	ND	35.20	59.40
33. *A. puer*	20	H	ND	29.10	46.90
34. *Phyllotis xanthopygus*	20	G	D	61.70	62.50
35. *Sigmodon hispidus*	20	H	ND	80.00	103.03
36. *Phodopus sungorus sungorus*	22	G	D	37.20	47.99
37. *Cricetus cricetus*	20	H	ND	370.00	230.22
38. *Nannospalax leucodon*	20	H	ND	200.00	132.16
39. *Gerbillus allenbyi*	25	G	D	22.30	22.30
40. *G. dasyurus*	25	O	D	34.13	45.35
41. *G. henleyi*	25	G	D	9.10	28.50
42. *G. pyramidum*	25	G	D	31.90	32.24
43. *G. pusillus*	22	G	D	13.35	33.29
44. *Gerbillurus paeba*	20	O	D	27.90	69.19
45. *G. paeba*	25	O	D	26.69	57.25
46. *G. setzeri*	25	O	D	39.02	63.40

small eutherian mammals

Body mass (BMR; g)	BMR (kJ day^{-1})	ADMR (kJ g$^{-0.52}$ day^{-1})	ADMR/ ADMR$_{pred}$ [d]	BMR/ ADMR	Reference
20.70	26.36			0.49	French et al. (1976)
13.10	32.22			0.78	French et al. (1976)
17.10	26.56			0.58	French et al. (1976)
					French et al. (1976)
					French et al. (1976)
7.40	26.89			0.86	French et al. (1976)
7.90	26.78			0.47	French et al. (1976)
7.90	26.78			0.40	Grodzinski (1971)
7.90	26.78			0.42	Grodzinski (1971)
4.60	15.59			0.59	French et al. (1976)
101.50	47.07	7.87	0.92	0.47	Karasov (1981)
150.37	42.74	9.54	1.12	0.28	French et al. (1976)
224.00	119.79	9.79	1.15	0.53	Grodzinski (1971)
224.00	119.79	13.99	1.64	0.63	Grodzinski (1971)
224.00	119.79	11.99	1.40	0.78	French et al. (1976)
		9.46	1.12		Grodzinski (1971)
		10.04	1.18		French et al. (1976)
69.50	37.88	9.60	1.12	0.44	French et al. (1976)
		8.78	1.03		French et al. (1976)
9.00	9.07	8.76	1.02	0.30	French et al. (1976)
8.73	4.75	9.09	1.06	0.16	French et al. (1976)
19.20	15.93	8.09	0.94	0.39	Withers (1982)
20.90	12.00	8.94	1.05	0.33	French et al. (1976)
		7.72	0.90		French et al. (1976)
14.60	7.62	7.98	0.93	0.22	French et al. (1976)
9.00	11.04	8.75	1.02	0.40	French et al. (1976)
21.85	16.39	9.78	1.14	0.34	French et al. (1976)
19.09	17.29	8.20	0.95	0.46	French et al. (1976)
					Withers (1982)
		6.52	0.76		Ebensperger et al. (1990)
		7.60	0.88		Ebensperger et al. (1990)
42.30	27.74	7.51	0.88	0.57	Ebensperger et al. (1990)
27.00	23.84	9.32	1.09	0.52	Ebensperger et al. (1990)
		8.12	0.95		Ebensperger et al. (1990)
55.00	27.33	7.32	0.86	0.49	Ebensperger et al. (1990)
141.00	70.05	10.54	1.23	0.39	French et al. (1976)
33.20	25.92	7.31	0.86	0.60	Weiner (1987)
349.35	107.26	10.62	1.24	0.49	French et al. (1976)
176.93	63.02	8.40	0.98	0.54	French et al. (1976)
23.00	11.27	4.43	0.52	0.49	Degen and Kam (1991)
27.60	14.11	7.23	0.85	0.38	Degen and Kam (1991)
		9.04	1.06		
35.00	19.60	5.32	0.62	0.55	Degen and Kam (1991)
12.60	6.50	8.65	1.01	0.21	Buffenstein (1984)
32.50	15.05	10.40	1.21	0.19	Buffenstein (1984)
32.50	15.05	12.25	1.43	0.21	Downs and Perrin (1990)
46.10	17.79	9.43	1.10	0.24	Downs and Perrin (1990)

Table 7.9 (*contd.*)

Species	T^a (°C)	Diet[b]	Habitat[c]	Body mass (ADMR; g)	ADMR (kJ day^{-1})
47. *G. tytonis*	25	O	D	30.46	67.92
48. *Meriones crassus*	25	O	D	72.60	83.24
49. *M. unguiculatus*	20	O	D	70.90	99.89
50. *M. unguiculatus*	25	O	D	74.10	89.35
51. *Psammomys obesus*	25	H	D	127.60	106.26
52. *Clethrionomys gapperi*	20	H	ND	22.50	50.46
53. *C. glareolus*	20	H	ND	23.60	45.63
54. *C. rutilus*	10	H	ND	24.20	56.84
55. *C. rutilus*	15	H	ND	24.60	50.31
56. *C. rutilus*	20	H	ND	25.00	43.53
57. *Arvicola terrestris*	20	H	ND	102.00	89.54
58. *Microtus agrestis*	20	H	ND	21.00	43.66
59. *M. arvalis*	20	H	ND	22.43	45.85
60. *M. daghestanicus*	20	H	ND	12.98	35.09
61. *M. oeconomus*	10	H	ND	28.10	67.77
62. *M. oeconomus*	15	H	ND	25.50	56.82
63. *M. oeconomus*	20	H	ND	24.00	50.53
64. *M. pennsylvanicus*	20	H	ND	31.80	61.05
65. *M. pinetorum*	23	H	ND	25.00	51.88
66. *M. pinetorum*	20	H	ND	23.00	47.70
67. *Petromyscus collinus*	21	G	D	19.00	39.90
68. *Apodemus agrarius*	20	G	ND	20.97	49.23
69. *A. flavicollis*	20	G	ND	30.85	49.93
70. *A. sylvaticus*	20	G	ND	19.00	33.91
71. *Micromys minutus*	20	O	ND	8.70	31.77
72. *Acomys cahirinus*	25	O	D	47.40	49.01
73. *A. russatus*	25	O	D	44.03	45.55
74. *Aethomys namaquensis*	20	O	D	49.93	70.06
75. *Mus musculus*	20	O	ND	17.00	43.46
76. *M. spicilegus*	20	O	ND	14.00	33.63
77. *Myoxus glis*	20	O	ND	160.00	125.02
78. *Muscardinus avellanarius*	20	O	ND	21.00	41.53
79. *Dryomys nitedula*	20	O	ND	26.00	51.67
80. *Zapus hudsonicus*	20	G	ND	22.00	44.57
81. *Napaeozapus insignis*	20	O	ND	26.00	52.67
82. *Octodon degus*	20	H	ND	202.70	131.90
83. *Petromus typicus*	21	G	D	130.00	71.50

[a]T_a Air temperature of ADMR measurements. [b]Diet habits: O, omnivore; G, granivore; H, herbivore; I, insectivore. [c]Habitat: D, desert; ND, non-desert. [d]$ADMR_{pred}$: ADMR predicted, $8.55m_b^{0.520}$.

Body mass (BMR; g)	BMR (kJ day^{-1})	ADMR (kJ g$^{-0.52}$ day^{-1})	ADMR/ ADMR$_{pred}$^{d}	BMR/ ADMR	Reference
29.90	15.29	11.49	1.34	0.23	Downs and Perrin (1990)
120.10	68.35	8.96	1.05	0.50	
67.00	37.16	9.52	1.11	0.39	Ru-yung and Shao-liang (1984)
67.00	37.16	10.88	1.27	0.46	Ru-yung and Shao-liang (1984)
127.60	35.80	8.53	1.00	0.34	Degen et al. (1988)
23.53	24.42	9.99	1.17	0.46	French et al. (1976)
22.97	27.32	8.81	1.03	0.57	Drodz (1968); French et al. (1976)
28.00	37.14	10.83	1.27	0.56	Grodzinski (1971)
28.00	37.14	9.51	1.12	0.65	Grodzinski (1971)
28.00	37.14	8.12	0.95	0.76	French et al. (1976)
97.33	54.71	8.07	0.94	0.64	French et al. (1976)
22.30	32.83	8.96	1.04	0.71	French et al. (1976)
21.95	29.90	9.09	1.06	0.63	French et al. (1976)
13.00	23.58	9.25	1.08	0.67	Gebczynski (1964)
32.00	45.48	9.67	1.13	0.59	Grodzinski (1971)
32.00	45.48	11.95	1.40	0.64	Grodzinski (1971)
32.00	45.48	10.54	1.23	0.67	Gebczynski (1970)
37.93	36.76	10.09	1.18	0.50	French et al. (1976)
25.25	26.16	9.72	1.14	0.49	Lochmiller et al. (1982)
25.00	24.03	9.34	1.09	0.46	French et al. (1976)
		8.62	1.01		Withers et al. (1980)
21.00	17.12	10.11	1.18	0.35	Drodz (1968); Gorecki (1969)
33.20	53.88	8.34	0.98	0.99	Gebczynski (1966); Drodz (1968)
22.95	24.41	7.33	0.85	0.60	Grodzinski (1985)
8.71	21.72	10.31	1.20	0.68	Gorecki (1971)
42.00	22.34	6.59	0.77	0.51	Degen and Kam (1992) Kam and Degen (1991,1993)
51.07	19.76	6.36	0.74	0.37	Kam and Degen (1993)
48.40	20.78	9.16	1.07	0.29	Withers et al. (1980) Buffenstein (1985)
16.00	12.94	9.95	1.16	0.32	French et al. (1976)
		8.52	1.00		French et al. (1976)
43.20	8.92	1.04	0.37		Gebczynski et al. (1972)
	8.52	1.00			Gebczynski et al. (1972)
	9.50	1.11			Gebczynski et al. (1972)
21.65	8.93	1.04	0.36		French et al. (1976)
22.70	21.48	9.67	1.14	0.47	French et al. (1976)
206.00	92.86	8.32	0.97	0.69	Ebensperger et al. (1990)
		5.68	0.66		Withers et al. (1980)

°C and taking one value per species is (Fig. 7.10):

$$\text{ADMR (kJ d}^{-1}) = 8.55 \, m_b \, (g)^{0.520},$$
$(n = 63; \, SE_{y \cdot x} = 0.74; \, SE_a = 0.038; \, SE_b = 0.024; \, r^2_{adj} = 0.88; \, F = 472; \, P < 0.01).$

The difference between the ADMR and the BMR of an animal is due mainly to the heat increment of feeding for maintenance and some thermoregulatory costs. BMR values were available for 50 of the species in which ADMR has been measured. The linear regression equation for ADMR related to BMR takes the form:

$$\text{ADMR (kJ day}^{-1}) = 0.87 + 1.09 \, \text{BMR (kJ day}^{-1}),$$
$(n = 50; \, SE_{y \cdot x} = 0.581; \, SE_a = 0.166; \, SE_b = 0.185; \, r^2_{adj} = 0.40; \, F = 34.9; \, P < 0.01).$

The ratio BMR:ADMR, k_m, should be approximately the same for all small mammals feeding on similar diets. However, although the regression of the BMR:ADMR ratio related to body mass is not significant, there is quite a scatter of values within each diet and within desert and non-desert mammals (Fig. 7.11). The regression equation of the BMR:ADMR ratio related to body mass for all animals takes the form:

$$\text{BMR:ADMR ratio} = 0.443 + .0004 \, mb \, (g),$$
$(n = 50; \, SE_{y \cdot x} = 0.167; \, SE_a = 0.030; \, SE_b = 0.0003; \, r^2_{adj} = 0.008; \, F = 1.4; \, P < 0.23).$

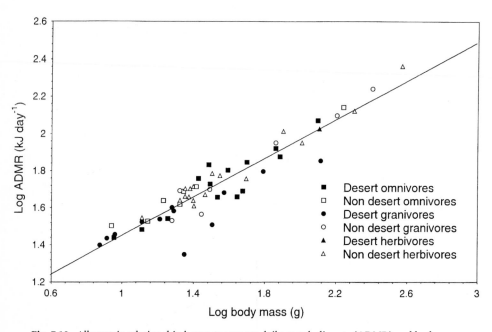

Fig. 7.10. Allometric relationship between average daily metabolic rate (ADMR) and body mass in small desert and non-desert eutherian mammal species

The slopes of the regression for all mammals, with each diet and for desert and non-desert mammals, were all close to zero (Table 7.10), indicating that the intercept was close to the efficiency of utilization of energy for maintenance (BMR/ADMR or k_m) for that group of animals. It was expected that the k_m for granivores would be higher than that for omnivores and herbivores – that is, the heat increment of consuming seeds would be lower than that for other diets, but according to analyses this was not the case. The k_m equalled 0.47 ± 0.21, 0.37 ± 12 and 0.56 ± 0.11 for granvores, omnivores and herbivores, respectively, and 0.37 ± 0.12 and 0.55 ± 0.16 for desert and non-desert mammals, respectively (Table 7.10). The mean value for herbivores was significantly higher than that for omnivores, whereas granivores did not differ from omnivores and herbivores. The value for non-desert rodents was higher than that for desert rodents.

The allometric regression of BMR (kJ day^{-1}) related to body mass (m_b; g) of these 50 species takes the form (Table 7.10):

$$BMR = 2.53 \; m_b^{0.634},$$

and the allometric regression of ADMR (kJ day^{-1}) related to body mass (m_b; g) of these same 50 rodents takes the form (Table 7.10):

$$ADMR = 8.54 \; m_b^{0.527}.$$

By dividing ADMR by BMR, the energetic costs of ADMR over BMR can be cal-

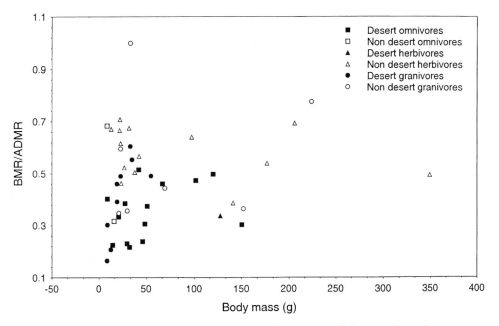

Fig. 7.11. Relationship between BMR/ADMR and body mass in small desert and non-desert eutherian mammal species. Regression analyses showed that the relationship was not significant for all animal species nor for desert and non-desert species nor for any of the dietary habits

Table 7.10. Coefficients of the linear regression equations taking the form $\log y = a + b \log$ body mass (m_b; g) where y is either average daily metabolic rate (ADMR; kJ day^{-1}) for small non-producing eutherian mammals or their basal metabolic rate (BMR; kJ day^{-1}). The linear regression equation, BMR/ADMR = a + b m_b, is also presented

y	n	a	b	S_a	S_b	$S_{y.x}$	r^2_{adj}	F	P
All animals									
ADMR	50	0.927	0.527	0.044	0.027	0.076	0.88	375.4	0.000
BMR	50	0.403	0.634	0.096	0.059	0.166	0.70	114.2	0.000
BMR/ADMR	50	0.443	0.000	0.031	0.000	0.167	0.01	1.4	0.243
20 °C									
ADMR	40	0.943	0.527	0.028	0.017	0.046	0.96	936.5	0.000
BMR	40	0.433	0.637	0.098	0.061	0.162	0.74	110.1	0.000
BMR/ADMR	40	0.460	0.000	0.035	0.000	0.173	0.00	1.1	0.292
25 °C									
ADMR	12	0.703	0.640	0.281	0.172	0.129	0.54	13.8	0.004
BMR	12	0.099	0.755	0.241	0.145	0.120	0.70	27.2	0.000
BMR/ADMR	12	0.390	0.000	0.069	0.001	0.126	0.00	0.0	0.893
Desert animals									
ADMR	24	0.902	0.524	0.085	0.055	0.093	0.79	89.9	0.000
BMR	24	0.172	0.710	0.102	0.065	0.114	0.84	118.2	0.000
BMR/ADMR	24	0.344	0.001	0.040	0.001	0.124	0.04	0.8	0.372
Non-desert animals									
ADMR	26	0.978	0.512	0.033	0.020	0.043	0.96	668.0	0.000
BMR	26	0.689	0.523	0.096	0.057	0.123	0.77	84.0	0.000
BMR/ADMR	26	0.545	0.000	0.040	0.000	0.161	0.00	0.1	0.813
Omnivores									
ADMR	18	1.014	0.477	0.081	0.053	0.069	0.83	82.0	0.000
BMR	18	0.469	0.549	0.157	0.100	0.148	0.63	29.9	0.000
BMR/ADMR	18	0.366	0.000	0.049	0.001	0.127	0.06	0.1	0.812
Granivores									
ADMR	16	0.860	0.548	0.092	0.060	0.098	0.84	82.7	0.000
BMR	16	0.132	0.797	0.167	0.109	0.165	0.78	53.7	0.000
BMR/ADMR	16	0.415	0.001	0.065	0.001	0.202	0.06	1.9	0.188
Herbivores									
ADMR	16	0.984	0.507	0.045	0.026	0.044	0.96	386.9	0.000
BMR	16	0.828	0.449	0.101	0.057	0.097	0.80	61.6	0.000
BMR/ADMR	16	0.582	0.000	0.036	0.000	0.110	0.05	0.8	0.391

n, sample size; S_a, S_b, $S_{y.x}$, SE of a, b and of estimate, respectively; r^2_{adj}, adjusted coefficient of determination; F, F-statistic for the linear terms; P, significance level of regression equation.

culated as:

$$ADMR/BMR = 3.34 \, m_b^{-0.107}.$$

However, this equation should be considered quite robust and used with caution as the allometric regression of ADMR/BMR related to body mass is not significant (Fig. 7.12):

$$ADMR/BMR = 3.05 \, m_b^{-0.086},$$
$$(n = 50; \, SE_{y \cdot x} = 0.174; \, SE_a = 0.100; \, SE_b = 0.062; \, r^2_{adj} = 0.02; \, F = 2.0; \, P < 0.17.$$

Of the 50 species of rodents used in the study with BMR measurements, the BMR of the desert species were lower than those of the non-desert species, as has been reported in other studies comparing desert with non-desert species, while values for omnivores and granivores were lower than those for herbivores. Similarly, the ADMR per $g^{0.522}$ of desert rodents (8.20 ± 1.50; $n = 29$) was significantly lower than that of non-desert rodents (9.05 ± 0.98; $n = 34$). Furthermore, granivores (8.04 ± 1.50; $n = 21$) had a lower ADMR than omnivores (8.94 ± 1.23; $n = 23$) and herbivores (9.00 ± 0.94; $n = 19$), which were not different from each other (Fig. 7.13). A scattergram of the residuals of the regression of ADMR related to body in desert and non-desert mammals according to dietary habits is presented in Fig. 7.14.

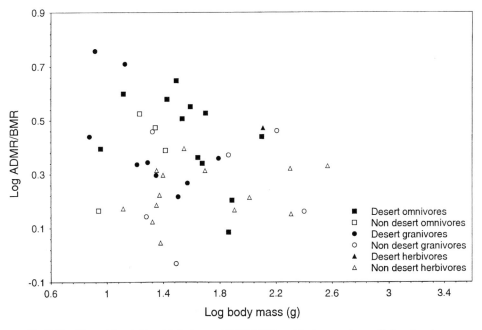

Fig. 7.12. Allometric relationship between ADMR/BMR and body mass in small desert and non-desert eutherian mammal species. Regression analyses showed that the relationship was not significant for all animal species nor for desert and non-desert species nor for any of the dietary habits

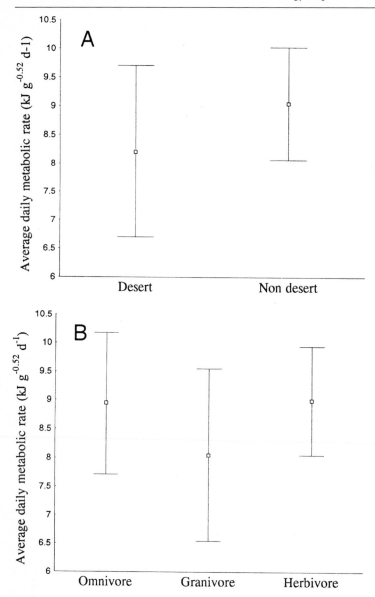

Fig. 7.13. Comparison of average daily metabolic rate in small eutherian mammals: **A** between desert and non-desert species and **B** among dietary habits

The statistics on the regression lines of BMR and ADMR related to body mass were also unexpected. BMR measurements are standardized among laboratories; ADMR measurements are not. In fact, the ADMR measurements used here were made at 20 °C and 25 °C. Yet the coefficient of determination (r^2_{adj}) of the allometric regression of BMR related to body mass was 0.70 and that of ADMR

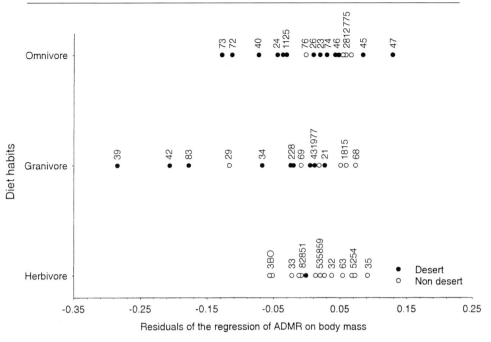

Fig. 7.14. Residuals of the regression of log ADMR on log body mass for small desert and non-desert mammalian species of different dietary habits. *Numbers* refer to species listed in Table 7.9

related to body mass was 0.88. That is, differences in body mass explained more of the variation for ADMR than for BMR and would indicate that factors other than body mass are more important in influencing BMR than ADMR. More significant though were the standard errors of the estimates $(S_{y \cdot x})$ for these two regressions: 0.166 for BMR related to body mass was considerably greater than 0.077 for ADMR related to body mass. Therefore, 68% of the measurements fall between 0.84 and 1.19 times the predicted value for ADMR and between 0.68 to 1.47 times the predicated value for BMR. That is, body mass proves to be a better predictor of ADMR than of BMR.

7.6
Field Metabolic Rate

The field metabolic rate (FMR) of an animal is its rate of heat production under free-living conditions. If the animal is maintaining its body mass, then it is considered to be in energy balance and the energy lost as heat production equals the metabolizable energy required for maintenance. If the animal is retaining energy (indicated by an increase in body mass), then the FMR includes heat production for maintenance and HIF_g; if the animal is mobilizing energy (indicated by a decrease in body mass), then FMR includes heat production from catabolism of tissue.

The number of BMR measurements available for homeotherms is large, and basically one method is used for its estimation. This is not the case for FMR, where there are few measurements but a large number of methods in use, making comparisons among animal species more difficult. For example, in collecting FMR from 47 bird species, Bennett and Harvey (1987) used data from studies in which ten methods had been used to measure FMR.

Measurements of time activity budgets have been used in many large mammals (Owen-Smith and Cooper 1989) and bird (Carmi-Winkler et al. 1987) studies. In this method different activities are timed and are assigned energy values derived in the laboratory and expressed as multiples of BMR (Goldstein 1988). This method is not usually a practical option for most small desert mammals as they are hidden throughout the day and difficult to observe at night.

7.6.1
BMR, ADMR and FMR

FMR has been estimated simply as multiples of metabolic rates measured in the laboratory (Gessaman 1973). For example, FMR has been expressed as two times BMR for small mammals (Golley 1960) and two to three times BMR (Robbins 1983) and three times RMR (Karasov 1981) for mammals in general. FMR has also been equated to ADMR in many studies and as two times ADMR (Robbins 1983). Ebensperber et al.(1990) measured ADMR in eight rodent species and concluded that, for rodents, FMR as predicted by Nagy (1987) can be estimated as 1.33 times ADMR.

These estimates are based on the premise that the energy costs of activities or for free-living are constant multiples of BMR, RMR and ADMR in different-sized animals. However, examination of the exponents to which body mass is raised in allometric relationships with BMR and FMR suggests that this is not the case. In a comparison of BMR and FMR in birds, marsupials and eutherian mammals, the exponent for BMR was greater than that for FMR in all cases (Koteja 1991; Degen and Kam 1995).

Koteja (1991) concluded that BMR is not a reliable index of energy expenditure for free-living animals. Indeed, Peterson et al. (1990) had studied the same three groups of animals as Koteja and found a large range of FMR to BMR ratios, most falling between 1.5 and 5. Furthermore, in 47 bird species, BMR was related to body mass to the power of 0.68 and FMR to the power of 0.61 (Bennett and Harvey 1987). As a result, these authors cautioned against the use of the FMR:BMR ratio in estimating the energy expenditure of free-living birds as it is not constant among differently sized species.

Degen and Kam (1995) also reported that the slope of BMR was significantly higher than that of FMR in birds, marsupials and eutherian mammals. However, they found that the FMR:BMR ratio was significantly correlated to body mass in the three groups (Table 7.11) and in each decreased as a power function of

Table 7.11. Body mass (m_b), basal metabolic rate (BMR), field metabolic rate (FMR) and FMR:BMR ratio for non-reproducing bird, marsupial and eutherian mammal species. FMR: BMR was calculated as mass-specific FMR divided by mass-specific BMR. All measurements of FMR were determined by the doubly labelled water method and of BMR by indirect calorimetry using O_2 consumption. (Degen and Kam 1995)

Species	m_b-BMR (g)	BMR (kJ day^{-1})	m_b-FMR (g)	FMR (kJ day^{-1})	FMR:BMR	Reference
Birds						
Calypte anna	4.8	10.0	4.5	31.8	3.39	1, 2
Estrilda troglodytes	6.5	10.9	6.7	57.4	5.11	1, 3
Auriparus flaviceps	6.6	13.4	6.6	30.0	2.24	4, 4
Acanthorhynchus tenuirostris	9.7	21.5	9.7	53.0	2.47	4, 4
Lanius ludovicianus	48.6	43.0	45.5	105.9	2.63	5, 5
Callipepla gambelii	140.9	49.4	143.4	90.8	1.83	6, 6
Ammoperdix heyi	172.0	86.2	187.7	160.3	1.70	7, 8
Alectoris chukar	412.0	145.9	440.2	306.2	1.96	7, 8
Eudyptula minor	900.0	383.5	1 076.0	986.1	2.15	9, 10
Dendragapus obscurus	1 150.0	417.5	1 131.0	657.0	1.60	11, 12
Marsupials						
Sminthopsis crassicaudata	14.1	9.0	16.6	68.8	6.49	13, 14
Antechinus stuartii	36.5	17.5	22.7	65.1	5.98	13, 15
Petaurus breviceps	128.1	42.7	121.0	169.0	4.19	13, 16
Gymnobelideus leadbeateri	166.3	48.9	125.0	226.0	6.15	17, 17
Pseudocheirus peregrinus	889.0	172.3	717.0	556.0	4.00	13, 18
Petauroides volans	1 000.0	205.0	1 018.0	532.1	2.55	19, 20
Setonix brachyurus	2 796.0	404.2	1 900.0	548.0	2.00	13, 21
Phascolarctos cinereus	4 765.0	498.3	9 300.0	2 040.0	2.10	13, 22
Macropus eugnii	4 878.0	665.3	4 380.0	1 150.0	1.93	13, 21
Eutherian mammals						
Macrotus californicus	11.7	6.7	12.9	22.8	3.09	23, 23
Mus musculus	15.0	12.1	13.0	39.8	3.80	13, 18
Peromyscus crinitus	16.1	11.9	13.4	39.3	3.97	13, 24
P. maniculatus	19.9	17.6	18.4	51.4	3.16	25, 25
P. leucopus	21.4	16.3	21.0	68.0	4.25	13, 26

Table 7.11 (contd.)

Species	m_b-BMR (g)	BMR (kJ day^{-1})	m_b-FMR (g)	FMR (kJ day^{-1})	FMR:BMR	Reference
Clethrionomys rutilus	28.0	37.1	16.0	57.6	2.72	13, 27
Gerbillus allenbyi	30.7	10.3	22.8	35.6	4.65	29, 29
Dipodomys merriami	38.0	20.7	34.7	51.3	2.71	13, 30
Praomys natalensis	41.5	15.4	57.3	86.6	4.07	31, 32
Microgale dobsoni	42.6	20.1	42.6	77.1	3.80	33, 33
M. talazaci	42.8	21.0	42.8	66.5	3.20	33, 33
Acomys cahirinus	44.0	22.3	38.3	51.8	2.67	13, 34
A. russatus	51.1	19.7	45.0	47.6	2.74	13, 34
Gerbillus pyramidum	53.0	33.2	31.8	45.3	2.27	35, 29
Dipodomys microps	57.0	32.1	57.9	101.2	3.10	13, 30
Sekeetamys calurus	59.5	22.0	41.2	44.0	2.89	30, 34
Tamias striatus	87.0	43.2	96.3	143.0	2.99	13, 36
Ammospermophilus leucurus	95.7	45.2	86.2	97.5	2.39	13, 37
Arvicola terrestris	97.5	54.5	85.8	118.9	2.48	13, 38
Psammomys obesus	127.6	35.8	165.6	146.3	3.15	39, 40
Thomomys bottae	143.0	57.9	103.8	130.3	3.10	13, 41
Lepus californicus	2 300.0	632.4	1 800.0	1 295.5	2.62	13, 42
Bradypus variegatus	3 790.0	330.6	4 347.0	615.1	1.62	13, 43
Alouatta palliata	4 670.0	990.2	7 332.5	2 582.5	1.66	13, 44

1. Bennett and Harvey (1987); 2. Powers and Nagy, (1988); 3. Weathers and Nagy (1984); 4. Weathers and Nagy (1984); 5. Weathers et al. (1984); 6. Goldstein and Nagy (1985); 7. Frumkin et al. (1986); 8. Kam et al. (1987); 9. Ellis (1984); 10. Costa et al. (1986); 11. Pekins et al. (1992); 12. Pekins et al. (1994); 13. Hayssen and Lacy (1985) and/or McNab (1988); 14. Nagy et al. (1988); 15. Nagy et al. (1978); 16. Nagy and Suckling (1985); 17. Smit et al. (1982); 18. Nagy (1987); 19. Foley (1987); 20. Foley et al. (1990); 21. Nagy et al. (1990); 22. Nagy and Martin (1985); 23. Bell et al. (1986); 24. Mullen (1971a); 25. Hayes (1989); 26. Munger and Karasov (1989); 27. Holleman et al. (1982); 28. Haim (1984); 29. Degen et al;. (1992); 30. Mullen (1971b); 31. Haim and Le Fourie (1980); 32. Green and Rowe-Rowe (1987); 33. Stephenson et al. (1994); 34. Degen et al. (1986); 35. Haim and Harari (1992); 36. Randolph (1980); 37. Karasov (1981); 38. Grenot et al. (1984); 39. Degen et al. (1988); 40. Degen et al. (1991); 41. Gettinger (1984); 42. Shoemaker et al. (1976); 43. Nagy and Montgomery (1980); 44. Nagy and Milton (1979)

body mass. The slope for marsupials (-0.216) was significantly steeper than that for birds (-0.114) and eutherian mammals (-0.109), and the elevation of the line for eutherian mammals was significantly higher than that for birds (Fig. 7.15).

Body mass explained about 45% of the variation in the FMR:BMR ratio in birds, 87% in marsupials and 55% eutherian mammals. Possibly this was a result of the marsupials used in the study, as they were more homogenous in their habitat than were the birds or eutherian mammals. Most marsupials (7 of 9)

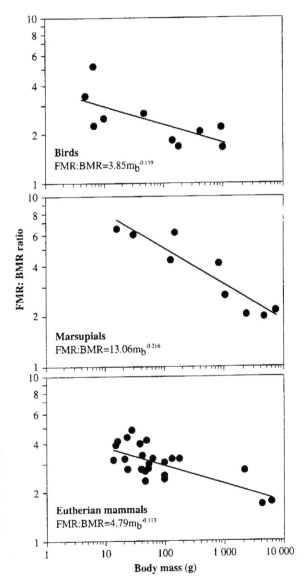

Fig. 7.15. Allometric relationships of the field metabolic rate to basal metabolic rate (FMR:BMR) ratio related to body mass in birds (*upper graph*), marsupials (*middle graph*) and eutherian mammals (*lower graph*). (Degen and Kam 1995)

Birds
FMR:BMR=$3.85 m_b^{-0.119}$

Marsupials
FMR:BMR=$13.06 m_b^{-0.216}$

Eutherian mammals
FMR:BMR=$4.79 m_b^{-0.113}$

FMR: BMR ratio

Body mass (g)

were from an eucalyptus forest, while the mammals and birds were from a much more diverse habitat (Nagy 1987).

The FMR:BMR ratio increases with a decrease in body mass, indicating that small animals expend proportionately more energy above BMR for free-living than do large animals. It can be reasoned that this is due primarily to thermoregulatory costs. Energy requirements for the maintenance of free-living animals include mainly: (1) BMR and heat increment of feeding for maintenance (HIF_m; processing of food and specific dynamic action of feeding); (2) activity costs such as locomotion, foraging and predator escape; and (3) thermoregulatory costs. This assumes no change in the energy content of the body and no costs for combatting diseases. Thermoregulatory costs in mammals scales to body mass to the power 0.50, and therefore it is proportionately, more expensive than BMR in smaller animals (Grodzniski and Weiner 1984). Smaller animals are more sensitive to changes in air temperatures than larger animals, which have a higher T_{lc} and narrower thermal neutral zone. Thus, if thermoregulatory costs are a substanial part of FMR for homeotherms, the FMR:BMR ration will decline with increasing body mass, if other costs such as HIF and locomotion remain the same relative to body mass.

The heat increment of feeding is related mainly to the diet and not to the size of an animal and is higher for animals consuming forage than concentrated feeds (Webster 1979). Smaller animals generally consume more concentrated diets than do larger animals (Demment and Van Soest 1985). Therefore, HIF as a proportion of maintenance energy should decline with a decrease in body mass. That is, HIF should decrease proportionately faster in smaller animals than in larger animals, and if it is related allometrically, the exponent of body mass should increase. Locomotory costs scale to 0.74 in bipedal birds and to 0.68 and 0.65 in marsupials and eutherian mammals, respectively (Paladino and King 1979). This suggests that activity costs and BMR scale to body mass in a similar way. However, this does not exclude the possibility that small animals may spend more time on costly activities, such as foraging, and that there may be differences in their behaviour, predator/prey relationships, food-seeking habits and so on that are dependent on body size.

7.6.2
Doubly Labelled Water Method

Measurements of CO_2 production (measure of FMR) and water influx have been achieved through the application of doubly labelled water (DLW), and its use is becoming more widespread. DLW contains ^2H (deuterium) or ^3H (tritium) instead of H, and ^{18}O instead of ^{16}O: ^3H is a radioactive isotope, the other two are not. The DLW method is attractive in that: (1) it allows measurements under near normal free-living conditions with little interference: (2) validation studies have shown that estimations are within 10% of actual measurements; and (3) it does not affect metabolic and behavioural responses, at least in mice (Speakman et al. 1991). However, there are two main limitations in its use, namely: (1) few laboratories are equipped for analysis of ^{18}O, which is difficult and expensive; and (2) the high cost of ^{18}O makes it expensive to measure large animals, and as a result, mainly small animals have been studied. Because of the importance of DLW in ecophysiological studies, it warrants a detailed description. Here, I will stress methods commonly used for small mammals (for more details, see Nagy 1983).

The first studies were initiated by Lifson et al. (1949, 1955). They reported that the concentration of ^{18}O in expired CO_2 is in equilibration with that in the animals body fluids, as ^{18}O rapidly equilibrates with CO_2 in the blood. This is due to the action of the enzyme carbonic anhydrase, which catalyses the reversible reaction of carbonic acid to H_2O and CO_2 (Maren 1967).

Assumptions Used in the DLW Method. Lifson and McClintock (1966), Nagy (1980) and Speakman (1988) summarized the theory involved in this method and discussed the potential errors associated with its use. Assumptions upon which the method is based are as follows: (1) the total body water volume within the animal remains constant during the period of measurement; (2) the rates of CO_2 production and water turnover remain constant throughout the period of measurement; (3) the isotopes $^{*}H$ and ^{18}O leave the body only in the form of H_2O and CO_2; (4) the isotopes $^{*}H$ and ^{18}O that leave the body are in the same concentration as these isotopes in body fluids, that is, there is no isotopic fractionation; (5) the isotopes $^{*}H$ and ^{18}O label only H_2O and CO_2, that is, they are not incorporated into non-aqueous molecules; (6) labelled or unlabelled H_2O and CO_2 does not enter the animal; and (7) the natural background abundances of the isotopes remain constant during the measurement period. All of these assumptions are violated to some extent, but the DLW method can still give accurate results if appropriate correction factors are applied to the data.

Procedure for Using DLW. The most common procedure for the use of DLW in small mammals is as follows (Fig. 7.16). A fluid sample, usually blood, is collected from the study animal to measure the background activity of isotopes. If the animals have never been injected with isotopes, it may not be necessary to take background samples from all of them; the mean of a few may suffice. DLW is then injected into the animal, either intramuscularly or intraperitoneally, and allowed to equilibrate with the body fluids. During this time, the animals do not have access to either food or water. The equilibration time depends mainly upon the size of the animals and takes up to 2 h in rats. After equilibration, the animal is weighed, a fluid sample, usually blood, is then collected, and the animal released. $^{*}H$ penetrates the total body water volume (TBWV) space. It also exchanges with some non-aqueous hydrogen, principally in fat and to a lesser extent in proteins and carbohydrates. As a result, the $^{*}H$ space, measured from the dilution of $^{*}H$ at equilibrium, overestimates the actual TBWV of an animal. The ^{18}O penetrates the body water pool, and at equilibrium, the concentration of ^{18}O in expired CO_2 and in water leaving the body equals the concentration in the body fluid.

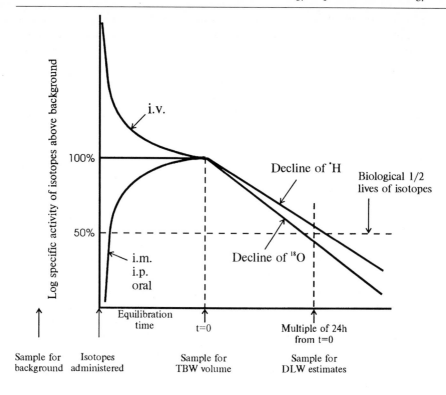

Fig. 7.16. Schedule for isotope administration and blood sampling when using the doubly labelled water method to estimate rates of water flux and CO_2 production

The animals are recaptured and weighed, another fluid sample is collected, and then they are released. Recaptures should be made in multiples of 24 after their release to avoid errors due to the large variability in energy expenditure within a day (Speakman and Racey 1988). Nagy (1983) suggested that recaptures should be made between one and two biological half-lives of the ^{18}O isotope when using the high level technique. Sampling earlier than this does not allow sufficient time for the isotopes to decline, and small errors in isotope analysis may result in large errors in energy flux estimates. Sampling later than this can, however, result in the specific activity of ^{18}O concentrations being close to background values, making the energy flux estimations unreliable.

Fluid sample are microdistilled (Wood et al. 1975; Nagy 1983), and the water obtained is analysed for concentrations of ^{18}O and *H. The diluiton of the isotopes in the animal provides an estimate of TBW (Nagy and Costa 1980; Degen et al.1981). In an animal administered with DLW, the specific activity of above-background *H in the body fluids declines exponentially over time. This is due to the uptake of unlabelled water into the water pool via drinking, preformed and metabolic water from food and from body tissue catabolism. There is a simultaneous output of isotope in water via evaporation, urine and faeces. The rate of decline of the specific activity of *H is a measure of water flux through an animal. The ^{18}O in a labelled animal is lost in the form of CO_2 as well as in water. Consequently, the rate of decline of above-background ^{18}O is faster than that of *H (Fig. 7.16). The difference in rate of decline of the two isotopes is due to CO_2 production, which is a measure of the metabolic rate.

Blood Sampling Schedule for DLW Measurements. The most common sampling schedule for DLW measurements is the collection of a sample for TBWV estimation at the time of isotope equilibration and then after a period of free-living. The log specific activities above background of the initial and final samples are used to calculate the fractional turnover rates of the isotopes. Another approach is to collect multiple samples following the equilibration of isotopes, and to use regression analysis to calculate the rate constants of the exponential declines of above background levels of *H and ^{18}O.

The advantages of each method are discussed in detail by Speakman (1990). In the analyses of isotopes, there are errors associated with each point as well as an error due to random variation in background specific activities. With multiple samples, therefore, the estimate of the decline rate of isotopes is unbiased by random analytical errors and background variations. Such is not the case with the two-point method because variations in background directly affect the specific activities of the isotopes. The can be particularly critical in analysis of the final sample if the specific activities of the isotopes are close to background values. The multiple sampling technique appears therefore to be more reliable though this need not necessarily be the case. In the two-point method, the decline in activity is measured between the initial and final samples, independently of what path the decline of isotope activity may have followed in between. In regression analysis, the equation line takes the best fit and may or may not pass through the final sample collected. If the decline in isotopes is constant, the lines pass through the final points and give an accurate estimation of energy and water fluxes during the period of measurement. Studies have shown, however, that the decline curves of the isotopes are often not smooth exponential lines but may be highly variable. This could be particularly problematical in free-living animals that show large differences in energy intake from day to day and could result in errors being recorded in energy and water fluxes during the period of measurement.

The option of choosing the sampling method is often not available in the case of small mammals where the taking of multiple samples would be unfeasible. In fact, removal of two blood samples may be harmful to some small animals, while constraining them until equilibration of the isotopes with body fluids has taken place may be stressful. In such cases, a single sampling technique can be used in which the animals are released immediately after being administered with isopotes. TBWVs of the animals are estimated either by isotope dilution in another population of animals of the same species, or by killing and desiccating a number of individuals. Values from these animals are then used to predict the TBWVs of the released animals. The initial specific activities of the isotopes can then be calculated from the known doses administered to the animals and their predicted TBWVs. Moreover, the ratio of isotopes at equilibration within the animals can be determined by analysing the injectate.

Calculations of CO_2 Production Using DLW. The difference in rates of decline between the two isotopes multiplied by the volume of the body space in which they turn over provides an estimate of the CO_2 production of the animal. This can be expressed as:

$$rCO_2 = 0.5\ N\ (K_o - K_d),$$

where rCO_2 is the rate of CO_2 production in mol CO_2 h^{-1}; N is the total body water pool (mol), and K_o and K_d are the rate constants of the exponential declines of above-background levels of ^{18}O and *H, respectively (Lifson and McClintock 1966). The constant 0.5 is used because the turnover of a water molecule is associated with half a molecule of CO_2: there is only one atom of oxygen in each molecule of water, compared with two atoms in each molecule of CO_2. The rate constant decline of ^{18}O, or the fractional turnover rate of the ^{18}O pool, can be calculated as:

$$k_o = (\ln\ ^{18}O_i - \ln\ ^{18}O_f)t,$$

where $^{18}O_i$ and $^{18}O_f$ are initial and final enrichments of ^{18}O (ppm) above the background enrichment of ^{18}O (ppm); and t is time between initial and final blood samples (h). The rate constant decline of *H, or the fractional turnover rate of the *H pool, can be calculated as:

$$k_d = (\ln\ {}^*H_i - \ln\ {}^*H_f)/t,$$

where $^*H_i$ and $^*H_f$ are initial and final enrichments of *H (dpm) above the background enrichment of *H (dpm).

The equation can be modified to take into account the physical fractionation of hydrogen at H_2O liquid to gas change ($f_1 = 0.93$), of oxygen at H_2O liquid to gas change ($f_2 = 0.99$), and of oxygen at CO_2 gas to H_2O change ($f_3 = 1.04$) and proportion of water loss that is fractionated (r_{gf}). These physical fractionation factors are determined at 25 °C. The equation then takes the form (Speakman 1990):

$$rCO_2 = 0.5\ [N(K_o - K_d) - (f_2 - f_1)r_{gf}]/f_3.$$

This equation could be written as:

$$rCO_2 = [N\ (K_o - K_d)/2.08] - 0.0144\ N_d k_d.$$

In this equation, one pool is used for both ^{18}O and *H spaces. However, these spaces do differ, and the equation can be modified to correct for this as follows (Coward et al. 1985a):

$$rCO_2 = N/2.08 \, (N_oK_o - N_dK_d) - 0.0144 \, N_dk_d,$$

where $N_o = {}^{18}O$ space (mol) and $N_d = {}^*H$ space (mol).

Initial studies on humans reported that the ^{18}O space is 1.01 of [TBWV], and $^*H$ space is approximately 1.04 TBW; that is, a ratio of 1.03 (1.04/1.01; Coward et al. 1985a, b). Using these values the equation can be represented as (Schoeller et al. 1986 b):

$$rCO_2 = [N/2.08 \times (1.01 \, K_o - 1.04 \, K_d)] - 0.0246 \, r_{gf},$$

where N is TBW $= [(N_o/1.01) + (N_d/1.04)]$; r_{gf} is the rate of water loss through routes subject to fractionation and is determined as $1.05 \, N \, (1.01 \, K_o - 1.04 \, K_d)$.

However, these spaces and, consequently, the ratios may vary. For example, Speakman et al. (1993b) further examined the $^*H$ to ^{18}O pool size ratio and in 161 individual measurements found a mean $^*H$ space of 1.0531 and an ^{18}O space of 1.01 for a ratio of 1.0427. These values were than substituted in the above equation.

The use of these equations have been discussed at length (Speakman 1987, 1988, 1990, 1993b; Speakman et al. 1993b). Speakman and Racey (1988) used bats to compare CO_2 production rates determined by DLW and by indirect calorimetry in which CO_2 production rates were computed using the equations of Lifson and McClintock (1966), Coward et al. (1985a) and Schoeller et al. (1986b). The mean algebraic differences between these equations and indirect calorimetry were +9.5, +5.1, and +3.4%, respectively. However, these means were not significantly different from each other because of the large standard errors associated with them. The mean absolute deviations between DLW and indirect calorimetry were 13.5, 13.7 and 14.5%, respectively, for the three equations.

Nagy (1980) modified the equations of Lifson and McClintock (1966) in which one TBW pool was used to yield ml CO_2 produced per g body mass per h. Most DLW studies on free-living animals have used this version. In an animal maintaining constant TBWV over the measurement period (between two blood samples and usually indicated by constant body mass), the equation reads:

$$CO_2 \text{ produced (ml g}^{-1} \text{ h}^{-1}) = [25.93W \ln({}^{18}O_i{}^*H_f/{}^{18}O_f{}^*H_i)]/Mt$$

where W is body water pool in ml; M is body mass in g; t is time days; i and f are initial and final, and the rest of the symbols as above. The conversion factor of 25.93 incorporates conversion factors from mmol to ml, days to hours, etc.

This equation was modified to account for an animal either gaining or losing TBWV over the measurement period. When TBWV in an animal changes linearly between the two blood samples, then the equation reads:

$$CO_2 \text{ produced (ml g}^{-1} \text{ h}^{-1}) = \frac{51.86(W_f - W_i) \ln ({}^{18}O_i{}^*H_f/{}^{18}O_f{}^*H_i)}{(M_i + M_f) \ln(W_f/W_i)t}.$$

And, if the change in TBWV is exponential, then the volume of CO_2 produced can be calculated as:

$$CO_2 \text{ produced (ml g}^{-1}\text{ h}^{-1}) = \frac{51.86 \ W_i \ ln(W_f/W_i)ln(^{18}O_i^*H_f/^{18}O_f^*H_i)}{(M_i+M_f) \ [1-(W_i/W_f)]t}.$$

It is generally assumed that the TBWV as a proportion of body mass remains constant. Very little error is engendered in using any of these equations if the change in TBW space is below 20% (Fig. 7.17).

Validation and Accuracy of DLW. Validation studies have shown that estimates of energy expenditure by DLW and, simultaneously, by other methods such as indirect calorimetry and metabolizable energy intake and energy balance are similar and generally range up to about 10% (Table 7.12). This indicates that the DLW method is accurate, and it is often used as the standard against which other methods are evaluated. Poppitt et al. (1993) suggested that the energy expenditure of lactating mammals can be overestimated when using DLW, due to an exchange of unlabelled CO_2 from the litter to the mother.

Interestingly, as pointed out by Speakman (1990), very little progress has been made in improving the accuracy of the DLW method since its initiation, in spite of the many modifications made over the years. However, there has been little, if any, room for improvement in accuracy as the algebraic error and mean deviation between CO_2 production estimated by DLW and CO_2 measured by (or calculated) in validation trials were relatively small in the early studies (Fig. 7.18).

Analysis of Isotopes. Tritium is a radioactive material with a radiological half-life of 12.3 years. It is a soft beta-emitter and can be measured with a liquid scintillation counter. This isotope is inexpensive and easy to analyse, but requires necessary permits for its use.

Deuterium is also relatively inexpensive, but it is difficult and expensive to measure its concentration. The natural abundance of deuterium is about 0.015 atom % excess (150 ppm). Concentrations as low as 0.017 atom% or 0.002 atom% excess above natural background can be measured reliably using an isotope

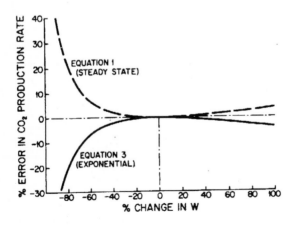

Fig. 7.17. Percentage error in calculated rates of CO_2 production due to use of equations for steady state or exponential change in body mass when change in body mass is actually linear. Errors are shown as a function of the percent by which body mass has changed during measurement period. (Nagy 1980)

Table 7.12. Summary of doubly labelled water validation studies done on vertebrates. (Speakman and Racey 1988)

Reference	Speicies	n individuals	n runs	Mean	Algebraic Error n+	Algebraic Error n−	Range Min	Range Max	Mean deviation	Calculation	Comparsion technique	Source of body water
Mammals.[c]												
Schoeller et al. (1986 b)	Human	6	6	+8.0	5	1	−7.0	+22.0	10.3	L+M	IC	Urine
	Human	3	3	+3.6	3	0	8.0	+9.0	3.6	L+M	IC	Urine
	Human	3	3	+.3	1	2	−7.0	+10.0	7.0	L+M	IC	Plasma
	Human	3	3	−16.0	0	3	−24.0	−5.0	16.0	L+M	IC	Saliva
	Human	3	3	+5.0	3	0	+3.3	+6.8	5.0	L+M	FI	Urine
	Human	3	3	+.8	2	1	−2.3	+4.2	2.3	L+M	FI	Plasma
	Human	3	3	−14.9	0	3	−20.2	−10.0	14.9	L+M	FI	Saliva
Jones et al. (1987)	Human infants	9	10	−.7	5	5	−12.0	+7.0	4.9	S	IC	Urine
Roberts et al. (1986)	Human infants	4	4	−1.4	1	3	−4.8	+5.8	4.3	C+P	IC	Urine
Schoeller et al. (1986a)	Human	5	5	+3.3	4	1	−6.0	+8.8	5.7	S	FI	Urine
Prentice et al. (1986)	Human (lean)	9	9	+10.2	6	3	−4.9	+40.4	11.5	?C+P	IC	Urine
	Human (obese)	7	7	+12.6	7	0	+4.9	+42.6	12.6	?C+P	IC	Urine
Prentice et al. (1985)	Human	12	14	+7.5	9	5	−5.3	+41.6	10.2	L+M	IC	Urine
Coward et al. (1985b)	Human	4	4	+2.4	3	1	−.9	+3.9	2.8	C+P	IC	Urine
Coward et al. (1985b)	Human infant	4	4	+2.1	4	0	+4.5	+10.8	6.9	C+P	IC	Urine
Schoeller and Webb (1984)	Human	5	5	+5.9	4	1	−6.5	+14.1	8.5	L+M	IC	Urine
Klein et al. (1984)	Human	1	1	+1.8	1	0	...	...	28.0	L+M	IC	Urine
Schoeller and Van Santen (1982)	Human	4	4	+2.1	3	1	−5.8	+7.1	5.8	L+M	FI	Urine
Westerterp et al. (1985)	Human	2	2	−2.5	1	1	−6.0	+1.0	3.5	L+M	IC	Urine
Gettinger (1983)	Pocket gopher (Thyomonys)	6	6	+3.7	4	2	−8.7	+14.5	6.8	L+M	IC	Blood
Karasov (1981)	Ground squirrel (Ammospermophilus)	7	7	+.8	?	?	−12.4	+17.2	?	L+M	FI	Blood
Little and Lifson (1975)	Chipmunk (Tamias)	2	2	+4.5	2	0	+1.0	+8.0	4.5	L+M	IC	Blood
Lifson et al. (1975)	Chipmunk (Tamias)	3	20	+4.9	15	5	−19.0	+18.0	5.8	L+M	IC	Urine/blood

Table 7.12 (*contd.*)

Reference	Speicies	n individuals	n runs	Mean	Algebraic Error		Range		Mean deviation	Calculation	Comparsion technique	Source of body water
					n+	n−	Min	Max				
Mullen (1970)	*Perognathus*	4	4	+.85	3	1	−9.2	+5.3	5.5	L+M	IC	Blood
Randolph (1980)	Chipmunk (*Tamias*)	6	6	−6.9	0	1	−13.3	−3.6	6.9	L+M	IC	Blood
Lifson and Lee (1961)	Rat	6	6	+1.0	4	1	−10.0	+8.0	4.0	L+M	Ic	Blood
Lee and Lifson (1960)	Rat	7	7	+2.0	4	2	−9.0	+7.0	4.4	L+M	IC	Blood
McClintock and Lifson (1958a)	Rat	8	8	+2.0	6	2	−2.0	+10.0	3.0	L+M	IC	Blood
McClintock and Lifson (1958b)	Rat	5	5	+.6	2	2	−2.0	+4.0	2.2	L+M	IC	Blood
McClintock and Lifson (1957)	Mice	7	7	−4.0	2	5	−12.0	+8.0	7.0	L+M	IC	Blood
Lifson et al. (1955)	Mice	15	15	−3.0	4	10	−21.0	+20.0	7.0	L+M	IC	Blood
Holleman et al. (1982)	Red-backed vole	5	5	+7.0	?	?	?	?	?	?L+M	IC	?Blood
Present study	Pipistrelle (*n*=7) and long-eared bats (*n*=2)	9	9	+9.5	6	4	−14.3	28.6	11.5	L+M	IC	Blood
		9	9	+5.1	5	5	−17.8	24.5	13.3	C+P	IC	Blood
		9	9	+3.5	5	5	−19.2	21.6	13.0	S	IC	Blood
Overall[c]		165	185	+3.1*	108	62	−19.0[b]	+42.6[b]	6.84*			
Birds:												
Buttemer et al. (1986)	Budgerigar	9	9	−.04	3	6	−5.2	+6.2	2.5	L+M	IC	Blood
Williams (1985)	Sparrows	11	13	−3.5	5	8	−17.0	+4.2	4.5	L+M	IC	Blood
	Starlings	4	4	−6.2	1	3	−21.6	+7.0	9.7	L+M	IC	Blood
	Sparrows	11	12	−3.1	6	6	−29.0	+16.1	9.8	L+M	FI	Blood
	Starlings	4	4	−2.8	1	3	−14.4	+8.0	6.8	L+M	FI	Blood
Williams and Nagy (1984)	Savannah sparrow	5	7	+6.5	6	1	−.2	+11.0	6.5	L+M	IC	Blood
				(−1.5)					(1.5)			

Goldstein and Nagy (1985)	Gambel's quail (Callipepla)	6	6	−6.0	1	5	−23.2	+15.5	11.1	L+M	FI	Blood
Hails and Bryant (1979)	House martin (Delichon)	4	4	+3.4	?	?	?	?	4.5	L+M	IC	Blood
Lefebvre (1964)	Pigeon (Columba)	10	10	+4.0 (−6.0)	7	3	−2.2	+48.5	6.6	L+M	IC	Blood
Masman (1986)	Kestrel (Falco)	8	8	+2.1	6	2	−5.1	+9.5	4.4	L+M	IC	Blood
Westerterp and Bryant (1984)	Sand martin (Riparia)	2	2	+4.4	2	0	+2.4	+6.4	4.4	L+M	IC	Blood
Reptiles:												
Nagy (1980)	Uta	?	?	−7.4	?	?	?	?	?	?L+M	?IC	?Blood
Nagy (1980)	Tortoise (Gopherus)	?	?	+2.2	?	?	?	?	?	?L+M	?IC	?Blood
Congdon et al. (1978)	Sceloparous sp.	4	4	13.2	2	2	−5.7	16.2	8.0	L+M	IC	Blood
Overall[d]		70	77	−2.11*	21	33	−29.0[b]	+48.5[b]	6.07*			

Note. Calculation refers to equation used: L+M, Lifson and McClintock (1966) equation and subsequent derivatives, e.g. Nagy (1980); C, Coward et al. (1985a) equation, with turnover of each isotope referred to its own dilution space and S, Schoeller et al. (1986b) equation. Comparison technique refers to method of simultaneous measurement: IC, indirect calorimetry, typically in openflow system, with stream gas analysis but also by gravimetric measurement of trapped expired CO_2; FI, food intake/energy balance.
[a] Weighted.
[b] Across all studies.
[c] Twenty-six studies, nine species.
[d] Nine studies, 12 species.

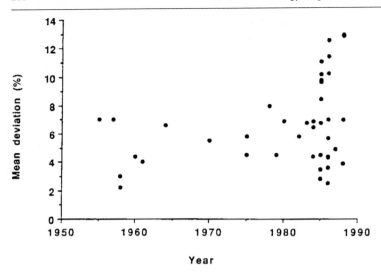

Fig. 7.18. Mean deviations for all validations of the doubly labelled water technique in comparison to indirect calorimetry, between the years 1955 and 1988. (Speakman 1990)

ratio mass spectrometer (IRMS; Schoeller et al. 1980). This procedure requires the conversion of H in distilled H_2O to H_2 in a vacuum line.

^{18}O is very expensive and both difficult and expensive to measure. The natural abundance of ^{18}O is approximately 0.2 atom% excess (2000 ppm). Two main methods are in use today for measuring ^{18}O concentrations.

1. High enrichment and proton activation analysis (PAA; Wood et al. 1975). In cyclotron-generated proton activation analysis, ^{18}O is converted to gamma-emitting ^{18}F following proton bombardment and the ^{18}F is counted in a gamma-counter. Only very small volumes of water (under 10 μl) are needed for analysis. This procedure requires a relatively high enrichment of isotope, that is levels in the fluids of animals to which they have been administered of about 6000 ppm or 4000 ppm above background. To achieve this, a dose of approximately 3.0 g of water containing 97 atom % ^{18}O per kg body mass must be administered. This high concentration has limited studies using this procedure mainly to small animals because of the great expense of ^{18}O. However, due to the large difference between enriched and natural abundances of ^{18}O in animals, small changes in natural abundances have little effect on the estimates of CO_2 production rates.

2. Low enrichment and gas IRMS (Speakman et al. 1993b). In this procedure water is converted to gas, preferably CO_2, and ionized at very low pressure. The most common method for the preparation of CO_2 from water for IRMS analysis is equilibration in which water and CO_2 are mixed at controlled temperatures and react as follows:

$$H_2^{18}O + C^{16}O_2 \rightarrow H_2^{16}O + C^{18}O_2.$$

The CO_2 is then purified and transferred to a mass spectrometer for analysis of the $^{18}O{:}^{16}O$ ratio to be used in the calculation of ^{18}O enrichment of the water sample. Very small volumes of water (under 5 µl) can be used for analysis. This procedure works with a relatively low enrichment of the isotope, about 100 ppm above background, but the analysis is more difficult and expensive than PAA. The low dose of ^{18}O required by this procedure allows measurements to be made on larger animals. Care must be taken, however, in measuring the natural abundance of ^{18}O in the animals; as the difference between enriched and natural abundances is low, small changes in natural abundance could result in substantial changes in excess concentrations. Furthermore, natural abundances in animals are not constant throughout the day.

Converting CO_2 Production Rates to Rates of Heat Production and Energy Intake. Rates of heat production, energy and food intakes, and metabolic water production can be estimated from rates of CO_2 production by the use of proper conversion factors, provided that the dietary composition is known and that the animals are maintaining their energy and water balances. Carbohydrates, fat, protein with urea as end product and protein with urate as end product produce, respectively, 20.8, 27.7, 23.1 and 24.8 J metabolizable energy per ml CO_2 (Table 7.1). In general, moreover, vegetation yields 21.7 J ml^{-1} CO_2, seeds yield 21.9 J ml^{-1} CO_2 and meat yields 25.7 J ml^{-1} CO_2 (Nagy 1983).

Dry matter intake can be calculated by dividing metabolizable energy intake by the metabolizable energy yield of the diet (Degen et al. 1986, 1991; Geffen et al. 1992a). Furthermore, in an animal which eats mainly two types of feed that differ somewhat in energy yield and water contribution (total of preformed and metabolic water) and does not drink water, the two dietary items can be estimated by the use of simultaneous equations (Degen et al. 1986; Geffen et al. 1992a). For example, in a study of *Acomys cahirinus*, *A. russatus* and *Sekeetamys calurus*, three symatric omnivorous desert rodents, it was found that insects and dry, mature vegetation were mainly consumed (Degen et al. 1986). It was assumed that dry, mature vegetation contained 0.111 g preformed water and 9.0 kJ metabolizable energy per g dry matter, and that 21.7 J energy and 0.637 µl metabolic water were produced per ml CO_2 production. For insects, the factors were 2.333 g preformed water and 20.3 kJ metabolizable energy per g dry matter, and 25.7 J energy and 0.660 µl metabolic water generated per ml CO_2 production. CO_2 production per gram dry matter intake can be calculated as:

insects: 20300 J g^{-1}/25.7 J (ml CO_2)$^{-1}$ = 789.9 ml CO_2 g^{-1}
vegetation: 9000 J g^{-1}/21.7 (ml CO_2)$^{-1}$ = 414.7 ml CO_2 g^{-1}.

The combined preformed and metabolic water produced per gram dry matter was calculated as:

insects: 2.333 ml g^{-1} + 0.521 ml g^{-1} = 2.854 ml g^{-1}
vegetation: 0.111 ml g^{-1} + 0.264 ml g^{-1} = 0.375 ml g^{-1}.

Then, combining these four equation and allowing X = g dry matter of insects consumed and $Y = g$ dry matter of vegetation consumed:

$789.9X + 414.7Y = ml \ CO_2$ production.

7.6.3
Calculation of Daily Energy Expenditure, Metabolizable Energy Intake and Energy Retention in Animals Not in Energy Balance

The above methods assume that the animals are in energy and water balance in order to estimate their energy intake. However, this is often not the case. For example, daily body mass changes in DLW studies range between 3.2% and 16.9% in large (3 kg) and small (5–164 g) birds (Costa and Prince 1987; Pettit et al. 1988; Gabrielsen et al. 1991; Williams 1993; Powers and Conley 1994), 1.5% and 4.3% in large (9.5 kg) and small (121 g) marsupials (Nagy et al. 1991; Nagy and Suckling 1985), and 2.0% and 14.8% in small eutherian mammals (13–230 g; Bell et al. 1986; Kenagy et al. 1990; Mutze et al. 1991). If an animal is in negative energy balance, then the mobilization of body energy, mainly in the form of lipid, will produce CO_2. This will be considered to arise from the catabolism of energy intake, and the energy intake will consequently be overestimated. If the animal is in positive energy balance, however, then some of the energy intake will be added to the energy reserves, and the energy intake will be underestimated. This can lead to a substantial error in estimating dry matter intake.

However, water influx and CO_2 production rate estimated from DLW, in conjunction with dietary habits and chemical composition of the diet, can be used to estimate total daily energy expenditure (DEE), metabolizable energy intake (MEI) and energy retention (ER; either negative or positive) of free-living animals. The model is based on the different water to energy ratios of each dietary component, and of body tissue reserves that are either mobilized or deposited by the animals (Kam and Degen, unpubl. data).

The total water influx (TWI) of an animal is composed of four components:

$$TWI = DW + PW + MW + CW, \tag{1}$$

where DW is drinking water; PW is preformed water from food; MW is metabolic water from metabolized food compounds; CW is metabolic water either from catabolism of body lipid tissue or from water produced in the process of transforming carbohydrates or proteins into lipid tissue.

Drinking water cannot be estimated from this model, and therefore Eqn (1) can be modified for DW-free total water influx (TWI_0) as:

$$TWI_0 = TWI - DW = (PW + MW) + CW = FW + CW, \tag{2}$$

where FW is total water yield from food items. Equation (2) can be used in cases where either $DW = 0$, as is the case for most small desert mammals, or where DW is determined. Animals in three states of energy balance can be defined: (1) for an animal is negative energy balance and mobilizing body energy reserves, mainly lipid tissue. In this case, CW equals 1.07 g water for each gram of fat catabolized, and hence $TWI_o > FW$; (2) for an animal maintaining energy balance. In this case, $CW = 0$ and $TWI_o = FW$; and (3) for an animal in positive energy balance and depositing part of its metabolizable energy intake as body energy reserves. In this case, MW is less than total dietary water yield, CW depends on dietary composition and biochemical pathways, and $WI_o < FW$.

Where there is no change in energy retention (case 2), calculations can be made as for animals in negative energy balance (case 1). The approach used for cases 1 and 3 is similar, but slightly different biochemical pathways are involved. Thus, the model differentiates between animals in positive and negative balance, and these are treated separately.

To generate the model, characteristics of the diet, namely, its dry matter (DM) content (DMC; $g\,g^{-1}$ DM), metabolizable energy yield (ME; $kJ\,g^{-1}$ DM), and its components, allow calculations of the $J\,ml^{-1}$ CO_2 production conversion factor and metabolic water yield. The conversion factors used are taken from Schmidt-Nielsen (1975a) and Nagy (1983). Here, catabolism of carbohydrate, fat and protein in mammals and protein in birds and reptiles produces 20.8, 27.7, 23.1 and 24.8 $J\,ml^{-1}$ CO_2, respectively, and 0.662, 0.754, 0.509 and 0.546 μl metabolic water ml^{-1} CO_2, respectively. The ratio of water yield to CO_2 production (R; ml water ml^{-1} CO_2 production) was calculated for PW (R_{pw}), MW (R_{mw}) and CW (R_{cw}).

Negative Energy Balance. For animals in negative energy balance, metabolizable energy intake (MEI) does not provide the total DEE; some is obtained by the mobilization of body energy reserves. Thus, CO_2 production from MEI is a fraction $(X_1; 0 < X_1 < 1)$ of the total CO_2 production (TCO_2) of an animal, and Eq. (2) can be modified as follows:

$$WI_o = X_1 TCO_2 (R_{pw} + R_{mw}) + (1 - X_1) TCO_2 R_{cw}. \tag{3}$$

Dividing Eq. (3) by TCO_2 results in:

$$WI_o/TCO_2 = R = X_1 (R_{pw} + R_{mw}) + (1 - X_1) R_{cw}. \tag{4}$$

That is, R is expressed as a combination of the estimated water yield in relation to CO_2 production of the diet and lipid tissue mobilized. Collecting Eq.(4) results in:

$$R = X_1 (R_{pw} + R_{mw} - R_{cw}) + R_{cw}. \tag{5}$$

Solving X_1 for an animal in negative energy balance yields:

$$X_1 = (R - R_{cw})/(R_{pw} + R_{mw} - R_{cw}). \tag{6}$$

Since X_1 stands for CO_2 production yield of MEI as a fraction of TCO_2, an estimate of metabolizable energy intake (MEI*; J) using the dietary energy conversion factor (E_d; $J\,ml^{-1}$ CO_2 production) is given by:

$$MEI^* = X_1 TCO_2 E_d, \tag{7}$$

and an estimate of energy retention (ER*; J), using a conversion factor for lipid tissue (E_f) of 27.7 J/ml CO_2 is given as:

$$ER^* = -(1-X_1)\, TCO_2\, E_f. \tag{8}$$

Total heat production of an animal or its energy expenditure (DEE* or FMR*, field metabolic rate, as is commonly used in many publications) is therefore given as:

$$DEE^* = MEI^* - ER^*. \tag{9}$$

For comparison, DEE in doubly labelled water studies is traditionally calculated from the dietary conversion factor as:

$$DEE = TCO_2\, E_d. \tag{10}$$

The ratio of DEE* to TCO_2 is the correct energy to total CO_2 conversion factor (E). It is higher than E_d in all negative energy balance cases, and the ratio E/E_d is a measure of the error in estimating DEE based on Eq. (10).

Positive Energy Balance. For animals in positive energy balance, MEI provides more than total DEE, and the remainder is deposited as body energy reserves. The biochemical pathways in these animals are more complex than for animals in negative energy balance and require slightly different analyses. In the following paragraphs, equations similar to those presented for the negative energy balance animals have the same integer with the letter "b" added.

Given (1) X_2 is the fraction of TCO_2 resulting from the total catabolism of the dietary components; (2) R'_{mw} is the metabolic water per ml CO_2 produced via the biochemical process of lipid deposition, and thus the quantity $[(1-X_2)TCO_2\, R'_{mw}]$ equals the total metabolic water produced in the process of fat deposition; and (3) E_{pl} is the metabolizable energy for lipid deposition per ml CO_2 produced, and the quantity $[(1-X_2)TCO_2\, E_{pl}]$ refers to the MEI deposited as lipid which, when multiplied by R_{pw} and divided by E_d, equals the preformed water consumed by the animal via its diet that is converted into lipid. Equation (2) can then be modified as follows:

$$WI_o = X_2\, TCO_2(R_{pw}+R_{mw})+(1-X_2)\, TCO_2\, [R'_{mw}+R_{pw}\, (F_{pl}/E_d)]. \tag{3b}$$

Dividing Eq. (3b) by TCO_2 results in:

$$WI_o/TCO_2 = R = X_2(R_{pw}+R_{mw})+(1-X_2)[R'_{mw}+R_{pw}\, (E_{pl}/E_d)], \tag{4b}$$

and collecting terms:

$$R = X_2\, [R_{pw}+R_{mw}-R'_{mw}-R_{pw}\, (E_{pl}/E_d)]+[R'_{mw}+R_{pw}\, (E_{pl}/E_d)]. \tag{5b}$$

R is expressed as a combination of the estimated water yield to CO_2 production of the metabolizable dietary components and the part that is used for lipid deposition. Thus, solving X_2 for an animal in positive energy balance yields:

$$X_2\, [R-R'_{mw}-R_{pw}\, (E_{pl}/E_d)]/[R_{pw}-R'_{mw}-R_{pw}\, (E_{pl}/E_d)]. \tag{6b}$$

Based on MEI, ER and DEE and given X_2, MEI* can be calculated as follows:

$$MEI^* = X_2 TCO_2 E_d + (1 - X_2) TCO_2 E_{pl}, \tag{7b}$$

and ER* as:

$$ER^* = (1 - X_2) TCO_2 E_{p2}, \tag{8b}$$

where E_{p2} is the energy stored as lipid per ml CO_2 in the biochemical process of lipid deposition. Thus, DEE* is calculated by difference [Eq. (9)], which varies from the traditional method of its calculation [Eq. (10)].

For simplicity and practical purposes, two diets are described in which mammals are in positive energy balance and are consuming either a herbivorous (mainly carbohydrate) diet or a carnivorous (mainly protein) diet. In each case both R'_{mw} and E_p are determined.

For a herbivore consuming a diet rich in carbohydrates, lipogenesis is considered as the main metabolic process. As suggested by Elia and Livesey (1988), this process yields a wide range of RQ values (1.9–9.6) and can be separated into two distinct stochiometric components, the oxidation of glucose and the conversion of unoxidized glucose into lipid. However, the unoxidized component which is converted into lipid is practically identical in all cases (Elia and Livesey 1988). Based on animal lipid (mainly dioleylpalmityltriglyceride) which yields 39.64 kJ g^{-1} and glucose which yields 15.56 kJ g^{-1}, the unoxidized lipogenesis component:

$$12.9\ C_6H_{12}O_6 \rightarrow 1.0\ C_{55}H_{102}O_6 + 22.5\ CO_2 + 26.5\ H_2O,$$

yields 67.94 J lipid tissue/ml CO_2 produced ($= E_{p2}$), or 72.17 J glucose that is deposited as lipid ml^{-1} CO_2 produced ($= E_{pl}$) and 0.953 µl H_2O ml^{-1} CO_2 ($= R'_{mw}$) is produced. The other part of the MEI that is required for the oxidized component of lipogenesis is considered to be part of total heat production as the glucose is oxidized completely at RQ = 1.

For a carnivore, the conversion of protein to lipid can be separated into two processes: the formation of glucose via gluconeogenesis and lipogenesis as described above. Gluconeogenic oxidation of 1 part by weight of Kleiber's standard protein to produce 0.6 parts by weight of glucose in mammals (with urea as the nitrogenous end product) could be described stochiometrically (Livesey and Elia 1988) as:

$$C_{100}H_{159}N_{26}O_{32}S_{0.7} + 60.18\ O_2 \rightarrow$$
$$13\ CON_2H_4 + 7.52\ C_6H_{12}O_6 + 41.88\ CO_2 + 7.68\ H_2O + 0.7\ H_2SO_4.$$

Based on the 20.1 kJ g^{-1} of metabolizable energy of protein catabolism in mammals (Gessaman and Nagy 1988) and combining this process with lipogenesis, 16.12 J lipid tissue is formed from protein/ml CO_2 produced ($= E_{p2}$), or 36.88 J protein is converted into fat ml^{-1} CO_2 produced ($= Epl$), and 0.338 µl metabolic water ml^{-1} CO_2 (R'_{mw}) is produced.

Birds, with uric acid as the main nitrogenous end product, have lower metabolizable energy yield from protein catabolism (18.4 kJ g^{-1}; Gessaman and

Nagy 1988) than mammals. Since the efficiency of use of metabolizable energy in birds is similar to that in mammals (Bondi 1982) although the metabolizable energy yield of protein in birds is about 92% that of mammals, the gluconeogenic oxidation of 1 part by weight of Kleiber's standard protein will yield approximately 0.55 parts by weight of glucose in birds and could thus be described stochiometrically as:

$$C_{100}H_{159}N_{26}O_{32}S_{0.7}+54.15\ O_2\rightarrow$$
$$6.5\ C_5H_4O_3N_4+6.9\ C_6\ H_{12}O_6+26.1\ CO_2+24.4\ H_2O+0.7\ H_2\ SO_4.$$

Analysis and Use of the Model. The water budget model [Eq. (2)] can be used separately for animals in negative [Eq.(3)] and positive [Eq. (3b)] energy balances. For animals that maintain their energy balance, either equation could be used. These equations were further analysed for the ratio of water influx to CO_2 production [R; Eqs.(5) and (5b)]. Based on either X_1 or X_2 for negative [Eq. (6)] and positive [Eq. (6b)] energy balances, MEI* [Eqs. (7) and (7b)], ER* [Eqs. (8) and (8b)] and DEE* [Eq. (9)] can be calculated.

The decision as to which equation route (either a or b) to use is based on the ratio (R) of the water influx to CO_2 production in any given measurement. Equations (4) and (4b) are in the form:

$$R = X(R_{pw}+R_{mw})+(1-X)Y. \tag{11}$$

where Y depends on the energy balance of the animal. R_{cw} is similar to R_{mw} but is generally smaller than $R_{pw}+R_{mw}$ (this depends mainly on dietary water content). Thus, for animals in negative energy balance [Eq. (4)], $Y<(R_{pw}+R_{mw})$, and R is smaller than the ratio of the dietary water yield to CO_2 production ($R_{pw}+R_{mw}$). In contrast, $Y>(R_{pw}+R_{mw})$ in animals in positive energy balance [Eq.(4b)] for the following reasons: E_{pl} (for either herbivorous or carnivorous diet) is much greater than E_d and thus $(E_{pl}/E_d)>>1$. This ratio is multiplied by R_{pw}, and the product is generally greater than R_{mw} (this, too, depends mainly on dietary water content). Thus, the overall Y ratio is greater than the dietary ratio.

From the above analysis, it can be seen that an R value smaller than $R_{pw}+R_{mw}$ is the result of an animal in negative ER, and an R value in excess of $R_{pw}+R_{mw}$ is the result of an animal in positive ER. Yet caution should be exercised in the case of specific diets in which the moisture content is very low (for example, extremely dry seeds), as a result of which R_{pw} could be smaller than R_{mw}. In these cases, careful examination of conversion factors based on previous analyses [Eqs.(4) and (4b)] and on Eq. (11) should reveal either negative or positive energy balance when $Y\neq R_{pw}+R_{mw}$ and , as a result, which equation route to use.

The model has three stages of calculations: (1) determine whether an animal is in negative or positive ER; (2) determine the corresponding X value based on either Eq. (6) or (6b); and (3) determine MEI* [Eqs.(7) and (7b)], ER* [Eqs. (8) and (8b)] and DEE* [Eq. (9)] using the X value. It is clear that the dietary habits of the animal are important in determining the range of possible errors in estimating its energy expenditure and energy intake. The larger the difference between E_d and E_p, the larger is the possible error in calculating energy expenditure in the traditional way [Eq. (10)]. This can be demonstrated by comparing DEE*/DEE ratios from R/R_d ratios for different diets. Based on this model, the diets of three species consuming either sucrose solution, leaves or insects (Table 7.13) were analysed (Fig. 7.19). Animals consuming mainly carbohydrate diets (leaves and sucrose solution) have a higher R/R_d ratio and a higher DEE*/DEE ratio than animals consuming mainly protein diets (insects). The DEE*/DEE ratio in animals consuming carbohydrate diets can be as high as 1.3 or as low as 0.3. This indicates the range of possible error in estimating energy expenditure based on Eq. (10).

Use of the Model and Sensitivity Analyses of Errors. The model was used to re-analyse the data from the literature of DLW studies in which animals did not drink water and measurements for individuals and diet charecteristics were given. To cover a wide range of body sizes and dietary habits, the studies included a small nectarivorous hummingbird, a small insectivorous bat, an insect-fed chick, a large folivorous marsupial and a fish-eating petrel (references in Table 7.13).

With all diets, a negative energy balance resulted in $R < R_d$ and a positive energy balance, in $R > R_d$. Data from references c, d and e (Table 7.13) each contained one case where the R ratio was beyond the model limits, and these three were not used. Differences between R_d and R reflected the magnitude of positive or negative body energy balances. As expected with negative energy balance, species consuming carbohydrate-rich diets (references a and d) had higher DEE*/DEE ratios than species consuming protein-rich diets (references b, c and e; Table 7.13). The highest ratio between energy expenditure using this model and the traditional model was found for the hummingbird, which averaged 1.18 and ranged between 1.02 and 1.32. For animals in positive energy balance, the herbivorous koala had the lowest ratio: it averaged 0.76 and ranged between 0.51 and 0.95. To indicate the magnitude of the non-balanced energy budget, the MEI*/DEE* ratio is presented. In four of the five species, most individuals were in negative energy balance. Metabolizable energy intake was as low as 43% of energy expenditure in negative energy balanced hummingbirds and as high as 245% of energy expenditure in positive balanced koalas (Table 7.13).

The model enables estimation of energy expenditure, metabolizable energy intake and energy retention when water influx, CO_2 production rate and diet are known. In cases where dietary habits and diet composition are not well described, the model could be used in assessing the possible range of errors for both energy expenditure and energy intake. It is assumed that both positive and negative energy retention are the result of deposition and mobilization of lipid. This can be inaccurate in some cases, for example, in young growing animals. Estimations should therefore be treated cautiously.

The accuracy of the model is dependent on estimates of water influx and CO_2 production rate, which are calculated from hydrogen and isotope enrichments

Table 7.13. Metabolizable energy (ME) and water content(WC) of diets consumed by five species of homeotherms from the literature. The ratio of the estimated daily energy expenditure (DEE*) based on the model in the text to the published DEE and the ratio of ME intake (MEI*) based on the model in the text to DEE* for animals in both negative and positive energy balances are included. Values are means ± SD. (Kam and Degen, unpublished data)

Species:	Humming Bird[a] (Calypte anna)	Wren chicks[b] (Troglodytes aedon)	Bat[c] (Macrotus californicus)	Koala[d] (Phascolarctos cinereus)	Wilson's storm petrel[e] (Oceanites oceanicus)
Diet	Sucrose solution	Crickets	Insects	Leaves	Krill
ME (kJ g^{-1} DMI)	16.0	16.4	20.5	10.5	18.3
E_d(J ml^{-1} CO$_2$)	20.8	25.7	25.7	21.8	25.8
WC (% fresh matter)	80	75	70	57	76
R_{pw}(µl ml^{-1} CO$_2$)	5.200	4.697	2.925	2.711	4.451
R_{mw}(µl ml^{-1} CO$_2$)	0.715	0.660	0.660	0.670	0.399
R_d(µl ml^{-1} CO$_2$)	5.915	5.357	3.585	3.381	4.849
Negative energy balance					
n	7	14	14	5	9
m_b(g)	4.49 ± 0.55	9.70 ± 0.93	13.0 ± 0.8	8300 ± 1943	43.3 ± 3.4
E(J ml^{-1} CO$_2$)	24.5 ± 2.4	26.1 ± 0.2	26.5 ± 0.5	22.9 ± 0.7	26.4 ± 0.2
R(µl ml^{-1} CO$_2$)	3.171 ± 1.794	4.478 ± 0.449	2.447 ± 0.718	2.912 ± 0.314	3.544 ± 0.494
DEE*/DEE	1.18 ± 0.12	1.01 ± 0.01	1.03 ± 0.02	1.05 ± 0.03	1.02 ± 0.01
MEI*/DEE*	0.43 ± 0.34	0.80 ± 0.10	0.58 ± 0.25	0.79 ± 0.14	0.67 ± 0.12
Positive energy balance					
n	1	2	1	11	3
m_b(g)	4.45	9.03 ± 0.69	13.2	9836 ± 1292	39.5 ± 1.5
E(J ml^{-1} CO$_2$)	19.4	26.1 ± 0.2	24.0	16.6 ± 3.5	26.1 ± 0.4
R(µl ml CO$_2$)	7.008	6.167 ± 0.419	3.836	5.330 ± 1.300	5.586 ± 7.909
DEE*/DEE	0.93	1.02 ± 0.01	0.95	0.76 ± 0.16	1.01 ± 0.01
MEI*/DEE*	1.29	1.15 ± 0.08	1.17	2.45 ± 1.21	1.13 ± 0.14

Taken and/or calculated from data in literature: E_d, metabolizable energy yield to CO$_2$ production; E_a, total water yield to CO$_2$ production; R_{pw}, preformed water to CO$_2$ production; R_{mw}, metabolic water to CO$_2$ production; R_d, total water yield to CO$_2$ production of the diet.
Calculated from this model: E, total heat production to total CO$_2$ production; R, total water influx to CO$_2$ production
References: [a]Powers and Nagy (1988). [b]Dykstra and Karasov (1993). [c]Bell et al (1986). [d]Nagy and Martin (1985). [e]Obst et al (1987).

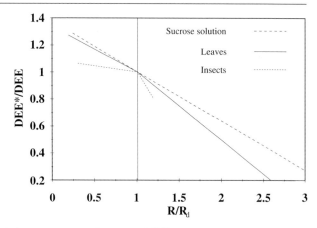

Fig. 7.19. Effect of R (the ratio between water influx to total CO_2 production) over R_d (dietary water yield to CO_2 production conversion factor) on DEE* (energy expenditure estimated using present model) over DEE (energy expenditure estimated in the traditional way) for three diets. The diets of sucrose solution, leaves and insects are based on diets in Table 7.13. (Kam and Degen, unpubl. data)

in the body fluids, and TBWV of the animal. The present model presumes that the DLW method yields error-free measurements of these parameters. However, errors in either water influx and/or CO_2 production rate can cause an error in the energy budget components. Most probable errors are in the hydrogen isotope turnover or in the TBWV estimate. By dividing water influx by CO_2 production rate, the TBWV is cancelled out, but an error in the hydrogen isotope turnover rate is magnified and results in an error in the R ratio. To assess such an error, the percent error in total water influx to CO_2 production ratio resulting from an analytical error of 1% in hydrogen isotope analysis (Nagy 1980) was determined. For R values above 3 μl water ml^{-1} CO_2, the most common range of water influx to CO_2 production ratio (Nagy and Peterson 1988), the error is below 1% and decreases gradually. The same holds true for X values, either X_1 [Eq. (6)] or X_2 [Eq. (6b)] which, in addition to the constants involved, depend entirely on the ratio R. However, if the CO_2 production rate contains an error, either through an error in the isotope turnover rate or in the TBWV estimate, this error will affect the energy budget components directly. As a result, MEI* = true MEI, ER* = δ true ER and DEE* = δ true DEE [Eqs. (7), (8), (7b), (8b) and (9)]. Thus, all components will be displaced equally by δ, and the error in determining DEE and DEE*/DEE ratio (Fig. 7.20) remains the same irrespective of δ.

A different type of error could arise where animals drink free water [Eq.(2)]. An error in DW estimate would cause an error in WI_o and, as a result, in the estimated total water to CO_2 production ratio, R* = εR. How does such an error affect the energy budget components? The positive energy balance state could be tested. For errors of 5.0, 7.5 and 10.0% (ε values of 1.050, 1.075 and 1.100,

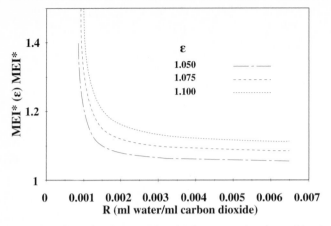

Fig. 7.20. Effect of underestimating drinking water (DW), resulting in an overestimation of the water influx to CO_2 ratio (R) by a value of ε on the estimated metabolizable energy intake (MEI*) for ε of 1.050, 1.075 and 1.100. (Kam and Degen, unpubl. data)

respectively), the ratio between X_1 calculated and X_1 without error is similar to ε values for R values greater than 0.002 (Fig. 7.21) – which is most likely for animals that are not fasting. As a result, in most cases MEI* is overestimated, and ER* is underestimated by approximately ε values. The ratio [DEE*(ε)/DEE*] can be taken to estimate the error in energy expenditure resulting from a 5% overestimation of WI_o ($\varepsilon = 1.05$) in three typical diets: sucrose solution, leaves and insects. Over most of the range of R values, [DEE*(ε)/DEE*] is less than 5%, resulting in an underestimation of DEE for animals in negative energy balance and an overestimation for animals in a positive energy balance (Fig. 7.21).

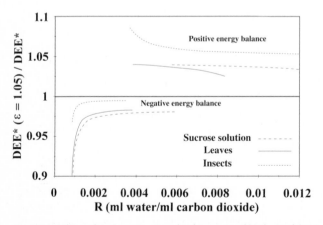

Fig. 7.21. Effect of underestimating drinking water (DW), resulting in an overestimation of water influx to CO_2 ratio (R) by an error of 5% ($\varepsilon = 1.05$), on the estimated daily energy expenditure (DEE*) for three diets. (Kam and Degen, unpubl. data)

The largest errors would occur with fasting animals that are drinking. Here, small errors in DW would result in large errors in DEE estimate.

7.6.4
Comparison of Field Metabolic Rates Among Mammals

Nagy (1994a) analysed FMR measurements in 44 species of eutherian mammals ranging in mass from the 7.3 g pipistrelle bat (*Pipistrellus pipistrellus*) to the 316000 g northern elephant seal (*Mirounga angustirostris*) and of 17 species of metatherians ranging in mass from the 33 g brown antechinus (*Antechinus stuartii*) to the 43900 g eastern grey kangaroo (*Macropus giganteus*). Most of the variation of FMR was explained by body mass and the infraclass of the animals. Allometric regression analyses revealed that the scaling of the FMR rate differed between eutherian and metatherian mammals. The regression equation for eutherian mammals was:

$$\text{FMR (kJ day}^{-1}) = 4.63 \; g^{0.762},$$

and that for marsupials was:

$$\text{FMR (kJ day}^{-1}) = 10.8 \; g^{0.582}.$$

In general, habitat and diet did not affect the FMR of mammals within each infraclass.

Eutherian mammals were separated into six groups according to habitat, and of these, only desert species ($n = 16$) had FMRs that differed from that predicted for eutherians with their body masses. The FMR for desert species averaged 78% (95% CI = 67–89%) of that predicted for mammals with their body masses. The other five groups did not have FMRs that differed from those predicted for eutherian mammals with their body masses. These included eutherians that inhabited (1) tundra (n = 3) in which the FMR averaged 162% (65–259%) of that predicted; (2) temperate forests (n = 8) in which the FMR averaged 123% (94–152%); (3) meadows (n = 6) in which the FMR averaged 119% (71–167%); (4) scrub (n = 5) in which the FMR averaged 115% (84–146%); and (5) sea (n = 6) in which the FMR averaged 132% (66–198%). Analyses for marsupials were carried out on groups inhabiting two habitats: scrub residents (n = 4) which averaged 82% (54–108%) of predicted FMR, and eucalyptus forest residents (n = 12) which averaged 102% (89–105%) of predicted FMR.

Analysis of dietary effects showed that, overall, no difference from the expected FMR was due to diet in either the eutherian or metatherian mammals. Of the eutherians, granivores (n = 4) averaged 79% (47–110%) of predicted FMR; insectivores and carnivores (n = 11) averaged 123% (90–156%) of predicted FMR; and omnivores (n = 15) and herbivores (n = 13) fell between these two extremes. Of marsupials, insectivores (n = 4), omnivores (n = 3) and herbivores (n = 9) all averaged close to 100% of predicted FMR.

However, as pointed out by Nagy (1994a), the small sample sizes may be part of the reason for failure to detect statistical differences in parameters influencing FMR. Furthermore, measurements were made in different seasons and on animals in different physiological states which could have had a bearing on the results. For example, the FMR of male springbok antelope (*Antidorcas marsupialis*) in rut was approximately three times greater than that of springbok during the dry season (Nagy and Knight 1994).

FMR was analysed in 28 small eutherian mammals, 24 of which were rodents (Degen, unpubl. data). Measurements were made on non-reproducing animals and also mainly during summer to minimize thermoregulary costs (Tables 7.14 and 7.15). The regression equation for all 28 mammals is (Fig. 7.22):

$$\text{FMR (kJ day}^{-1}) = 4.17 \ g^{0.734},$$
$$(n = 28; \ SE_{y \cdot x} = 0.148; \ SE_a = 0.086; \ SE_b = 0.046; \ r^2_{adj} = 0.90; \ F = 253.2; \ P < 0.01).$$

When eutherian mammals were compared per $g^{0.734}$, the FMR of desert species (3.74 ± 0.98; n = 18) was significantly lower than that of non-desert species (5.58 ± 2.14; n = 10). According to dietary habits, the FMR of granivorous (3.27 ± 0.70; n = 6) mammals tended to be the lowest, but were not significantly different from those of omnivorous (4.34 ± 1.22; n=12), herbivorous (5.78 ± 2.65; n = 7) and carnivorous (4.32 ± 1.19; n = 3) mammals. A scattergram of the residuals of the regression of FMR related to body mass in desert and non-desert mammals according to dietary habitats is presented in Fig. 7.23.

The regression equation for the 24 rodents takes the form:

$$\text{FMR (kJ day}^{-1}) = 6.22 \ g^{0.622},$$
$$(n = 24; \ SE_{y \cdot x} = 0.150; \ SE_a = 0.138; \ SE_b = 0.084; \ r^2_{adj} = 0.71; \ F = 54.6; \ P < 0.01).$$

When compared per $g^{0.662}$, the FMR of desert rodent species (5.25 ± 1.14; n = 14) was significantly lower than that of non-desert species (8.51 ± 3.15; n = 10). In addition, granivorous species (4.67 ± 0.69; n=6) tended to have the lowest FMRs; granivorous and omnivorous (6.40 ± 1.52; n = 12) species had significantly lower FMRs than herbivorous species (8.97 ± 4.10; n = 6; Fig. 7.24).

7.6.5
Seasonal Field Metabolic Rates

From body mass considerations, it could be predicted that smaller mammals would be more vulnerable to low air temperatures than larger animals. As a result, smaller animals should theoretically respond proportionately more to changes in air temperature than larger animals. That is, the increase in metabolic rates in the winter compared with summer should be greater in smaller animals than in larger animals. An allometric relationship between winter FMR:summer FMR ratio and body mass in desert eutherian mammals, using regression analysis of the log-log transformed data, confirms this (Degen et al. unpubl. data). Non-desert species are not included in the analysis (Table 7.16). The ratio decreases with an increase in body mass, and equals one for an

Table 7.14. Body mass, field metabolic rate (FMR), diet and habitat of small eutherian mammals in which FMR was measured using doubly labelled water. Where possible, measurements were made in summer. Predicted FMR is from the equation for rodents only

Species	Body mass (g)	FMR (kJ day^{-1})	Diet[a]	Habitat[b]	FMR pred[c]	FMR/FMR pred	FMR (kJ g$^{-0.622}$ day^{-1})	Reference
Rodentia								
1. *Gerbillus henleyi*	9.7	22.2	G	D	25.6	0.87	5.40	Degen et al. (unpubl.)
2. *Mus musculus*	13.0	39.8	O	ND	30.7	1.30	8.07	Nagy (1987)
3. *Peromyscus crinitus*	13.4	39.3	O	D	31.3	1.26	7.28	Mullen (1971a)
4. *Mus domesticus*	15.5	45.2	O	ND	34.2	1.32	8.22	Mutze et al. (1992)
5. *Clethrionomys rutilus*	16.0	57.6	H	ND	34.9	1.65	10.27	Holleman et al. (1982)
6. *Perognathus formosus*	16.3	25.4	G	D	35.3	0.72	4.48	Mullen and Chew (1973)
7. *Peromyscus leucopus*	19.4	36.6	O	ND	39.3	0.93	5.79	Randolph (1980)
8. *Gerbillus allenbyi*	22.8	35.6	G	D	43.5	0.82	5.09	Degen et al. (1992)
9. *G. pyramidum*	31.8	45.3	G	D	53.5	0.85	5.27	Degen et al. (1992)
10. *Pseudomys albocinereus*	32.6	62.2	O	D	54.3	1.14	7.12	Nagy (1987)
11. *Dipodomys merriami*	33.2	35.9	G	D	55.0	0.65	4.06	Nagy and Gruchacz (1994)
12. *Acomys cahirinus*	38.3	51.8	O	D	60.1	0.86	5.37	Degen et al. (1986)
13. *Sekeetamys calurus*	41.2	44.0	O	D	62.9	0.70	4.36	Degen et al. (1986)
14. *Acomys russatus*	45.0	47.6	O	D	66.4	0.72	4.46	Degen et al. (1986)
15. *Dipodomys microps*	55.0	64.5	O	D	75.2	0.86	5.34	Mullen (1971b)
16. *Lemmus trimucronatus*	55.2	201.0	H	ND	75.4	2.67	16.59	Peterson et al. (1976)
17. *Praomys natalensis*	57.3	86.6	O	ND	77.2	1.12	6.98	Green and Rowe-Rowe (1987)
18. *Meriones crassus*	77.6	55.7	G	D	93.2	0.60	3.72	Degen et al. (unpubl.)
19. *Ammospermophilus leucurus*	86.0	79.3	O	D	99.3	0.80	4.97	Karasov (1981)
20. *Arvicola terrestris*	89.7	88.7	H	ND	102.0	0.87	5.41	Grenot et al. (1984)
21. *Tamias striatus*	96.3	143.0	O	ND	106.6	1.34	8.35	Randolph (1980)
22. *Thomomys bottae*	103.8	130.3	H	ND	111.7	1.17	7.26	Gettinger (1984)
23. *Psammomys obesus*	165.6	146.3	H	D	149.3	0.98	6.10	Degen et al. (1991)
24. *Spermophilus saturatus*	239.4	248.2	H	ND	187.8	1.32	8.23	Kenagy et al. (1989a)
Chiroptera								
25. *Macrotus californicus*	13.3	22.8	C	D				Bell et al. (1986)
Lagomorpha								
26. *Lepus californicus*	1800.0	1175.0	H	D				Nagy et al. (1976)
Carnivora								
27. *Vulpes cana*	955.8	617.0	C	D				Geffen et al. (1992a)
28. *V. Velox*	1990.0	1488.0	C	D				Miller et al. (unpubl. data).

[a] Diet habits: O, omnivore; G, granivore; H, herbivore; C, carnivore; N, nectarivore. [b] Habitat: D, desert; ND, non-desert. [c] FMR$_{pred}$; FMR predicted, 6.22 m$_{b}$$^{0.622}$.

Table 7.15. Coefficients of the linear regression equations taking the form $\log y = a + b \log$ body mass (m_b; g) where y is the field metabolic rate (FMR; kJ d^{-1}) for small non-reproducing eutherian mammals. Where possible, measurements were made in summer

y	n	a	b	S_a	S_b	$S_{y.x}$	r^2_{adj}	F	P
All animals	28	0.618	0.733	0.086	0.046	0.148	0.90	253.2	0.000
Desert	18	0.479	0.779	0.071	0.037	0.107	0.96	442.9	0.000
Non-desert	10	0.943	0.602	0.188	0.110	0.144	0.76	30.0	0.001
Omnivores	12	1.011	0.481	0.160	0.102	0.102	0.66	22.3	0.001
Granivores	6	0.888	0.464	0.105	0.073	0.050	0.89	40.2	0.003
Herbivores	7	0.966	0.609	0.260	0.119	0.183	0.81	26.4	0.004
Carnivores	3	0.417	0.818	0.138	0.052	0.087	0.99	245.0	0.001
Non-rodents	4	0.420	0.816	0.097	0.034	0.062	1.00	564.1	0.002
Rodents	24	0.794	0.622	0.138	0.084	0.150	0.70	54.6	0.000
Desert	14	0.802	0.564	0.121	0.076	0.091	0.81	54.8	0.000
Non-desert	10	0.943	0.602	0.188	0.110	0.144	0.76	30.0	0.001
Omnivores	12	1.011	0.481	0.160	0.102	0.102	0.66	22.3	0.001
Granivores	6	0.888	0.464	0.105	0.073	0.050	0.89	40.2	0.003
Herbivores	6	1.310	0.419	0.366	0.187	0.173	0.45	5.0	0.088

n, sample size; Sa, Sb, Sy.x, SE of a, of b, and of estimate, respectively; r^2_{adj}, adjusted coefficient of determination; F, F-statistic for the linear term; P, significance level of regression equation.

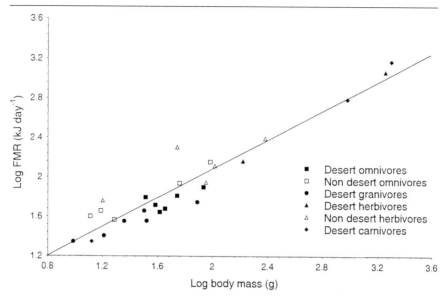

Fig. 7.22. Allometric relationship between field metabolic rate (FMR) and body mass for small desert and non-desert eutherian mammals

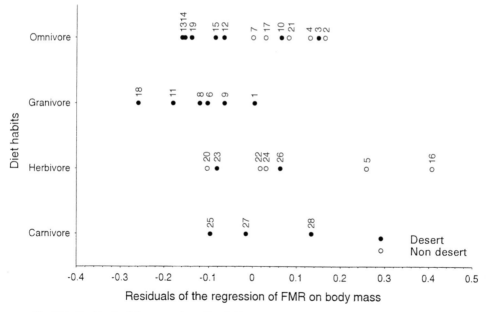

Fig. 7.23. Residuals of the regression of log field metabolic rate (FMR) on log body mass for small desert and non-desert mammalian species of different dietary habits. *Numbers* refer to species listed in Table 7.14

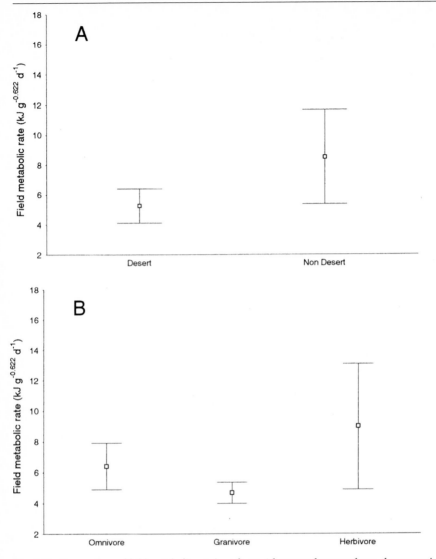

Fig. 7.24. Comparison of field metabolic rate in rodents: **A** between desert and non-desert species and **B** among dietary habits

animal of 427 g. The regression equation takes the form (Fig. 7.25):

log (winter FMR: summer FMR ratio) = 0.384 – 0.144 log m_b,
(n = 10; $SE_{y·x}$ = 0.080; SE_a = 0.073; SE_b = 0.032; r^2_{adj} = 0.68; F = 19.9; $P < 0.02$).

Table 7.16. Body mass (m_b), field metabolic rate (FMR), diet and habitat of small eutherian mammals in which FMR was measured using doubly labelled water

Species	Diet[a]	Habitat[b]	Summer m_b (g)	Summer FMR (kJ g^{-1} day^{-1})	Winter m_b (g)	Winter FMR (kJ g^{-1} day^{-1})	Mean m_b (g)	Ratio measured	Ratio predicted	Ratio meas/pred	Reference
Gerbillus henleyi	G	D	9.7	2.29	8.8	3.47	9.3	1.52	1.76	0.86	Unpublished data
Mus musculus	O	ND	18.3	2.86	13.1	2.98	15.7	1.04	1.63	0.64	Mutze et al. (1991)
Clethrionomys rutilus	H	ND	15.9	3.78	13.6	4.34	14.8	1.15	1.64	0.70	Holleman et al. (1982)
Perognathus formosus	G	D	16.3	1.56	18.0	2.94	17.2	1.88	1.61	1.17	Mullen (1970)
Dipodomys merriami	G	D	33.2	1.08	33.1	1.77	33.2	1.64	1.46	1.12	Nagy and Gruchacz (1994)
D. microps	O	D	55.0	1.17	60.3	1.69	57.7	1.44	1.35	1.07	Mullen (1971)
Meriones crassus	G	D	77.6	0.83	60.7	1.27	69.2	1.53	1.31	1.17	Unpublished data
Ammospermophilus leucurus	O	D	79.9	0.99	96.1	0.82	88.0	0.83	1.27	0.65	Karasov (1981)
Thomomys bottae	H	ND	99.4	1.28	108.0	1.19	103.7	0.93	1.24	0.75	Gettinger (1984)
Psammomys obesus	H	D	165.4	0.88	175.7	1.05	170.6	1.19	1.15	1.03	Degen et al. (1991)
Vulpes cana	C	D	900.9	0.66	1015.0	0.63	958.0	0.95	0.90	1.06	Geffen et al. (1992a)
Lepus californicus	H	D	1800.0	0.79	1800.0	0.65	1800.0	0.82	0.82	1.00	Shoemaker et al. (1976)
Vulpes velox	C	D	2200.0	0.95	1990.0	0.75	2095.0	0.79	0.80	0.98	Miller et al. (unpubl.)

[a]Diet G, granivore; H, herbivore; O, omnivore; C, carnivore, [b]Habitat: D, desert; ND, non-desert; FMR ratio predicted = $2.31m_b$ (g)$^{-0.144}$.

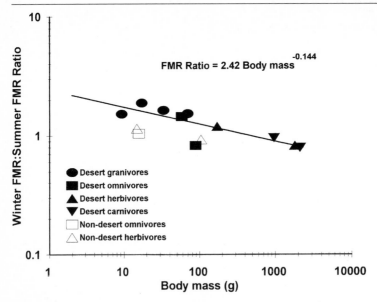

Fig. 7.25. Allometric relationship between the ratio of winter to summer field metabolic rate (FMR) related to body mass in eutherian mammals. Only desert species were used in the regression equation

7.6.6
Water Influx to FMR Ratio

Small desert mammals that do not drink water and consume a dry diet such as seeds obtain relatively little water from their food and have a low water influx (ml day^{-1}) in relation to their field metabolic rate (kJ day^{-1}) ratio (ml kJ^{-1}). In contrast, mammals consuming a relatively moist diet, such as green vegetation, obtain more water from their diet and have a higher WI:FMR ratio (Fig. 7.26). Thus the WI:FMR ratio could be indicative of the diet of an animal provided that it is maintaining its energy and water balances. For example, in a monthly study of water influx and the FMR of the kangaroo rat *Dipodomys merriami*, the lowest WI:FMR ratio was 0.049 ml kJ^{-1} in December when the diet consisted of 99% seeds, and the highest WI:FMR ratio was 0.121 ml kJ^{-1} in March when the diet was 90% vegetation and no seeds were eaten (Fig. 7.27; Nagy and Gruchacz 1994).

The WI:FMR ratio can be relatively low in desert animals that are well adapted to conserve water. Nagy and Gruchacz (1994) calculated that a WI:FMR ratio of approximately 0.047 ml kJ^{-1} is the lowest possible ratio for a rodent consuming a diet of only seeds such as husked oats that have a preformed water content of 2.9% of fresh matter and a digestibility of 0.906 g dry matter per dry matter intake. This calculated ratio is composed of approximately 0.030 ml kJ^{-1} from metabolic water formed by the oxidation of carbohydrates, lipids and proteins, 0.015 ml kJ^{-1} obtained through water vapour exchange, and 0.002 ml kJ^{-1}

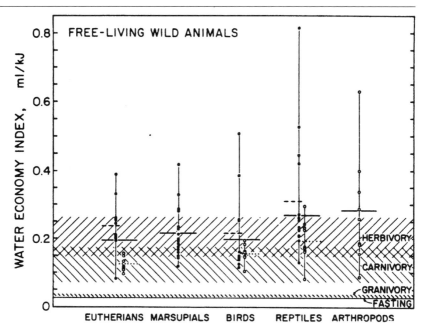

Fig. 7.26. Water economy index (WEI; the amount of ml water used per kJ of energy metabolized by animals) for air-breathing animals maintaining constant body mass. *Solid horizontal lines* show means for each taxon; *dotted and dashed horizontal lines* indicate means for desert (o) and non-desert (●) animals, respectively, within taxa. *Cross-hatched areas* indicate expected values of WEI for non-drinking animals maintaining water and energy balances on a diet of leaves (62–72% water content, to represent the food of a herbivore), animal matter (*Tenebrio* larvae to whole vertebrates, to represent the food of a carnivore) or mature seeds (0–10% water, for a granivore), and *"fasting" line* indicates the WEI for fat catabolism. (Nagy and Peterson 1988)

from preformed water. Some rodents have a WI:FMR ratio close to this minimum value (Table 7.17). In addition to the ratio of 0.049 ml kJ^{-1} for *Dipodomys merriami* (body mass = 33 g), a desert heteromyid (Nagy and Gruchacz 1994), a ratio of 0.045 ml kJ^{-1} was determined for *Perognathus formosus* (Mullen 1970), a small heteromyid (body mass = 16.3g) restricted to arid regions (Schmidly et al. 1993) which forages mainly for seeds. Furthermore 0.048 ml kJ^{-1} has been calculated for *Gerbillus henleyi* (Degen, Khokhlova and Kam, unpubl. data), a small gerbillid (body mass = 10 g) that inhabits deserts with an annual precipitation of 30–100 mm and also forages for seeds (Khokhlova et al. 1994). In contrast, herbivores consuming relatively moist vegetation have a high WI:FMR ratio. The lowest ratios for *Psammomys obesus,* a large desert gerbil which is strictly herbivorous, is 0.205 ml kJ^{-1}. This ratio was measured during the summer (Degen et al. 1991). The WI:FMR ratios of *Acomys cahirinus, A. russatus* and *Sekeetamys calurus,* all desert omnivores, were 0.098, 0.119 and 0.134 ml kJ^{-1}, respectively (Degen et al. 1986). In general, these values fell to levels be-

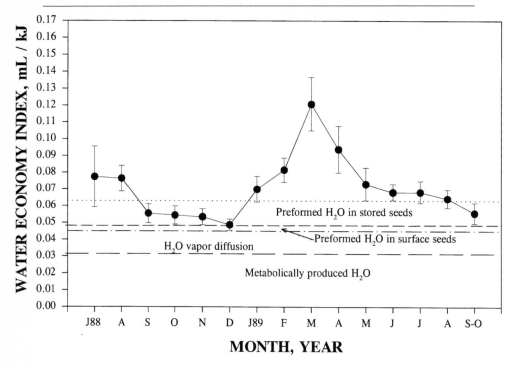

Fig. 7.27. Water economy index of free-living Merriam's kangaroo rats (*Dipodomys merriami*) throughout a 15-month period in the Mojave desert. *Horizontal lines* represent (begining at the bottom): (1) only metabolic water (ml kJ^{-1}) produced from a diet of seeds; (2) metabolic water plus the water vapour exchange expected given the high humidities in burrows; (3) metabolic plus vapour plus preformed water in air-dried seeds on the soil surface; and (4) metabolic plus vapour plus preformed water in seeds stored in high humidity air (as in burrows). (Nagy and Gruchacz 1994)

tween those of granivores and herbivores, which is consistent with the relative moisture contents of their diets.

In eutherian mammals, there is some overlap in the WI:FMR ratios among diets (Table 7.17). Ratios for *Gerbillus allenbyi* and *G. pyramidum*, two granivorous gerbillidae, were 0.078 and 0.071 ml kJ^{-1}, respectively, in the summer (Degen et al. 1992). These are higher than the minimum values calculated for a dry seed diet. This indicates that these rodent species were either consuming seeds with a high content of preformed water or that, in addition to seeds, the animals were consuming considerable amounts of green vegetation. In contrast, the desert jack rabbit, a strict herbivore, had a relatively low ratio of 0.057–0.058 ml kJ^{-1} (Nagy et al. 1976), indicating that it was consuming a diet of very dry vegetation.

Nagy and Peterson (1988) used the term "water economy index" (WEI) for the WI:FMR ratio as it "indicates the amount of water used per unit of 'living' an animals does, it is independent of the rate at which an animal uses energy,

Table 7.17. Body mass, diet, water influx (WI), field metabolic rate (FMR), and WI:FMR in eutherian mammals under different free-living conditions. All measurements were made using doubly labelled water

Species	Body mass (g)	Diet[a]	Habitat[b]	Condition	WI (ml day⁻¹)	FMR (kJ day⁻¹)	WI:FMR (ml kJ⁻¹)	Reference
Rodentia								
Gerbillus henleyi	9.7	G	D	Summer	1.09	22.48	0.048	Degen et al. (unpubl.)
	8.8	G	D	Winter	1.59	30.70	0.052	Nagy and Peterson (1988)
Mus musculus	13.0	O	ND	Autumn	3.30	39.80	0.083	Holleman et al. (1982)
Clethrionomys rutilus	13.6	H	ND	Winter	3.60	59.00	0.061	
	18.3	H	ND	Spring	4.90	52.80	0.093	
	15.9	H	ND	Summer	5.70	60.20	0.095	
	16.1	H	ND	Autumn	5.80	58.50	0.099	Mullen (1970)
Perognathus formosus	16.3	G	D	Summer	1.13	25.35	0.045	
	18.0	G	D	Autumn	2.23	41.20	0.054	Degen et al. (1992)
Gerbillus allenbyi	22.8	G	D	Summer	2.78	35.60	0.078	
G. Pyramidum	31.8	G	D	Summer	3.24	45.30	0.072	Nagy and Gruchacz (1994)
Dipodomys merriami	32.6	G	D	Winter, 99% seeds	2.85	58.47	0.049	
	36.9	G	D	Spring, 0% seeds	6.99	57.82	0.121	
	32.5	G	D	Autumn, 97% seeds	1.95	35.92	0.054	
	34.6	G	D	Summer, 95% seeds	2.21	34.60	0.064	Nagy and Peterson (1988)
Pseudomys albocinereus	32.6	O	D	Autumn	7.76	62.20	0.125	Degen et al. (1986)
Acomys cahirinus	38.3	O	D	Spring, dry	5.06	51.80	0.098	Degen et al. (1986)
Sekeetamys calurus	41.2	O	D	Spring	5.89	44.00	0.134	Degen et al. (1986)
Acomys russatus	45.0	O	D	Spring, dry	5.65	47.60	0.119	Mullen (1971b)
Dipodomys microps	51.9	O	D	Spring	6.50	34.50	0.188	
	61.0	O	D	Summer	7.60	102.00	0.075	
	55.1	O	D	Autumn	10.70	137.00	0.078	Green and Rowe-Rowe (1987)
Praomys natalensis	55.0	O	D	Winter	6.20	64.50	0.096	Degen et al. (unpubl.)
	57.3	O	ND	Summer	12.23	86.60	0.142	
Meriones crassus	77.6	G	D	Summer	3.81	55.70	0.068	
	60.7	G	D	Winter	8.21	74.30	0.110	
Arvicola terrestris	84.4	H	ND	Spring	62.90	149.00	0.422	Grenot et al. (1984)
	93.6	H	ND	Summer	78.30	88.70	0.883	

Table 7.17 (contd.)

Species	Body mass (g)	Diet[a]	Habitat[b]	Condition	WI (ml day^{-1})	FMR (kJ day^{-1})	WI:FMR (ml kJ^{-1})	Reference
Ammospermophilus leucurus	90.0	O	D	Spring	18.00	114.00	0.158	Karasov (1983)
	79.9	O	D	Summer	11.90	79.30	0.150	
	82.1	O	D	Autumn	9.03	79.60	0.113	
	96.1	O	D	Winter	8.65	79.00	0.109	
Thomomys bottae	99.4	H	ND	Summer	26.00	127.00	0.205	Gettinger (1984)
	108.0	H	ND	Winter	27.00	128.00	0.211	
	104.0	H	ND	Spring	53.00	136.00	0.390	
Psammomys obesus	165.6	H	D	Summer adults	30.15	146.80	0.205	Degen et al. (1991)
	171.6	H	D	Winter adults	41.60	184.50	0.225	
	80.6	H	D	Winter juveniles	28.40	91.80	0.309	
Chiroptera								
Macrotus californicus	12.6	C	D	Spring	3.00	22.20	0.135	Bell et al. (1986)
	13.3	C	D	Winter	1.80	20.80	0.087	
Eptesicus fucus	20.8	I	ND	Pregnant	8.47	47.60	0.178	Kurta et al. (1990)
	17.4	I	ND	Lactating	17.13	75.30	0.227	
Lagomorpha								
Lepus californicus	1 800.0	H	D	Summer	82.40	1420.00	0.058	Nagy et al. (1976)
	1 800.0	H	D	Winter	67.30	1180.00	0.057	
Xenarthra								
Bradypus varigatus	3 830.0	H	ND	Lactating females	154.00	590.00	0.261	Nagy and Montgomery (1980)
	4 320.0	H	ND	Non-lactating females	125.00	490.00	0.255	
	4 450.0	H	ND	Males	181.00	739.00	0.245	
Primates								
Alouatta palliata	6 500.0	H	ND	Dry season	767.00	2308.00	0.332	Nagy and Milton (1979)
Artiodactyla								
Antidorcas marsupialis	42 100.0	H	D	Hot, dry, drinking water	3180.00	13100.00	0.243	Nagy and Knight (1994)
	36 800.0	H	D	Hot, dry, no drinking water	1600.00	12800.00	0.125	
	45 500.0	H	D	Hot, wet, drinking water	4180.00	31800.00	0.131	
	40 600.0	H	D	Cold, dry, drinking water	2480.00	16200.00	0.153	

		Diet[a]	Habitat[b]					Reference
Odocoileus hemionus	46 800.0	H	D	Hot, wet, drinking water, pre-rut	3840.00	26900.00	0.143	Nagy and Peterson (1988)
	45 000.0	H	D	Hot, wet, drinking water, rut	4930.00	43700.00	0.113	
	39 975.0	H	ND	Spring females	4550.00	23400.00	0.194	
	67 100.0	H	ND	Spring males	8050.00	40000.00	0.201	
Carnivora								
Vulpes cana	900.9	C	D	Summer	102.96	593.28	0.174	Geffen et al. (1992a)
V. velox	1 015.0	C	D	Winter	82.76	642.54	0.129	
	2 200.0	C	D	Summer	289.00	2079.00	0.139	Miller et al. (unpubl.)
	1 990.0	C	D	Winter	205.00	1488.00	0.138	
Callorihinus ursinus	30 900.0	C	Sea	Lactating females	5690.00	21800.00	0.261	Costa and Gentry (1986)
Arctocephalus gazella	32 000.0	C	Sea	Foraging	5464.00	26320.90	0.208	Costa et al. (1989)
Zalophus californianus	82 800.0	C	Sea	Lactating females	7620.00	38400.00	0.198	Costa (1984)
	75 200.0	C	Sea	Lactating females	7640.00	50200.00	0.152	

[a]Diet: O, omnivore; G, granivore; H, hebrivore; C, carnivore; I, insectivore.
[b]Habitat. D, desert; ND, non-desert.

and can be used to compare directly the water economy of animals in different taxa and of different body masses." These authors found that WEI values were significantly lower in desert eutherians than in non-desert eutherians. No correlation was found between the WEI values and the body mass of the mammals; but all mammals larger than 10 kg have a WEI above 0.175, and only mammals under 100 g have WEI values lower than 0.150. Nonetheless, some small mammals show relatively high WEI values. This led Nagy and Peterson (1988) to conclude that "small size may be a prerequisite for having a low WEI, but small size does not mandate water conservation."

CHAPTER 8

Reproduction

> "Reproduction in mammals is a complicated process that must occur in harmony
> with existing dietary, physical and social conditions. To this end natural selection
> has provided the mammal with a rich variety of signaling systems, each of which
> couples environmental variation of some kind with appropriate neuroendocrine re-
> sponses."
>
> (Bronson 1985)

8.1
Timing of Reproduction

Three strategies have been described in the timing of reproductive activity. In
the first, there is wide-open opportunism, and no predictors of any kind are
employed. In extreme forms, males are sexually prepared at all times and "females breed either seasonally or continually depending upon moment-to-mo-
ment energetic and nutritional considerations". In the second, seasonal plant
compounds are used as predictors for a period of increased food availability
(Bronson 1985).

Small mammals usually employ the first option, that is, one of opportunism.
These animals are relatively short-lived and have few body energy reserves.
Such reserves could only last for a short period without food input and small
animals therefore have to be assured of a steady energy supply for successful
reproduction. Thus, in the wild house mouse (*Mus musculus*), the onset of
breeding is determined by the availability of food (Bomford 1987a). In addition, protein intake may be a limiting factor in reproduction, as has been
demonstrated in the house mouse (Bomford 1987b) and the cotton rat
(*Sigmodon hispidus*; Randolph et al. 1995).

Small desert mammals are particularly vulnerable to unpredictable energy
and nutrient supplies which put a constraint on reproduction. In consequence,
reproductive activity is often linked to seasonal rainfall and the subsequent emergence of green vegetation. Indeed, positive correlations between seasonal rainfall and seasonality of breeding have been reported in African (Field 1975; Perrin
1980; Perrin and Swanepoel 1987), Australian (Breed 1990), Asian (Rogovin 1985)
and North American desert rodents (Kenagy and Bartholomew 1985; Randall
1993). In northern Africa, *Meriones libycus* begins reproductive activity following rainfall (Daly and Daly 1975b). Further evidence of the importance of water
was demonstrated by the pygmy jerboas, *Salpingotus crassicaida* and
Cardiocranius paradoxus, in the Gobi Desert which stop breeding during droughts
(Rogovin 1985). Moreover, populations of free-ranging rabbits decreased in arid
areas of Australia during droughts when no free water was available and the water

content of the vegetation declined below about 60%. Lactation of the females was affected, as were all aspects of their reproduction (Richards 1979).

However, as pointed out by several authors (Sadleir 1969; Perrin and Swanopoel 1987), this does not necessarily mean that rainfall directly initiates oestrus and spermatogenesis but that the reproductive activity may be a response to rainfall and the resultant increase in energy and nutrient supply. "In general, a wide variety of desert rodents breed towards the end of the rainy season and in the few months afterwards, regardless of when the rain occurs" (Randall 1994). This is possible, as these rodents have the ability of being flexible and opportunistic in their breeding. For example, in many species of heteromyids, the females are polyoestrus and are able to reproduce throughout the year in response to rainfall, green vegetation growth and seed production (Beatley 1969; Bradley and Mauer 1971). For example, Merriam's kangaroo rat (*Dipodomys merriami*) remains on standby and increases its body mass and reproduces sporadically according to the unpredictable availability of annuals at any time winter through early autumn (Kenagy 1973a). Reproduction in these species can be curtailed or cease completely if there is little rain and no green vegetation (Beatley 1969; Randall 1993; 1994).

Green vegetation may be the cue triggering reproductive activity. Many desert rodents shift their diet from seeds to one that includes a higher proportion of green vegetation prior to breeding, and this may be related to the increased amount of preformed water in the diet and perhaps to the stimulatory effects of the emerging vegetation (Bradley and Mauer 1971; Reichman and Van De Graaff 1975; Negus and Berger 1977). Accelerated reproductive activity in response to green vegetation has been recorded in the Indian desert gerbils *Tatera indica* and *Meriones hurrianae* (Prakash 1975). In addition, two breeding peaks which coincided with new vegetation growth in spring and late summer were reported in *Dipodomys merriami* (Reynolds 1960).

Desert rodents which feed on the green leaves of perennials are less dependent on rainfall for reproduction. *Dipodomys microps* consumes the chenopod *Atriplex confertifolia*. Its reproductive season is short, well defined and consistent from year to year (Kenagy 1973b). Fat sand rats (*Psammomys obesus*) consume chenopods, including *Atriplex* spp., and their litters are produced throughout the year according to Daly and Daly (1975a). These authors concluded that "the leaf diet of perennial shrubs should free *P. obesus* from the seasonal cycles of food availability experienced by granivores..."

The importance of the availability of water alone has been demonstrated in the reproduction of several desert rodents. Christian (1979b) provided permanent sources of drinking water in the Namib Desert and monitored the reproductive parameters of three rodent species. Female *Gerbillurus paeba* and *Rhabdomys pumilio* provided with supplementary water had a greater number of pregnancies and a higher proportion of lactating individuals than females which were not provided with water. Furthermore, females with additional water extended their breeding season further into the hot, dry portion of the year than controls. The effect of increased water availability was most pronounced during periods of the year when water stress was greatest. The enhanced reproductive response to supplementary water resulted in a marked increase in the number of *R. pumilio* and in a slight increase in *G. paeba*. Similarly, reproduction in *Peromyscus truei* was stimulated when artificial sources of water were provided (Bradford 1975). In contrast to these three species, *Desmodillus auricularis* showed no consistent response to the provision of water Christian (1979b). *D. auricularis* is physiologically more capable of conserving body water than *G. paeba* and *R. pumilio* and is therefore less dependent on an external supply. Furthermore, reproduction and population growth in *D. auricularis* are relatively aseasonal and continue consistently throughout the year; *G. paeba* and *R. pumilio* exhibit seasonally restricted breeding patterns and rapid population growth in response to favourable conditions.

The Namib desert rodents *Petromys collinus, Aethomys namaquensis* and *Petromus typicus* exhibit reproductive patterns which are unlike those described above (Withers 1983).These species breed in the summer before any significant seasonal rain falls; but they become reproductively active after an increase in the frequency of advective fogs. Their reproductive pattern is highly seasonal and of short duration, resulting in low reproductive potential but low annual mortality. Withers (1983) concluded that "it appears that a low reproductive potential and high annual survival are adaptations, or preadaptations, for the successful exploitation of desert niches by small mammals".

8.2
Water Intake, Milk Production and Pup Growth

Early work on the milk composition of *Dipodomys merriami* showed that water made up about 50% of total milk, fat averaged 23.5% and non-fat solids 26.1% (Kooyman 1963). This percentage of water is much lower than that of most eutherian mammals, including large desert mammals such as the camel (Degen et al. 1987) and the collared peccary, whose milk contains 84% water (Sowles et al. 1961; quoted by Kooyman 1963). Kooyman (1963) suggested that the low water content of the milk would be advantageous to the mother for water conservation and that the young may not need a large supply of water because the "newborn rat probably produces a concentrated urine similar to that of its parent." However, if this is true, it would suggest that water recycling from the pups to the mother (see Sect. 8.3.1) is relatively unimportant in *D. merriami* as the lactating females would have to excrete almost all the urine water that they consume.

The same conclusion that the low water content of mammal milk is an adaptation to the desert environment has been reached by a number of other authors. It was found to be the case when Meyerson-McCormick et al. (1990) analysed the milk of the punare (*Thrichomys apereoides*; Echimydae), a hystricomorph rodent that occurs in the xeric, rocky shrubland of Brazil and Paraguay and produces precocial young. *Thrichomys* are the only echimyids to inhabit a xeric environment. Milk of the punare is characterized by an extremely high fat content, ranging between 30% on day 2 of lactation to 21% on day 21, and a low water content. This composition is even more unusual when one considers that the females of species with precocial young usually produce relatively dilute milk with a high sugar content whereas those with altricial young produce relatively concentrated milk (Martin 1984). Meyerson-McCormick et al. (1990) concluded that "the unusual milk composition of this tropical rodent is one of several adaptations correlated with reproduction in a xeric environment" and that "high fat milk is an adaptation for year round reproduction. By increasing the fat portion of the milk, the dam decreases the aqueous portion, thereby reducing stress on maternal water reserves."

Kunz et al. (1995) also suggested that milk composition might be related to water availability after they had analysed the milk composition of three species of insectivorous bats that are similar in body mass (approximately 7–13 g during lactation) but differ in their access to water. *Tadarida brasilienses* (Molossidae) usually inhabits caves that are located in arid and semi-arid regions and has "seldom been observed feeding or drinking over bodies of water." In contrast, *Myotis lucifugus* and *M. velifer* (Vespertilionidae) inhibit caves that are "usually located near bodies of water, where bats feed and drink upon departing from their roost at dusk." The mean peak fat and water content of *Tadarida brasiliensis* milk were 25.8 and 63.5%, respectively, of *M. lucifugus* 15.8% and 72.9%, respectively, and of *M. velifer* 19.9 and 67.9% respectively. These results led the authors to conclude that "relatively high fat and low water levels in *T. brasiliensis* milk may reflect the limited access that lactating females have to free water, as well as needs to minimize mass of stored milk during long foraging trips. Conversely, lower fat concentrations and higher water levels in milk in *M. lucifugous* and *M. velifer* may relate to the propensity for colonies of these two species to roost and forage near bodies of water."

However, the high dry matter content of the milk of *Dipodomys merriami* and *Thrichomys aperoides* does not appear to be typical of all desert rodents. In a study of Australian desert mice and *Mus musculus*, offered water ad libitum, total solids in the milk of *Notomys alexis* averaged 21–33%, of *N. cervinus* 24–34%, of *N. mitchelli* 24–33%, of *Pseudomys australis* 25–30% and of *M. musculus* 31–41% (Baverstock et al. 1976). In general, the proportion of total solids increased with lactation time. This is then a consequence of an increase in the proportion of the fat content, since protein and sugars remained constant. Although the solid contents of the milk of these desert-adapted rodents were relatively high compared with those of other eutherian mammals, they were not exceptionally so. In fact, *M. musculus*, the lease desert-adapted of the five species, produced the most concentrated milk. Baverstock et al. (1976) suggested that the milk concentration was related to the rate of development of the young. Young of *M. musculus* open their eyes at 12 days post-partum, *P. australis* at 16 days, and the three species of *Notomys* at 20 days. Milk solids, particularily in the first 5 days of lactation, are proportionally higher in species in which the rate of development of the young is faster. The high total solid content of *D. merriami* milk can also be explained in this context since the pups open their eyes at 11 days.

Restricting the water intake of lactating females led to an increase in the milk concentration of *N. alexis, N. cervinus, P. australis* and *M. musculus*, but not of *N. mitchelli*. The increase in concentration was due to a greater proportion of fat; protein and sugar increases were negligible (Baverstock et al. 1976). However, these authors suggested that the "change in milk composition with water-restriction is a physiological consequence of water-restriction rather than a homeostatic response to save water." They based their arguments on two main points: (1) urine recycling from the pup to the dam (Sect. 8.3.1) is very efficient in all these rodent species, and there should therefore be no need to produce concentrated milk; and (2) if lactating females were attempting to save water, they would increase the fat, protein and sugar in the same proportions. Water restriction of lactating females also resulted in decreased milk production in five desert rodent species and, consequently, depressed the growth rate of their suckling young (Baverstock and Watts 1975; Baverstock et al. 1976). Decreased growth rate and increased mortality were also observed in *Peromyscus maniculatus* pups following water restriction of the lactating females (Porter and Busch 1978).

Water intake by pregnant and lactating fat sand rats was shown to have an effect on the birth mass and subsequent growth of their pups (Kam and Degen 1994). *Psammomys obesus*, a herbivorous desert rodent, is able to thrive and reproduce while consuming only the chenopod *Atriplex halimus*, a food low in energy yield and high in electrolyte and water content (Degen 1988; Kam and Degen 1993b; 1994). The water content of this plant differs greatly from season to season and can range between 40 and 90% of fresh matter (Degen et al. 1990). In a laboratory study, pregnant and lactating females were offered only *A. halimus* differing in water content varying between 60 and 85%. Litter size explained about 30% of pup body mass at birth. The relationship of birth mass per pup (m_{b-o}; g) to litter size (L) can be described by the non-linear regression equation (Kam and Degen 1994):

$$m_{b-o} = 8.05\ L^{-0.137},$$

where n = 38; SE of coefficient and exponent = 0.36 and 0.34, respectively; and $r^2 = 0.30$. However, when the effect of water content (W; proportion of fresh matter) of the *A. halimus* was included in the regression equation, 45% of the variation in pup body mass at birth was explained. The regression equation took the form (Kam and Degen 1994):

$$m_{b-o} = 9.11\ L^{-0.150} W^{0.427},$$

where n = 38; SE of coefficient and exponents of L and W = 0.51, 0.03 and 0.14, respectively, and $r^2_{adj} = 0.45$. For a decrease of 1% in water content of *A. halimus*, there was a decrease of about 0.5% in m_{b-o}.

Similarly, the growth rate of the *P. obesus* pups was affected both by the litter size (Fig. 8.1) and by the water content of the *A. halimus*. The effect of age (T; days), litter size and *A. halimus* water content (W) on m_b of growing fat sand rats was analysed using a modified Gompertz model for litters of three, four and six young (87 individuals from 21 liters) until 60 days of age. The equation

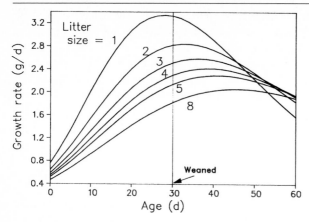

Fig. 8.1. Effect of litter size (one, two, three, four, five and eight pups) on the growth rate of fat sand rat (*Psammomys obesus*) pups from birth to 60 days of age. Pups were weaned at 30 days. Until 15 days of age, only lactating females consumed saltbush (*Atriplex halimus*) and thereafter, both females and their pups consumed the plant. Curves were fitted by the Gompertz equation for a water content of 75% of *A. halimus* leaves. (Kam and Degen 1993a)

took the form (Kam and Degen 1994):

$$m_b = 189.7 \, \text{Exp}[-3.82 \, \text{Exp}(-0.076T \, W^{1.60}L^{-0.233})]$$

where SE of parameters in order of presentation = 4.6, 0.03, 0.002, 0.05, 0.010; and $r^2 = 0.98$.

To demonstrate the effect of the water content of *A. halimus* on the growth rate of the pups, a simulation using the Gompertz model was generated for a range of 60–85% in the water content of the fresh matter, and for a litter size of four young that were weaned at 30 days of age (Fig. 8.2). The maximum difference in body mass per pup due to differing dietary water content reached 12 g at 15 days of age – this was 47% of the higher body mass – and 60 g at 60 days of age, which represented 43% of the higher body mass. When the effects of both litter size and water content on the growth rate of pups weaned at 30 days were examined, the maximum growth rate fell within the lactating period and close to weaning when the water content of the leaves was high (85% of fresh matter): the growth rate of weaned fat sand rats decreased rapidly. When leaves low in water content (60% of fresh matter) were consumed, there was an increase in growth rate towards weaning but at a much slower rate, which resulted in a reduced body mass of the weaned pup (Fig. 8.3).

8.3
Water Budget During Pregnancy and Lactation

A large number of studies have examined the energy costs of pregnancy and lactation; but few have measured water requirements during these physiologi-

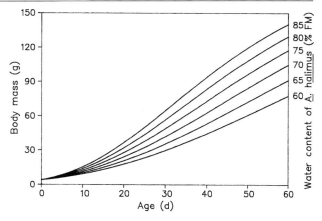

Fig. 8.2. Effect of water content of saltbush (*Atriplex halimus*) leaves on the body mass of fat sand rat (*Psammomys obesus*) pups from birth to 60 days of age. Pups were weaned at 30 days. Until 15 days of age, only lactating females consumed *A. halimus*, and thereafter, both females and their pups consumed the plant. Curves were fitted by the Gompertz equation for a litter size of four pups. (Kam and Degen 1993a)

Fig. 8.3. Effect of water content of saltbush (*Atriplex halimus*) leaves (*three upper curves are 85% and three lower curves are 60% water*) on the growth rate of fat sand rat (*Psammomys obesus*) pups from birth to 60 days of age. Pups were weaned at 30 days. Until 15 days of age, only lactating females consumed *A. halimus* and thereafter, both females and their pups consumed the plant. Curves were fitted by the Gompertz equation for litter sizes of one, two and three pups. (Kam and Degen 1993a)

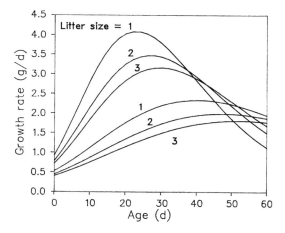

cal states. Water requirements increase as a result of two main factors: (1) evaporative water loss increases in consequence of increased energy requirements and heat production; and (2) milk production during lactation.

The water budget of control, pregnant, lactating and post-lactating *Acomys cahirinus* females was measured. These omnivorous rodents inhabit rocky desert areas. The females were offered a balanced rodent ration and water ad libitum. Each female gave birth to one or two pups, but there was no difference in the water intake or output of the females in relation to the different litter sizes. In non-pregnant females, the total water intake averaged 5.26 ml d^{-1} or 106.5 ml kg^{-1}d^{-1}. During pregnancy, total water intake increased by 38% on an absolute basis and by 21% on a mass-specific basis while during lactation these increases were 59% and 54%, respectively. Most of this increased water intake was due to the increased amounts of water drunk. Urine water increased substantially during both pregnancy and lactation. Milk production averaged 0.93 ml d^{-1} (Table 8.1).

A comparative study of water turnover during lactation was carried out with three rodent species inhabiting areas of different aridity (Oswald et al. 1993): red-backed voles (*Clethrionomys gapperi*) from mesic areas, white-footed mice (*Peromyscus leucopus*) from a variety of habitats including moist and arid areas, and two species of gerbils (*Gerbillus allenbyi* and *G. pyramidum*) from extremely dry areas. Water and either rodent chow (for voles and white-footed mice) or millet seeds (for gerbils) were provided ad libitum, and water turnover was measured from days 3–15 of lactation in the voles and white-footed mice and from days 11–19 in the gerbils. A complete water budget was established for the voles and white-footed mice.

Much larger proportional increases in total water intake were found in these rodents than in *A. cahirinus*. Water turnover was 21.7 ml d^{-1} in non-reproductive females voles and increased by 71–131% during lactation; 8.4 ml d^{-1} in non-reproductive female white-footed mice and increased by 58–129% during lactation; 2.7 ml d^{-1} in non-reproductive female gerbils and increased by 52–36% during lactation. Water turnover not only increased in females during lactation compared with the non-reproductive state, but it also increased with time during lactation. Drinking water accounted for 87–90% of total water turnover in voles, 75–78% in white-footed mice and only 42–55% in gerbils. That is, in the first two rodent species, drinking water contributed to almost all the increased water intake but did so less in the gerbils (Fig. 8.4). Apparently, the big difference in water intake among these rodents lay in the recycling of water from the pups to the mothers (see Sect. 8.3.1).

Table 8.1. Water intake and output in *Acomys cahirinus* females with litters of one or two pups before and during pregnancy, during lactation and after weaning. No difference was found between females with one or two pups, and the data were combined

	Before pregnancy	During pregnancy	During lactation	After weaning
Water intake	5.26 ± 1.39[a]	7.27 ± 0.98[b]	8.37 ± 0.64[c]	6.28 ± 0.67[ab]
Preformed water	0.32 ± 0.08[a]	0.37 ± 0.09[b]	0.42 ± 0.10[c]	0.35 ± 0.09[ab]
Metabolic water	1.47 ± 0.36[a]	1.69 ± 0.41[b]	1.91 ± 0.47[c]	1.59 ± 0.39[ab]
Water drunk	3.47 ± 0.92[a]	5.21 ± 0.70[b]	6.04 ± 0.46[c]	4.34 ± 0.19[ab]
Water output	5.26 ± 1.39[a]	6.89 ± 0.92[b]	8.97 ± 0.68[c]	6.13 ± 0.65[ab]
Urine water	0.87 ± 0.51[a]	1.93 ± 0.41[b]	1.88 ± 0.43[b]	1.22 ± 0.22[ab]
Faecal water	0.54 ± 0.21[a]	0.82 ± 0.31[b]	0.70 ± 0.27[ab]	0.77 ± 0.29
Evaporative water	3.85 ± 0.83	4.11 ± 0.55[a]	5.86 ± 0.44[b]	4.14 ± 0.44[a]
Milk water	0.00	0.00	0.92	0.00

Values are means ± SD (n = 19), and those with different letters within rows differ significantly from each other.

*Includes water intake plus or minus water available due to changes in body mass.

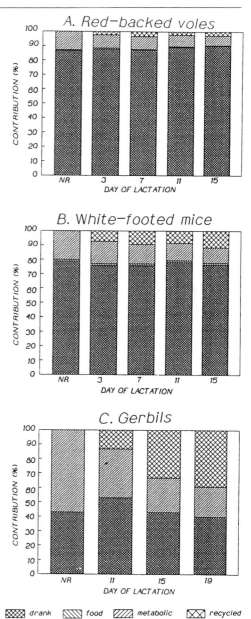

Fig. 8.4. Relative contributions of sources of water intake for **A** red-backed voles, **B** white-footed mice and **C** gerbils. *NR* Non-reproductive females. (Oswald et al. 1993)

Water intake during lactation was also determined in five xeric-adapted Australian rodent species which were offered water ad libitum (Baverstock and Watts 1975). In contrast to the previous results, water intake in lactating females was, in general, only slightly higher than that of non-lactating females. The water intake of lactating *Notomys cervinus* females was similar to that of non-lactating females, while the water intake of lactating *Pseudomys australis* was about twice that of non-lactating females. The water intake of lactating *N. alexis* females with 3 and 4 pups was similar to that of non-lactating females. With a litter of 5 young, however, water intake increased in the lactating females, but only after 10 days post-partum. In *N. mitchelli*, the water intake of lactating females nursing litters of 4 pups was only slightly higher than that of non-lactacting females. *Mus musculus* females nursing three to five young showed a slight increase in water intake over that of non-lactating females; females nursing litters of six pups showed a greater increase.

In rodents, evaporative water loss accounts for most of the water output during pregnancy and lactation. In *Acomys cahirinus*, these losses were 60–65% of total water loss during these physiological states. (unpubl. data). Evaporative water loss accounted for 74–79% of the total water turnover of lactating white-footed mice and for 43–49% of total water turnover of voles (Oswald et al. 1993). However, evaporative water loss as a proportion of total water loss does not increase during lactation, at least in *A. cahirinus* (Table 8.1).

Urinary water loss increased from 16% of total water output in non-reproductive *A. cahirinus* to 28% in pregnant females and 21% in lactating females. Absolute urine output in white-footed mice increased from 0.5 g d^{-1} in non-reproductive females to 2.6 g d^{-1} in lactating females, and in voles from 9.8 g d^{-1} in non-reproductive females to 24.1 g d^{-1} in lactating females. Although in absolute terms the increases were substantial, in relative terms they were slight. Urine output was not measured in gerbils, but lactating gerbils had a urine osmotic pressure of more than twice that of non-reproductive females. In contrast, no difference in urine osmotic pressure between lactating and non-reproductive females was measured in the white-footed mice and voles. This would suggest that the gerbils were conserving urinary water to a greater extent than the other rodent species.

Faecal water loss is relatively low in non-reproductive females. Although in absolute terms faecal water loss increases during lactation, when compared with non-reproductive females it decreases in relative terms (Table 8.1).

8.3.1
Recycling of Water During Lactation

Most detailed studies on water recycling from pups to lactating females have been carried out on laboratory rats. Pups obtain water and electrolytes via milk from their mothers, while lactating female rats reclaim water and electrolytes from their pups when they lick the anogenital area of the pups, stimulating urination, and then consume the urine (Fig. 8.5). Dams also consume the faeces of the young (Gubernick and Alberts 1983, 1985). Even when dams were offered water ad libitum, the combined fluid from urine and faeces consumed from a litter of eight 10-day-old pups can amount to 20 ml d^{-1}. This may represent 65–70% of milk production (Friedman and Bruno 1976; Gubernick and Alberts 1983). From 15-day-old pups, this fluid can amount to 40 ml d^{-1} which

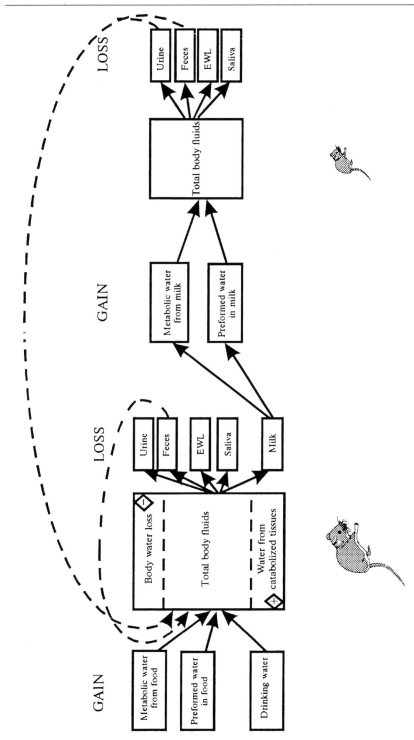

Fig. 8.5. Sources of water input and output in the lactating female and her pup, indicating recycling of urine and faecal water from the pup to the mother (*EWL* evaporative water loss)

presumably is greater than that used in milk production, although this was not measured. Thereafter, water reclaimed from the pups decreased rapidly and equalled about 5 ml d^{-1} from 30-day-old pups (Gubernick and Alberts 1983). Water and electrolytes obtained in this manner are important in balancing extracellular fluid volume and total body water in the lactating female, since female rats deprived of pup urine were significantly more hypovolemic than females deprived only of water (Friedman et al. 1981).

Time spent in anogenital licking is controlled by the uptake of fluid from the pups and by the salt needs, in particular of NaCl, of the dams. When dams were offered a 0.15 M solution of NaCl, anogenital licking of the pups decreased significantly by 19%, but there was no significant change in non-anogenital licking, that is, in the licking of other areas of the body. This indicates that only anogenital licking was affected by the intake of NaCl solution by the dams. Furthermore, when dams were offered a 0.08 M solution of KCl, there was no significant change in either anogenital or non-anogenital licking. These results were then interpreted to mean that there is "evidence for a sodium specific mechanism that modulates anogenital licking of pups by rat dams" (Gubernick and Alberts 1983).

The recycling of water should be of particular importance to the water budget of small mammals inhabiting xeric areas. The urine osmotic concentration of the suckling young is much lower than that of the urine the mother is capable of producing, and therefore the solute consumption in urine can be excreted by the mother in a much smaller volume of water. This leaves free water available to the lactating female. For example, lactating *Notomys alexis* females are capable of concentrating their urine to about five times that of their pups' urine. (Baverstock and Green 1975).

In the study of Oswald et al. (1993) described above, the absolute volume of recycled water increased during lactation and was similar among the three rodent species: 0.9–1.3 g d^{-1} for the voles, 1.0–2.2 g d^{-1} for the mice and 0.5–2.3 g d^{-1} for the gerbils. However, in relative terms, there was a great difference among them since reclaimed water contributed 2.1–3.2% of total water turnover in the mesic voles, 7.2–11.5% in mice and 13.0–39.0% in the xeric gerbils (Fig. 8.4.). The contribution of recycled water towards milk production was not, however, determined from these results. Oswald et al. (1993) concluded that water "recycling is a more important source of water for xeric than mesic species".

Baverstock et al. (1979) estimated that recycling of water from the pups of *Mus musculus* accounts for 16% of the total water intake of lactating females under conditions of ad libitum drinking water. The measurement was made on day 11 of lactation. The recycled water in this species was estimated to be 79% of the water lost by the pups, which equalled 38% of the water secreted in the milk (Baverstock and Green 1975). From computer simulation, it was estimated that the recycled water in *Mus musculus* equals about 50% of the volume of water secreted in the milk (Baverstock and Elhay 1981). A value of greater than 50% was measured for guinea pigs (*Cavia porcellus*) by Baverstock and Elhay (1979). Measurements of *Notomys alexis* showed that this desert species recycled only 58% of the water lost by the pups, which equalled 32% of milk secreted by the females (Baverstock and Green 1975).

From studies done thus far, it is difficult to assess the value of recycling water from pups on the water balance of lactating dams, in particular among desert rodents. Desensitized virgin female rats lick the anogenital area of pups and consume the urine. Moreover, anogenital licking and urine consumption is practised by male California mice (*Peromyscus californicus*), although the time spent licking and the volume of urine consumed by the males are less than those by lactating females (Gubernick and Alberts 1987). In all the above studies, water was available ad libitum, and therefore there was no real need to conserve water to balance water budgets. Perhaps if water were not available, as is often the case with desert gerbils, the differences noted among the species in the study of Oswald et al. (1993) would have been greater.

Baverstock and Green (1975) suggested that the recycling of urine and faecal water may be a behavioural pattern that has evolved for nest hygiene. For example, rodents, whether desert-adapted or not, usually recycle urine whereas lactating rabbits do not, and this may be related to the reduced maternal behaviour of rabbits. Thus, in desert species, recycling of water during lactation may be a "by-product of a behaviour that has evolved for some other purpose".

In addition to the transfer of fluids from the pups to the lactating mother, Friedman and Bruno (1976) also observed the transfer of water vapour between littermates and stated that this may be significant. However, Baverstock and Green (1976) calculated that the volume of water received by a rat pup in a litter of eight was only about 0.12 ml d $^{-1}$ by pulmocutaneous exchange of water. They concluded that the "net 'saving' in water to the entire litter is therefore about 1 ml/day. This seems small compared with the 21 ml of urine per day consumed by the mother."

8.4
Energy Requirements for Pregnancy and Lactation

Female eutherian mammals typically have relatively lengthy gestation and short lactation periods, whereas metatherian females reverse this and have short gestation and lengthy lactation periods. There is still a question as to which strategy is energetically more costly (Harvey 1986).

Energy requirements during pregnancy include: (1) production of fetal, uterine, placental and mammary tissue; (2) heat increment costs incurred in the production and growth of tissues; and (3) increased maintenance costs associated with these new tissues (Gittleman and Thompson 1988). Costs of pregnancy are usually determined by the difference in caloric intakes between pregnant and non-reproducing females. This method has usually revealed little increase in energy intake in small pregnant mammals. For example, little or no increase in energy intake (Slonaker 1925 and Wang 1925; quoted by Gittleman and Thompson 1988) or an increase of only 10% was found for the pregnant rat (*Rattus norvegicus*; Morrison 1953) and the pregnant desert spiny mouse (*Acomy cahirinus*; unpublished data), and increases of 18–25% were found for the bank vole (Kaczmarski 1966), European common vole (Migula 1969), white mouse (Myrcha et al. 1969) and hispid cotton rats (*Sigmodon hispidus*; Mattingly and McClure 1982).

Small mammals usually consume enough food to provide energy for the support and maintenance of the pregnant female, and for fetal growth and development. In fact, many females can consume food in excess of maintenance requirements for pregnancy and store energy that can be used during lactation. This has been demonstrated in the Djungarian hamster (*Phodopus sungorus sungorus*; Weiner 1987) and the cotton rat (Randolph et al. 1995). Females that do not consume enough energy generally use energy reserves to maintain near normal fetal development. However, this is not always the case for in white rats, "maternal nutrient stores are not mobilized for fetal utilization even when fetal growth is markedly impaired" (Lederman and Rosso 1981).

The method of estimating pregnancy costs by difference in caloric intake can result in errors, as pointed out by Gittleman and Thompson (1988). Some species may reduce the energy costs of activities not related to reproduction, perhaps by reducing the time spent on these activities, and allocating this energy to gestation (Racey 1982). Thus, the energy costs of pregnancy would be underestimated, and this could be part of the explanation of the low increases in energy intake during pregnancy. In order to assess pregnancy costs better, a "combination of caloric consumption, respirometry, and time budgets of behaviour, before and during gestation (reproduction) is a requisite for assessment of the allocation of energy to maintenance, . . . and net production during gestation" (Gittleman and Thompson 1988).

Lactation is energetically much more demanding than pregnancy (Millar 1977). In fact, lactation is considered to be the most energy-expensive phase of reproduction for the female mammal and, as a consequence, is accompanied with a great increase in energy expenditure. Moreover, energy requirements increase throughout lactation "so that late lactation is the most energetically critical period of the breeding cycle" (Millar 1977).

Most of the increase in energy expenditure of the lactating female is a consequence of the energy required for milk production. This includes both the energy content of the milk and the heat increment of feeding for milk production (Kam and Degen 1993b). In general, small lactating homeothermic mammals can meet these additional energy demands by increasing their food consumption (Gittleman and Thompson 1988); there is usually little difference in digestive efficiencies between reproducing and non-reproducing individuals (Kasczmarski 1966; Migula 1969; Randolph et al. 1977; Millar 1978; Lochmiller et al. 1982; Poppitt et al. 1994).

The increase in nutritional energy intake by the lactating female above that of non-pregnant, non-lactating females rodents usually ranges between 65 and 200% (Kaszmarski 1966; Migula 1969; Randolph et al. 1977; Millar 1978; Mattingly and McClure 1982; Glazier 1985; Gittleman and Thompson 1988;

Poppitt et al. 1994). However, in small rodents and other small mammals, this increase may be even higher and can reach 400% in the house mouse, *M. musculus* (Konig and Markl 1987), and even 800% in a small shrew, *Sorex coronatus* (Genoud and Vogel 1990). Moreover, high increases in energy intake during lactation have been reported in mammals consuming a diet that has a low efficiency of energy utilization. Energy increases greater than 200% were found in the lactating fat sand rat, a herbivore that consumes low energy chenopods (Kam and Degen 1993b), and also in the lactating red panda (*Ailurus fulgens*), a folivore that consumes low energy bamboo leaves (Gittleman 1988).

Unequivocal results have been reported regarding the effect of litter size on energy intake in lactating females. Lactating fat sand rats increased their metabolizable energy intake over the maintenance requirements of non-reproducing females by up to 85% when nursing 1–3 pups, by up to 118% when nursing 4–5 pups, and up to 138% when nursing 6–8 pups (Kam and Degen 1993b). Similarly, an increase in energy intake with an increase in litter size has been reported in the pine vole (Lochmiller et al. 1982) and *Peromyscus leucopus* (Millar 1978). In contrast, no difference in energy intake was noted in lactating *Acomys cahirinus* females with different litter sizes. Females nursing 1 pup increased their metabolizable energy intake by 46%, while females nursing 2 pups increased metabolizable energy intake by 44% (unpubl. data).

8.4.1
Energy Partition in Lactating Female

Energy intake by the lactating female is partitioned into the energy required for maintenance and that for milk production. When offered food ad libitum, some small lactating mammals such as the pine vole (*Microtus pinetorum*; Lochmiller et al. 1982) and golden-mantled ground squirrel (*Spermophilus saturatus*; Kenagy et al. 1989b) remain approximately in energy balance. Others such as the lesser hedgehog tenrec (*Echinops telfairi*; Poppitt et al. 1994) can add to their energy reserves, while still others such as the Djungarian hamster cannot assimilate enough energy and must mobilize energy reserves (Weiner 1987).

If energy intake does not cover the requirements for maintenance and lactation of the dam, several options are available, including the following:

1. Total body energy of the female is maintained and milk production reduced. This results in a slower growth rate of the pups and maintenance of body mass of the female. Such a strategy appears to be used by Norway rats which "apparently do not monitor and defend a maximal pup growth rate. Rather, rat dams seem to continue to defend their own homeostatis, and by doing so, allow the young to grow and survive under a wide variety of circumstances" (Leon and Woodside 1983).

2. Milk production is maintained near normal, and the energy shortage is offset by the mobilization of the body reserves of the lactating female. This re-

sults in fast growth of the pups and a reduced body mass of the female. Such a strategy is used by a number of species that do not have the potential to consume enough nutritional energy to support lactation.

3. A combination of points (1) and (2) which results in a slower growth rate of the pups and a reduced body mass of the female. This strategy has been demonstrated by cotton rats from Kansas and Tennessee (Mattingly and McClure 1985) and is the one most commonly used. The extent of maternal tissue mobilization may depend upon whether there is concurrent pregnancy and lactation and on the energy allocations to each (Oswald and McClure 1990).

Furthermore, Millar (1977) suggested three ways in which mammals might respond to a restricted energy intake during reproduction: (1) to reduce the size of the litter and, in this way, maintain the weaning size and age of the individual offspring; (2) to reduce weaning size but retain litter size and weaning age; and (3) to increase weaning age and maintain litter and weaning sizes and, in this way, decrease daily energy requirements. Energy requirements of the lactating female could be reduced further by thermolability, a metabolic compensation used by a number of mammals, including the brown long-eared bat (*Plecotus auritus*; Speakman and Racey 1987). Moreover, reduced body temperature, in addition to a reduced growth rate of the pups, as noted in cotton rats, can reduce energy demands on lactating females with restricted energy intake (Mattingly and McClure 1985).

The total energy in milk is dependent upon the energy intake and the changes in the energy retention (negative or positive) of the lactating female (Kam and Degen 1993b). The importance of energy and water intakes in reproduction was examined by Porter and Busch (1978), who used fractional factorial analysis to determine factors affecting lactating *Peromyscus maniculatus* with regard to the growth and weaning success of their offspring. Of the parameters examined, available food and water were found to be more important than seven other variables. There were no survivors when the females consumed less than 50% of the food or less than 20% of the water of the ad libitum intakes of lactating females (Fig. 8.6). The importance of body reserves in reproduction was demonstrated in a study on Kansas cotton rats (*Sigmodon hispidus*; KS*h*) by Mattingly and McClure (1985). These authors found that "growth and survivorship of KS*h* young were largely determined by the body mass of the dam, indicating the potentially significant role of body size for small mammals that experience periodic food shortages during lactation".

Change in body mass of the growing pup is a good indication of milk production, and change in body mass of the lactating female is a good indication of change in energy retention by the female. The effect of nutrient energy intake and body mass change of lactating females on the growth rate of the pups were examined in *Acomys cahirinus* nursing either one or two pups (unpubl. data). The growth rate of 2 pups together did not differ from that of a single pup, and therefore the data were combined. The linear regression of change in

Fig. 8.6. Three-dimensional response surface, with centre replicate variable levels except for food and water, showing that sensitivity to food deprivation is about twice as great as to water deprivation as judged by the minimum percentage of the "free access" ration that will result in some young surviving. Each point on the surface represents the mean of a minimum of 4 litters and in one case 27 litters. SD are omitted for viewing clarity, but they range from a low of ±0.28 to a high of ±3.5 g in the "valley" area, which represents 90% of the "free access" food consumption. Typical SD are about ±1.0 g. (Porter and Busch 1978)

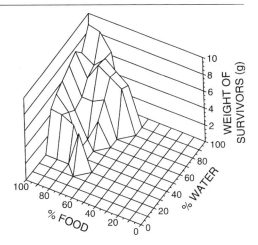

the body mass of the pup (Δm_b; g d^{-1}) on the metabolizable energy intake of the lactating female (LMEI; kJ·kg$^{0.75}$·d^{-1}) takes the form:

$$\Delta m_b = -0.5701 + 0.0016 \text{ LMEI},$$

where $n = 18$; $S_{y \cdot x} = 0.206$; $r^2 = 0.48$; $P < 0.01$.

The effect of the relative increase in metabolic energy intake during lactation, that is, metabolic energy intake during lactation/metabolic energy intake before pregnancy (LMEI/BPMEI), on the growth rate of the pup can be described by the regression equation.

$$\Delta m_b = -0.6312 + 0.7456 \text{ LMEI/BPMEI},$$

where $n = 18$; $S_{y \cdot x} = 0.210$; $r^2 = 0.47$; $P < 0.01$.

The effect of the absolute increase in metabolic energy intake during lactation (LMEI $-$ BPMEI; kJ kg$^{-0.75}$d^{-1}) on the growth rate of the pup can be described by the regression equation:

$$\Delta m_b = 0.0955 + 0.0018 \text{ LMEI} - \text{BPMEI},$$

where $n = 18$; $S_{y \cdot x} = 0.200$; $r^2 = 0.51$; $P < 0.01$.

The linear regression of change in body mass of the pup (Δm_b; g d^{-1}) on the change in body mass of the lactating female (M_b; d^{-1}) takes the form:

$$\Delta m_b = -0.1587 + 0.5486 \, \Delta M_B,$$

where $n = 18$; $S_{y \cdot x} = 0.210$; $r^2 = 0.46$; $P < 0.01$.

Therefore, either metabolizable energy intake (expressed as total intake, relative intake or absolute increased intake) or change in the body mass of the lactating female explained about 50% of the change in body mass of the growing pup.

The effect of two independent variables on Δm_b took the following forms:

$\Delta m_b = -0.6925 + 0.0009 \text{ LMEI} + 0.3932 \text{ LMEI/BPMEI}.$

where $n = 18$; $S_{y\cdot x} = 0.205$; $r^2_{adj} = 0.44$; $P < 0.01$;

$\Delta m_b = -0.6898 + 0.0014 \text{ LMEI} + 0.4717 \Delta M_B,$

where $n = 18$; $S_{y\cdot x} = 0.128$; $r^2_{adj} = 0.78$; $P < 0.01$;

$\Delta m_b = -0.6486 + 0.5968 \text{ LMEI/BPMEI} + 0.4333 \Delta M_B,$

where $n = 18$; $S_{y\cdot x} = 0.150$; $r^2_{adj} = 0.70$; $P < 0.01$.

$\Delta m_b = -0.0832 + 0.0015 \text{ (LMEI} - \text{BPMEI)} + 0.4432 \Delta M_B,$

where $n = 18$; $S_{y\cdot x} = 0.135$; $r^2_{adj} = 0.76$; $P < 0.01$.

Therefore, metabolizable energy intake (expressed as total intake, relative intake or absolute increased intake) and change in the body mass of the lactating female explained about 75% of the change in body mass of the growing pup.

8.5
Reproduction, Metabolic Rate and Body Size

Body size has an effect on metabolic rate (Sect. 7.4) and on reproductive rate. Large body size is associated with low reproductive rate and long life, while small body size is associated with high reproductive rate and short life (Millar and Hickling 1991). Moreover, the adult body size of mammals is positively correlated with the birth mass and growth rate of the young and weaning body mass. It is not correlated with litter size and time to weaning and is negatively correlated with reproductive effort. Reproductive effort is measured as the "amount of energy females must acquire, relative to her own maintenance requirements, in order to wean her offspring successfully" (Millar 1977).

The exponential population growth constant (r_m) can be defined as: $r_m = (\ln R_0)/T$, where $\ln R_0$ is the natural logarithm of the net rate of reproduction per generation, and T is the generation time (McNab 1980b and references cited therein). It follows then that the maximal intrinsic constant of population growth (r_{max}) occurs when R_0 is maximal and T is minimal. The r_m is influenced by the interaction of body mass and the rate of metabolism of the animal. Furthermore, according to McNab (1980b, 1984, 1986a, 1987), in mammals of similar body mass, those with a higher metabolic rate have a higher reproductive rate than those with a lower metabolic rate: that is, r_m is positively correlated with BMR. An increase in metabolic rate is associated with a decrease in gestation period, a decrease in time from conception to weaning, an increase in the post-natal growth rate and an increase in fecundity. Several examples are cited to support this trend. In eutherian mammals, microtine rodents and lagomorphs have a relatively high metabolic rate and a large r_m constant, cricetines have an intermediate rate of metabolism and intermediate r_m, while heteromyids and fossorial rodents have a low metabolic rate and a low r_m. Marsupials have a low BMR and a low rate of reproduction, and no correlation exists in them between BMR and reproductive rate (McNab 1986b). Moreover, mammals with high rates of metabolism, such as snowshoe hares (*Lepus americanus*) and lemmings *(Lemmus* spp.), tend to have larger periodic population fluctuations, whereas mammals with low metabolic rates, such as the heteromyids, tend to have stable populations with small fluctuations in numbers (McNab 1980b). It is theorized that a relationship between BMR and reproductive rate exists, since animals with high metabolic rates can increase their rates of maternal and fetal tissue biosynthesis and produce milk at a faster rate than mammals with low metabolic rates. In consequence, they can support a high rate of reproduction.

This would indicate that desert mammals with their relatively low BMR have lower reproductive rates than their non-desert counterparts with higher BMRs. McNab (1986a) also theorized that since food habits influence BMRs, then food habits must also be related to reproductive rate. Small mammals which feed on herbs have high metabolic rates, and those which feed on seeds have low metabolic rates. These dietary habits permit high and low reproductive rates, respectively.

Henneman (1983, 1984a) analysed data on the BMR and reproductive rate of mammals and also arrived at the conclusion that the maximal intrinsic rate of increase in the population of a species is positively correlated with its BMR. However, he found many other confounding factors exist that affect this relationship. For example, altricial eutherian mammals have a higher r_m than precocial eutherian mammals, in particular when they have small body masses (Henneman 1984b).

However, the positive relationship between r_m and BMR reported in some studies is not clear and has been disputed vigourously. Hayssen (1984) reanalysed the data set of Henneman (1983) and found that the correlation between the intrinsic rate of increase and the BMR was not statistically significant. She further argued that "there is no theoretical reason to expect a high correlation between *basal* metabolic rate and a population's maximum rate of increase".

Derting (1989) also questioned the validity of the relationship between BMR and reproductive rate in mammals, citing interspecific and intraspecific relationships among mammals to support his view. Comparisons among microtine rodents revealed that species with a low metabolism were associated with high reproductive rates, and in cotton rats, BMR was not correlated with production rate. Nonetheless, when subcutaneous capsules containing L-thyroxine (T_4) were implanted in cotton rats in order to increase their metabolic rate, females with elevated metabolic rates showed increased production rates. This led Derting (1989) to conclude "that if food energy is unlimited, increases in BMR are associated with increased production rate, at least on a short-term basis. Therefore, the hypothesis that increased rates of basal metabolism are associated with increased rates of production cannot be rejected at the individual level." However, the question of causality remained unanswered and two possible explanations were presented for the positive interspecific correlation between BMR and reproductive rate: (1) BMR may be related to rates of tissue biosynthesis resulting in enhanced reproduction; and (2) selection is acting on reproductive traits and therefore on BMR through production rates.

The relationship between BMR and reproductive rate was also rejected by Harvey et al. (1991) who examined the effects in mammals of BMR on 22 life-history variables when the analyses removed the effects of body mass. Analyses were undertaken for all mammals combined, for eutherian mammals only, and for metatherian mammals only. They were calculated at all taxonomic levels, that is, from species within genera and upward. The authors concluded that "there is no empirical evidence for the claim that basal or active daily metabolic rates contribute to taxonomic variation in mammalian or eutherian life histories when body-size effects are controlled for. There is also no empirical evidence for the claim that dietary differences contribute to taxonomic variation in mammalian or eutherian metabolic rates when body-size effects are controlled for."

The longitudinal measurements of energy expenditure during reproduction in a metatherian mammal, the South American short-tailed opossum (*Monodelphis domestica, Marsupialia*: Didelphidae) and in two eutherian mammals, the elephant shrew (*Elephantus rufenscens,* Macroscelidea: Macroscelididae) and the tenrec (*Echinops telfairi,* Insectivora: Tenrecidae), helped to clarify the relationship between metabolic rate and reproductive rate. These three species of diverse phylogeny are characterized by relatively low BMRs: 64%, 82% and only 28%, respectively, of what would be expected according to the Kleiber equation. All three species were able to elevate their metabolism for prolonged periods during gestation and lactation to levels similar to those of other eutherian mammals. (Thompson and Nicoll 1986). In the tenrec, all of the additional energetic cost could be met by an increase in energy intake. During pregnancy and lactation, females are able to add to their energy reserves. The females also remain homeothermic during pregnancy and lactation, which presumably ensures rapid rates of growth for the fetus and subsequently of the offspring. This is unlike the non-reproductive state, for this insectivore is heterothermic and thereby saves energy that would otherwise be used for thermoregulation (Poppitt et al. 1994).

It is possible, however, that the three species selected for the longitudinal measurements of energy requirements during reproduction were not representative of mammalian species in general. At least *Didelphis* and *Tenrec* are notable for their low metabolic rates and high fecundity rates (McNab 1986b, 1987). Further studies on a number of mammalian species are required to determine whether any relationship exists between the metabolic rates of reproducing and non-reproducing females.

References

Abbott KD (1974) Ecotypic and racial variation in the water and energy metabolism of *Peromyscus maniculatus* from the western United States and Baja California, Mexico. PhD Diss, Univ California, Irvine, 155 pp

Abramsky Z (1983) Experiments on seed predation by rodents and ants in the Israeli desert. Oecologia (Berl) 57: 328–332

Abramsky Z, Brand S, Rosenzweig ML (1985a) Geographical ecology of gerbilline rodents in sand dune habitats of Israel. J Biogeogr 12: 363–372

Abramsky Z, Rosenzweig ML, Brand S (1985b) Habitat selection of Israel desert rodents: comparison of a traditional and a new method of analysis. Oikos 45: 79–88

Adolph EF (1947) Blood changes in dehydration chap 10. In: Adolph EF (ed) Physiology of man in the desert. Interscience, New York, pp 160–171

Aitchison CW (1987) Winter energy requirements of soricine shrews. Mammal Rev 17: 25–38

Altschuler EM, Nagle RB, Braun EJ, Lindstedt SL Krutzsch PH (1979) Morphological study of the desert heteromyid kidney with emphasis on the genus *Perognathus*. Anat Rec 194: 461–468

Anand RS, Parker AH, Parker HR (1966) Total body water and water turnover in sheep. Am J Vet Res 27: 899–902

Asa CS, Wallace MP (1990) Diet and activity pattern of the Sechuran desert fox (*Dusicyon sechurae*). J Mammal 71: 69–72

Aschoff J (1981) Thermal conductance in mammals and birds: its dependence on body size and circadian phase. Comp Biochem Physiol 69A: 611–619

Austin PJ, Suchar LA, Robbins CT, Hagerman AE (1989) Tannin–binding proteins in saliva of deer and their absence in saliva of sheep and cattle. J Chem Ecol 15: 1335–1347

Baddouri K, Butlen D, Imbert-Teboul M, Bouffant F Le, Marchetti J, Chabardes D, Morel F (1984) Plasma antidiuretic hormone levels and kidney responsiveness to vasopressin in the jerboa, *Jaculus orientalis*. Gen Comp Endocrinol 54: 203–215

Bakken GS (1976) A heat transfer analysis of animals: unifying concepts and the application of metabolism chamber data to field ecology. J Theor Biol 60: 337–384

Bakken GS (1992) Measurement and application of operative and standard operative temperatures in ecology. Am Zool 32: 194–216

Banin D (1990) Effect of photoperiod on bioenergetics of *Microtus guentheri*. MSc Thesis, Technion, Haifa (Hebrew with English)

Banin D, Haim A, Arad Z (1994) Metabolism and thermoregulation in the Levant vole, *Microtus guentheri*: the role of photoperiodicity. J Therm Biol 19: 55–62

Bar Y, Abramsky Z, Gutterman Y (1984) Diet of gerbilline rodents in the Israeli desert. J Arid Environ 7: 371–376

Bartholomew GA (1987) Interspecific comparison as a tool for ecological physiologists. In: Feder ME, Bennett AF, Burggren WW, Huey RB (eds) New directions in ecological physiology. Cambridge University Press, Cambridge, pp 11–37

Bassett JE (1982) Habitat aridity and intraspecific differences in the urine concentrating ability of insectivorous bats. Comp Biochem Physiol 72A: 703–708

Bassett JE (1986) Habitat aridity and urine concentrating ability of nearctic, insectivorous bats. Comp Biochem Physiol 83A: 125–131

Baverstock PR, Elhay S (1979) Water-balance of small lactating rodents. IV. Rates of milk production in Australian rodents and the guinea pig. Comp Biochem Physiol 63A: 241–246

Baverstock PF, Elhay S (1981) Water-balance of small lactating rodents. III. Estimates of milk production and water recycling in lactating *Mus musculus* under various water regimes. J Math Biol 13: 1–22

Baverstock PR, Green B (1975) Water recycling in lactation. Science 187: 657–658

Baverstock PR, Green B (1976) Exchange of water during lactation. Science 191: 420

Baverstock PR, Watts CHS (1975) Water-balance of small lactating rodents. 1. *Ad libitum* water intakes and effects of water restriction on growth of young. Comp Biochem Physiol 50A: 819–825

Baverstock PR, Spencer L, Pollard C (1976) Water-balance of small lactating rodents. II. Concentration and composition of milk of females on *ad libitum* and restricted water intakes. Comp Biochem Physiol 53A: 47–52

Baverstock PR, Watts CHS, Spencer L (1979) Water-balance of small lactating rodents. V. The total water-balance picture of the mother-young unit. Comp Biochem Physiol 63A: 247–252

Beatley JC (1969) Dependence of desert rodents on winter annuals and precipitation. Ecology 50: 721–724

Begovic S, Dusic E, Sacirbegovic A, Tafro A (1978) Examination of variations of tannase activity in ruminal content and mucosa of goats on oak leaf diet and during intraruminal administration of 3 and 10% tannic acid. Veterinaria (Sarajevo) 27: 445–457

Bell AW, Thompson GE, Findlay JD (1974) The contribution of the shivering hind leg to the metabolic response to cold of the young ox (*Bos taurus*). Pfluegers Arch Eur J Physiol 346: 314–350

Bell GP, Bartholomew GA, Nagy KA (1986) The roles of energetics, water economy, foraging behavior, and geothermal refugia in the distribution of the bat, *Macrotus californicus*. J Comp Physiol B 156: 441–450

Benjamin RW, Koenig R, Becker K (1993) Body composition of young sheep and goats determined by the tritium dilution technique. J Agric Sci (Camb) 121: 399–408

Bennett AF (1987) The accomplishments of ecological physiology. In: Feder ME, Bennett AF, Burggren WW, Huey RB (eds) New directions in ecological physiology. Cambridge University Press, Cambridge, pp 1–8

Bennett AF, Ruben JA (1979) Endothermy and activity in vertebrates. Science (Wash DC) 206: 649–654

Bennett AF, Huey RB, John-Adler H, Nagy KA (1984) The parasol tail and thermoregulatory behaviour of the Cape ground squirrel, *Xerus inauris*. Physiol Zool 57: 57–62

Bennett NC, Jarvis JUM, Davies KC (1988) Daily and seasonal temperatures in the burrows of African rodent moles. S Afr J Zool 23: 189–195

Bennett PM, Harvey PH (1987) Active and resting metabolism in birds: allometry, phylogeny and ecology. J Zool (Lond) 213: 327–363

Bentley PJ, Schmidt-Nielsen K (1967) The role of the kidney in water balance of the echidna. Comp Biochem Physiol 20: 285–290

Bergeron J-M, Jodoin L (1987) Defining "high quality" food resources for herbivores: the case for meadow voles (*Microtus pennsylvanicus*). Oecologia (Berl) 71: 510–517

Berman SL (1985) Convergent evolution in the hindlimb of bipedal rodents. Zool Syst Evolutionsforsch 23: 1–80

Beuchat CA (1990a) Body size, medullary thickness, and urine concentrating ability in mammals. Am J Physiol 258: R298–R308

Beuchat CA (1990b) Metabolism and the scaling of urine concentrating ability in mammals: resolution of a paradox? J Theor Biol 143: 113–122

Beuchat CA (1996) Structure and concentrating ability of the mammalian kidney: correlations with habitat. Am J Physiol 271: R157–R179

Bintz GL, Roesbery HW (1978) Evaporative water loss by control and starved laboratory rats and *Spermophilus richardsoni*. Comp Biochem Physiol 59A: 275–278

Blake BH (1977) The effects of kidney structure and the annual cycle on water requirements in golden-mantled ground squirrels and chipmunks. Comp Biochem Physiol 58A; 413–419

Blix AS, Johnsen HK (1983) Aspects of heat exchange in resting reindeer. J Physiol 340: 445–454

Block W, Vannier G (1994) What is ecophysiology? Two perspectives. Acta Oecol (Berl) 15: 5–12

Bomford M (1987a) Food and reproduction of wild house mice I. A field experiment to examine the effect of food availability and food quality on the onset of breeding. Aust Wildl Res 14: 197–206

Bomford M (1987b) Food and reproduction of wild house mice III. Experiments on the breeding performance of caged house mice fed rice-based diets. Aust Wildl Res 14: 207–218

Bondi A (1982) Nutrition and animal productivity. In: Rechcigl M (ed) Handbook of agricultural productivity, vol II. CRC Press, Boca Raton, pp 195–215

Borut A, Shkolnik A (1974) Physiological adaptations to the desert environment. In: Robertshaw D (ed) MTP international review of science, physiology: series 1. Environmental physiology. Butterworth, London, pp 185–229

Borut A, Horowitz M, Castel M (1972) Blood volume regulation in the spiny mouse: capillary permeability changes due to dehydration. Symp Zool Soc Lond 31: 175–189

Borut A, Haim A, Castel M (1978) Non-shivering thermogenesis and implication of the thyroid in cold labile and cold resistant populations of the golden spiny mouse (*Acomys russatus*). In: Givardial L, Seydoux J (eds) Effectors of thermogenesis. Experientia Suppl, Bitkhäuser, Basel, pp 219–227

Boulière F (1975) Mammals, small and large: the ecological implications of size. In: Golley FB, Petrusewicz K, Ryszkowski L (eds) Small mammals: their productivity and population dynamics. Cambridge University Press, Cambridge, pp 1–8

Boyer DC (1987) Effect of rodents on plant recruitment and production in the dune area of the Namib Desert. MSc Thesis, Univ Natal, Pietermaritzberg, S Africa

Bozinovic F, Contreras LC (1990) Basal rate of metabolism and temperature regulation of two desert herbivorous octodonid rodents: *Octomys mimax* and *Tympanoctomys barrerea*. Oecologia (Berl) 84: 567–570

Bozinovic F, Rosenmann M (1988) Daily torpor in *Calomys musculinus*, a South American rodent. J Mammal 69: 150–152

Bozinovic F, Rosenmann M (1989) Maximum metabolic rate of rodents; physiological and ecological consequences on distributional limits. Funct Ecol 3: 173–181

Bradford DF (1974) Water stress of free-living *Peromyscus truei*. Ecology 55: 1407–1414

Bradford DF (1975) The effects of an artificial water supply on free-living *Peromyscus truei*. J Mammal 56: 705–707

Bradley SR, Deavers DR (1980) A re-examination of the relationship between thermal conductance and body weight in mammals. Comp Biochem Physiol 65A: 465–476

Bradley WG, Mauer RA (1971) Reproduction and food habits of Merriam's kangaroo rat, *Dipodomys merriami*. J Mammal 52: 497–507

Bradshaw SD, Bradshaw J, Lachiver F (1976) Taux de renouvellement d'eau et balance hydrique chez deux rongers desertiques, *Meriones shawii et Meriones libycus*, etudies dans leur environnement naturel en Tunisie. C R Acad Sci Paris D282: 481–484

Brand S, Abramsky Z (1987) Body masses of gerbilline rodents in sandy habitats of Israel. J Arid Environ 12: 247–253

Braun EJ (1985) Comparative aspects of the urinary concentrating process. Renal Physiol 8: 249–260

Breed WG (1990) Comparative studies on the timing of reproduction and foetal number in six species of Australian conlurine rodents (Muridae: Hydromyinae). J Zool (Lond) 221: 1–10

Breton L, Thomas DW, Bergeron J-M (1989) Digestive tract characteristics for meadow voles (*Microtus pennsylvanicus*) fed diets containing phenols and tannins. Mammalia 53: 656–659

Brody S (1945) Bioenergetics and growth. Reinhold, New York

Brody S, Proctor RC (1932) Growth and development with special reference to domestic animals. Further investigations of surface area in energy metabolism. Mo Agric Exp Stn Res Bull 116: 1–20

Bronson FH (1985) Mammalian reproduction: an ecological perspective. Biol Reprod 32: 1–26

Brooker B, Withers P (1994) Kidney structure and renal indices of dasyurid marsupials. Aust J Zool 42: 163–176

Brown JH, Bartholomew GA (1969) Periodicity and energetics of torpor in the kangaroo mouse, *Microdipodops pallidus*. Ecology 50: 705–709

Brown JH, Reichman OJ, Davidson DW (1979) Granivory in desert ecosystems. Annu Rev Ecol Syst 10: 201–227

Brown JS, Kotler BP, Smith RJ, Wirtz WO (1988) The effects of owl predation on the foraging behavior of heteromyid rodents. Oecologia (Berl) 76: 408–415

Brownfield MS, Wunder BA (1976) Relative medullary index: a new structural index for estimating urinary concentrating capacity of mammals. Comp Biochem Physiol 55A: 69–75

Brylski P (1993) The evolutionary morphology of heteromyids. In: Genoways HH, Brown JH (eds) Biology of the Heteromyidae, Spec Publ No 10. American Society of Mammalogists, Brigham Young University, Provo Utah, pp 357–385

Buffenstein R (1984) The importance of microhabit in thermoregulation and thermal conductance in two Namib rodents - a crevice dweller, *Aethomys namaquensis*, and a burrow dweller, *Gerbillurus paeba*. J Therm Biol 9: 235–241

Buffenstein R (1985) The effect of a high fibre diet on energy and water balance in two Namib desert rodents. J Comp Physiol B 155: 211–218

Buffenstein R, Jarvis JUM (1985) Thermoregulation and metabolism in the smallest African gerbil, *Gerbillus pusillus*. J Zool (Lond) 205: 107–205

Buffenstein R, Campbell WE, Jarvis JUM (1985) Identification of crystalline allantoin in the urine of African Cricetidae (Rodentia) and its role in their water economy. J Comp Physiol B 155: 493–499

Buret A, Hardin J, Olson ME, Gall DG (1993) Adaptation of the small intestine in desert-dwelling animals: morphology, ultrastructure and electrolyte transport in the jejunum of rabbits, rats, gerbils and sand rats. Comp Biochem Physiol 105A: 157–163

Burns TW (1956) Endocrine factors in the water metabolism of a desert mammal, *Gerbillus gerbillus*. Endocrinology 58: 243–254

Buttemer WA, Hayworth AM, Weathers WW, Nagy KA (1986) Time-budget estimates of avian energy expenditure: physiological and meteorological conditions. Physiol Zool 59: 35–44

Byman D (1985) Thermoregulatory behavior of a diurnal small mammal, the Wyoming ground squirrel (*Spermophilus elegans*). Physiol Zool 58: 705–718

Calder WA (1984) Size, function and life history. Harvard University Press, Cambridge

Calder WA, Braun EJ (1983) Scaling of osmotic regulation in mammals and birds. Am J Physiol 244: R601–R606

Calow P (1987) Towards a definition of functional ecology. Funct Ecol 1: 57–61

Canguilhem B, Marx C (1973) Regulation of the body weight of the European hamster during the annual cycle. Pfluegers Arch Eur J Physiol 338: 125–131

Caputa M (1979) Temperature gradients in the nasal cavity of the rabbit. J Therm Biol 4: 283–286

Carmi-Winkler N, Degen AA, Pinshow B (1987) Seasonal time-energy budgets of free-living chukars in the Negev Desert. Condor 89: 594–601

Carpenter RE (1966) A comparison of thermoregulation and water metabolism in the kangaroo rats *Dipodomys agilis* and *Dipodomys merriami*. Univ Calif Publ Zool 78: 1–36

Carpenter RE (1969) Structure and function of the kidney and the water balance of desert bats. Physiol Zool 42: 288–302

Cassuto Y (1968) Metabolic adaptations to chronic heat exposure in the golden hamster. Am J Physiol 214: 1147–1151

Castel M, Borut A, Haines H (1974) Blood titres of vasopressin in various murids (Mammalian: Rodentia). Isr J Zool 23: 208–209

Chappell MA (1984) Maximum oxygen consumption during exercise and cold exposure in deer mice. Respir Physiol 55: 367–377

Chappell MA, Bartholomew GA (1981a) Activity and thermoregulation of the antelope ground squirrel *Ammospermophilus leucurus* in winter and summer. Physiol Zool 54: 215–223

Chappell MA, Bartholomew GA (1981b) Standard operative temperatures and thermal energetics of the antelope ground squirrel *Ammospermophilus leucurus*. Physiol Zool 54: 81–93

Chew RM (1955) The skin and respiratory water losses of *Peromyscus maniculatus sonoriensis*. Ecology 36: 463–467

Chew RM, Chew AE (1970) Energy relationships of the mammals of a desert shrub (*Larrea tridentata*) community. Ecol Monogr 40: 1–21

Chew RM, Dammann AE (1961) Evaporative water loss of small vertebrates, as measured with an infrared analyzer. Science 133: 384–385

Choshniak I, Yahav S (1987) Can desert rodents better utilize low quality roughage than their non-desert kindred? J Arid Environ 12: 241–246

Choshniak I, Brosh A, Shkolnik A (1988) Productivity of Bedouin goats: coping with shortages of water and adequate food. In: Isotope aided studies on livestock productivity in Mediterranean and North African countries. International Atomic Energy Agency, Vienna, pp 47–64

Christian DP (1979a) Physiological correlates of demographic patterns in three sympatric Namib desert rodents. Physiol Zool 52: 329–339

Christian DP (1979b) Comparative demography of three Namib desert rodents: responses to the provision of supplementary water. J Mammal 60: 679–690

Christian DP, Matson JO, Rosenberg SG (1978) Comparative water balance in two species of *Liomys* (Rodentia: Heteromyidae). Comp Biochem Physiol 61A: 589–594

Clausen TP, Provenza FD, Burritt EA, Reichardt PB, Bryant JP (1990) Ecological implications of condensed tannin structure: a case study. J Chem Ecol 16: 2381–2392

Cock MJW (1978) The assessment of preference. J Anim Ecol 47: 805–816

Cole P (1953) Further observations on the conditioning of respiratory air. J Laryngol Otol 67: 669–681

Collins BG, Bradshaw SD (1973) Studies on the metabolism, thermoregulation and evaporative water loss of two species of Australian rats, *Rattus villosissimus* and *Rattus rattus*. Physiol Zool 46: 1–21

Collins JC, Pilkington TC, Schmidt-Nielsen K (1971) A model of respiratory heat transfer in a small mammal. Biophys J 11: 886–914

Congdon JD, King WW, Nagy KA (1978) Validation of the HTO[18] method for determination of CO_2 production of lizards (genus *Sceloporous*). Copeia 1978: 369–362

Conley KE, Porter WP (1985) Heat loss regulation: role of appendages and torso in the deer mouse and the white rabbit. J Comp Physiol B 155: 423–431

Cooke BD (1982) Reduction of food intake and other physiological responses to a restriction of drinking water in captive wild rabbits, *Oryctolagus cuniculus* (L.). Aust Wildl Res 9: 247–252

Costa DP (1984) Assessment of the impact of the California sea lion and northern elephant seal on commercial fisheries. California Sea Grant College Program Biennial Report, 1982–84. Institute of Marine Resources, University of California, San Diego

Costa DP, Gentry RL (1986) Free-ranging energetics of northern fur seals. In: Gentry RL, Kooyman GL (eds) Fur seals: maternal strategies on land and at sea. Princeton University Press, Princeton, pp 79–101

Costa DP, Prince PA (1987) Foraging energetics of grey-headed albatrosses *Diomedea chrysostoma* at Bird Island, South Georgia. Ibis 129: 149–158

Costa DP, Dann P, Disher W (1986) Energy requirements of free ranging little penguins, *Eudyptula minor*. Comp Biochem Physiol 85A; 135–138

Costa DP, Croxall JP, Duck CD (1989) Foraging energetics of Antarctic fur seals in relation to changes in prey availability. Ecology 70: 596–606

Coward WA, Prentice AM, Murgatroyd PR, Davies HL, Cole TJ, Sawyer M, Goldberg GR, Halliday D, MacNamara (1985a) Measurement of CO_2 and water production rates in man using ^2H, ^{18}O labelled H_2O: comparisons between calorimeter and isotope values. In: van Es AJH (ed) Human energy metabolism: physical activity and energy expenditure measurements in epidemiological research based upon direct and indirect calorimetry. Eur Nutr Rep 5, CIP-gegevens Koninklijke Bibliotheek, The Hague

Coward WA, Roberts SB, Prentice AM, Lucas A (1985b) The ^2H, ^{18}O method for energy expenditure measurements – clinical possibilities, necessary assumptions and limitations. In: Proc 7th Congr/Eur Soc Enteral and Parenteral Nutr, Munich, 1985

Daly M, Daly S (1975a) Behavior of *Psammomys obesus* (Rodentia: Gerbillinae) in the Algerian Sahara. Z Tierpsychol 37: 298–321

Daly M, Daly S (1975b) Socio-ecology of Saharan gerbils, especially *Meriones libycus*. Mammalia 39: 289–311

Dark J, Zucker I, Wade GN (1983) Photoperiodic regulation of body mass, food intake, and reproduction in meadow voles. Am J Physiol 245: R334–R338

Dawson TJ, Hulbert AJ (1970) Standard metabolism, body temperature, and surface areas of Australian marsupials. Am J Physiol 218: 1233–1238

Dawson WR (1955) The relation of oxygen consumption to temperature in desert rodents. J Mammal 36: 543–553

Deavers DR, Hudson JW (1979) Water metabolism and estimated field water budgets in rodents (*Clethrionomys gapperi* and *Peromyscus leucopus*) and an insectivore (*Blarina brevicauda*) inhabiting the same mesic environment. Physiol Zool 52: 137–152

Degen AA (1977) Responses to dehydration in native fat-tailed Awassi and imported German mutton merino sheep. Physiol Zool 50: 284–293

Degen AA (1988) Ash and electrolyte intakes of the fat sand rat, *Psammomys obesus*, consuming saltbush, *Atriplex halimus*, containing different water content. Physiol Zool 61: 137–141

Degen AA, Kam M (1991) Average daily metabolic rate of gerbils of two species: *Gerbillus pyramidum* and *Gerbillus allenbyi*. J Zool (Lond) 223: 143–149

Degen AA, Kam M (1995) Scaling of field metabolic rate to basal metabolic rate ratio in homeotherms. Ecoscience 2: 48–54

Degen AA, Young BA (1984) Effects of ingestion of warm, cold and frozen water on heat balance in cattle. Can J Anim Sci 64: 73–80

Degen AA, Pinshow B, Alkon PU, Arnon H (1981) Tritiated water for estimating total body water and water turnover rate in birds. J Appl Physiol 51: 1183–1188

Degen AA, Kam M, Hazan A, Nagy KA (1986) Energy expenditure and water flux in three sympatric desert rodents. J Anim Ecol 55: 421–429

Degen AA, Elias E, Kam M (1987) A preliminary report on the energy intake and growth rate of early-weaned camel (*Camelus dromedarius*) calves. Anim Prod 45: 301–306

Degen AA, Kam M, Jurgrau D (1988) Energy requirements of fat sand rats (*Psammomys obesus*) and their efficiency of utilization of the saltbush *Atriplex halimus* for maintenance. J Zool (Lond) 215: 443–452

Degen AA, Pinshow BP, Ilan M (1990) Seasonal water flux, urine and plasma osmotic concentrations in free-living fat sand rats feeding solely on saltbush. J Arid Environ 18: 59–66

Degen AA, Hazan A, Kam M, Nagy KA (1991) Seasonal water influx and energy expenditure of free-living fat sand rats. J Mammal 72: 652–657

Degen AA, Pinshow B, Kam M (1992) Field metabolic rates and water influxes of two sympatric Gerbillidae: *Gerbillus allenbyi* and *G. pyramidum*. Oecologia (Berl) 90: 586–590

Dehnel A (1949) Studies on the genus *Sorex* L. Ann Univ Mariae Curie-Sklodowska Sec C Biol 4: 17–102

Demment MW, Van Soest PJ (1985) A nutritional explanation for body-size patterns of ruminant and nonruminant herbivores. Am Nat 125: 641–672

Denny MJS, Dawson TJ (1975) Effects of dehydration on bodywater distribution in desert kangaroos. Am J Physiol 229: 251–254

Derting TL (1989) Metabolism and food availability as regulators of production in juvenile cotton rats. Ecology 70: 587–595

Dick-Peddie WA (1991) Semiarid and arid lands: a worldwide scope. In: Skujins J (ed) Semiarid lands and deserts: soil resources and reclamation. Marcel Dekker, New York, pp 3–32

Dippenaar NJ, Rautenbach IL (1986) Morphometrics and karyology of the southern African species of the genus *Acomys* I. Geoffroy Saint-Hilaire, 1838 (Rodentia: Muridae). Ann Transvaal Mus 34: 129–183

Dolph CI, Braun HA, Pfeiffer EW (1962) The effect of vasopressin upon urine concentration in *Aplodontia rufa* (Sewellel) and the rabbit. Physiol Zool 35: 263–269

Downs CT, Perrin MR (1989a) An investigation of the macro- and micro-environments of four *Gerbillurus* species. Cimbebasia 11: 41–54

Downs CT, Perrin MR (1989b) Thermal parameters of four species of *Gerbillurus*. J Therm Biol 15: 291–300

Downs CT, Perrin MR (1990) Field water-turnover rates of three *Gerbillurus* species. J Arid Environ 19: 199–208

Drabek CM (1973) Home range daily activity of the round-tailed ground squirrel, *Spermophilus tereticaudus neglectus*. Am Midl Nat 89: 287–293

Drodz A (1968) Digestibility and assimilation of natural foods in small rodents. Acta Theriol 13: 367–389

Du Plessis A, Erasmus T, Kerley GIH (1989) Thermoregulatory patterns of two sympatric rodents: *Otomys unisulcatus* and *Parotomys brantsii*. Comp Biochem Physiol 94A: 215–220

Du Plessis A, Kerley GIH, Deo Winter PE (1992) Refuge microclimates of rodents: a surface nesting *Otomys unisulcatus* and a burrowing *Parotomys brantsii*. Acta Theriol 37: 351–358

Dykstra CR, Karasov WH (1993) Daily energy expenditure by nestling house wrens. Condor 95: 1028–1030

Ebensperger L, Bozinovic F, Rosenmann M (1990) Average daily metabolic rate as predictor of energy expenditure in free-ranging rodents. Rev Chil Hist Nat 63: 83–89

Edwards BA, Donaldson K, Simpson AP (1983) Water balance and protein intake in the Mongolian gerbil (*Meriones unguiculatus*). Comp Biochem Physiol 76A: 807–815

Edwards RM, Haines H (1978) Effects of water vapor pressure and temperature on evaporative water loss in *Peromyscus maniculatus* and *Mus musculus*. J Comp Physiol 128: 177–184

Eisenberg JF (1975) The behavior patterns of desert rodents. In: Prakash I, Ghosh PK (eds) Rodents in desert environments. Dr W Junk, The Hague, pp 189–224

Elgar MA, Harvey PH (1987) Basal metabolic rates in mammals: allometry, phylogeny and ecology. Funct Ecol 1: 25–36

El Husseini M, Haggag G (1974) Antidiuretic hormone and water conservation in desert rodents. Comp Biochem Physiol 47A: 374–350

Elia M, Livesey G (1988) Theory and validity of indirect calorimetry during net lipid synthesis. Am J Clin Nutr 47: 591–607

Ellis HI (1984) Energetics of free-ranging seabirds. In: Whittow GC, Rahn H (eds) Seabird energetics. Plenum Press. New York, pp 203–234

Ellison GTH (1993) Evidence of climatic adaptation in spontaneous torpor among pouched mice *Saccostomus campestris* from southern Africa. Acta Theriol 38: 49–59

Ellison GTH, Bronner GN, Taylor PJ (1993) Is the annual cycle in body weight of pouched mice (*Saccostomus campestris*) the result of seasonal changes in adult size or population structure? J Zool (Lond) 229: 545–551

Emberger (1942) Un project d'une classification des climats, du point de vue phytogeographique. Bull Soc Hist Nat Toulouse 77: 97–124

Emmans GC (1994) Effective energy: a concept of energy utilization applied across species. Br J Nutr 71: 801–821

Eriksson L, Valtonen M, Makela J (1984) Water and electrolyte balance in male mink (*Mustela vison*) on varying dietary NaCl intake. Acta Physiol Scand Suppl 537: 59–64

Evenari M (1981) Ecology of the Negev Desert, a critical review of our knowledge. In: Shuval H (ed) Developments in arid zone ecology and environmental quality. Balaban ISS, Philadelphia, pp 1–33

Evenari M, Shanan L, Tadmor N (1971) The Negev – the challenge of a desert. Harvard University Press, Cambridge

Feder ME (1987) The analysis of physiological diversity: the prospects for pattern documentation and general questions in ecological physiology. In: Feder ME, Bennett AF, Burggren WW, Huey RB (eds) New directions in ecological physiology. Cambridge University Press, Cambridge, pp 38–75

Feder ME, Block BA (1991) On the future of animal physiological ecology. Funct Ecol 5: 136–144

Feder ME, Bennett AF, Burggren WW, Huey RB (eds) (1987) New directions in ecological physiology. Cambridge University Press, Cambridge

Feist DD, Feist CF (1986) Effects of cold, short day and melatonin on thermogenesis, body weight and reproductive organs in Alaskan red-backed voles. J Comp Physiol 156B: 741–746

Feist DD, Morrison PR (1981) Seasonal changes in metabolic capacity and norepinephrine thermogenesis in the Alaskan red-backed voles: environmental cues and annual differences. Comp Biochem Physiol 69A: 697–700

Field AC (1975) Seasonal changes in reproduction, diet and body composition in two equatorial rodents. East Afr Wildl J 13: 221–235

Fielden LJ, Perrin MR, Hickman GC (1990a) Water metabolism in the Namib desert golden mole, *Eremitalpa granti namibensis* (Chrysochloridae). Comp Biochem Physiol 96A: 227–234

Fielden LJ, Waggoner JP, Perrin MR, Hickman GC (1990b) Thermoregulation in the Namib golden mole, *Eremitalpa granti namibensis* (Chrysochloridae). J Arid Environ 18: 221–237

Finch VA, Dmi'el R, Boxman R, Shkolnik A, Taylor CR (1980) Why black coats in hot deserts? Effects of coat color and heat exchange of wild and domestic goats. Physiol Zool 53: 19–25

Fleming TH (1977) Response of two species of tropical heteromyid rodents to reduced food and water availability. J Mammal 58: 102–106

Fokin IM (1978) Features of jerboas' locomotion. Bull Moscow Soc Nat Sect Biol 68, 5: 22–28 (in Russian)

Foley WJ (1987) Digestion and energy metabolism in a small arboreal marsupial, the greater gilder (*Petauroides volans*), fed high terpene *Eucalyptus foliage*. J Comp Physiol 157: 355–362

Foley WJ Kehl JC, Nagy KA, Kaplan R, Borsboom AC (1990) Energy and water metabolism in free-living greater gliders *Petauroides volans*. Aust J Zool 38: 1–9

Folkow LP, Blix AS (1987) Nasal heat exchange in gray seals. Am J Physiol 253: R883–R889

Forman GL, Phillips CJ (1988) Histological variation in the proximal colon of heteromyid and cricetid rodents. J Mammal 69: 144–149

Forman GL, Phillips CJ (1993) The proximal colon of heteromyid rodents: possible morpho-physiological correlates to enhanced water conservation. In: Genoways HH, Brown JH (eds) Biology of the Heteromyidae, Spec Publ No 10 American Society of Mammalogists, Brigham Young University, Provo, Utah, pp 491–508

Fraguedakis-Tsolis SE, Chandropoulos BP, Nikoletopoulos, NP (1993) On the phylogeny of the genus *Acomys* (Mammalia: Rodentia). Z Saeugetierkd 58: 240–243

Frank CL (1988a) The influence of moisture content on seed selection by kangaroo rats. J Mammal 69: 353–357

Frank CL (1988b) The relationship of water content, seed selection, and the water requirements of a heteromyid rodent. Physiol Zool 61: 527–534

Frank CL (1988c) Diet selection by a heteromyid rodent: role of net metabolic water production Ecology 69: 1943–1951

French AR (1976) Selection of high temperatures for hibernation by the pocket mouse. *Perognathus longimembris*: ecological advantages and energetic consequences. Ecology 57: 185–191

French AR (1977a) Circannual rhythmicity and entrainment of surface activity in the hibernator, *Perognathus longimembris*. J Mammal 58: 37–43

French AR (1977b) Periodicity of recurrent hypothermia during hibernation in the pocket mouse, *Perognathus longimembris*. J Comp Physiol 115: 87–90

French AR (1989) Seasonal variation in the use of torpor by pallid kangaroo mice, *Microdidpodops pallidus*. J Mammal 839–842

French AR (1993) Physiological ecology of the Heteromyidae: economics of energy and water utilization. In: Genoways HH, Brown JH (eds) Biology of the Heteromyidae, Spec Publ No 10. American Society of Mammalogists, Brighan Young University, Provo, Utah, pp 509–538

French NR, Grant WE, Grodzinski W, Swift DM (1976) Small mammal energetics in grassland ecosystems. Ecol Monogr 46: 201–220

Friedman MI, Bruno JP (1976) Exchange of water during lactation. Science 191: 409–410

Friedman MI, Bruno JP, Alberts JR (1981) Physiological and behavioral consequences in rats of water recycling during lactation. J Comp Phychol 95: 26–35

Frumkin R, Pinshow B, Weinstein Y (1986) Metabolic heat production and evaporative heat loss in desert phasianids: chukar and sand partridge. Physiol Zool 59: 592–605

Furstenburg D, van Hoven W (1994) Condensed tannin as anti-defoliate agent against browsing by giraffe (*Giraffa camelopardalis*) in the Kruger National Park. Comp Biochem Physiol 107A: 425–431

Gabrielsen GW, Taylor JRE, Konarzewski M, Mehlum F (1991) Field and laboratory metabolism and thermoregulation in dovekies (*Alle alle*). Auk 108: 71–78

Gebczynska Z (1970) Bioenergetics of a root vole population. Acta Theriol 15: 33–66

Gebczynska Z, Gebczynski M (1965) Oxygen consumption in two species of water-shrews. Acta Theriol 10: 209–214

Gebczynski M (1964) Effect of light and temperature on the 24-hour rhythm in *Pitymys subterraneus* (de Sel.-Long.). Acta Theriol 9: 125–137

Gebczynski M (1965) Seasonal and age changes in the metabolism and activity of *Sorex araneus* Linnaeus 1758. Acta Theriol 10: 303–331

Gebczynski M (1966) The daily energy requirement of the yellow-necked field mouse in different seasons. Acta Theriol 11: 391–398

Gebczynski M (1971) The rate of metabolism of the lesser shrew. Acta Theriol 20: 329–339

Gebczynski M, Gorecki A, Drodz A (1972) Metabolism, food assimilation and bioenergetics of three species of dormice (Gliridae). Acta Theriol 17: 271–294

Geffen E, Degen AA, Kam M, Hefner R, Nagy KA (1992a) Daily energy expenditure and water flux of free-living Blanford's fox (*Vulpes cana*), a small desert carnivore. J Anim Ecol 61: 611–617

Geffen E, Hefner R, Macdonald DW, Ucko M (1992b) Diet and foraging behaviour of Blanford's fox, *Vulpes cana*, in Israel. J Mammal 73: 395–402

Geffen E, Mercure A, Girman DJ, Macdonald DW, Wayne RK (1992c) Phylogenetic relationships of the fox-like canids: mitochondrial DNA restriction fragment, site and cytochrome *b* sequence analyses. J Zool (Lond) 228: 27–39

Gehr P, Mwangi DK, Ammann A, Maloiy GMO, Taylor CR, Weibel ER (1981) Design of the mammalian respiratory system. V. Scaling morphometric pulmonary diffusing capacity to body mass; wild and domestic mammals. Respir Physiol 44: 61–86

Geiser F (1994) Hibernation and daily torpor in marsupials: a review. Aust J Zool 42: 1–16

Geist V (1987) Bergmann's rule is invalid. Can J Zool 65: 1035–1038

Geist V (1990) Bergmann's rule is invalid; a reply to J.D.Paterson. Can J Zool 68: 1613–1615

Geluso KN (1975) Urine concentration cycles of insectivorous bats in the laboratory. J Comp Physiol 99: 309–319

Geluso KN (1978) Urine concentrating ability and renal structure of insectivorous bats. J Mammal 59: 312–323

Genoud M (1985) Ecological energetics of two European shrews: *Crocidura russula* and *Sorex coronatus* (Soricidae: Mammalia). J Zool (Lond) 207: 63–85

Genoud M (1988) Energetic strategies of shrews: ecological constraints and evolutionary implications. Mammal Rev 18: 173–193

Genoud M, Vogel P (1990) Energy requirements during reproduction and reproductive effort in shrews (Soricidae). J Zool (Lond) 220: 41–60

Genoways HH, Brown JH (eds) (1993) Biology of the Heteromyidae, Spec Publ No 10. American Society of Mammalogists, Brigham Young University, Provo, Utah

Gessaman JA (ed) (1973) Ecological energetics of homeotherms: a view compatible with ecological modeling. Utah State University Monogr Ser, No 20

Gessaman JA, Nagy KA (1988) Energy metabolism: errors in gas-exchange conversion factors. Physiol Zool 61: 507–513

Gettinger RD (1983) Use of doubly-labelled water ($^3HH^{18}O$) for determination of H_2O flux and CO_2 production by a mammal in a humid environment. Oecologia (Berl) 59: 54–57

Gettinger RD (1984) Energy and water metabolism of free-ranging pocket gophers, *Thomomys bottae*. Ecology 65: 740–751

Gettinger RD, Arnold P, Wunder BA, Ralph CL (1986) Seasonal effects of temperature and photoperiod on thermogenesis and body mass of the kangaroo rat (*Dipodomys ordii*). In: Heller HC, Musacchia XJ, Wang LCH (eds) Living in the cold. Elsevier, New York, pp 505–509

Getz LL (1968) Relationships between ambient temperature and respiratory water loss of small mammals. Comp Biochem Physiol 24: 335–342

Ghobrial LI (1970) The water relations of the desert antelope *Gazella dorcas dorcas*. Physiol Zool 43: 249–256

Ghobrial LI, Hodieb ASK (1973) Climate and seasonal variations in the breeding of the desert jerboa, *Jaculus jaculus,* in the Sudan. J Reprod Fert Suppl 19: 221–233

Ghosh PK, Goyal SP, Prakash I (1979) Metabolism and ecophysiology of Rajasthan desert rodents. Thermoregulation at a moderately low temperature (21 °C) during winter. J Arid Environ 2: 77–83

Gittleman JL (1988) Behavioral energetics of lactation in a herbivorous carnivore, the red panda (*Ailurus fulgens*). Ethology 79: 13–24

Gittleman JL, Thompson SD (1988) Energy allocation in mammalian reproduction. Am Zool 28; 863–875

Glazier DS (1985) Relationship between metabolic rate and energy expenditure for lactation in *Peromyscus*. Comp Biochem Physiol 80A: 587–590

Goldstein DL (1988) Estimates of daily energy expenditure in birds: the time energy budget as an integrator of laboratory and field studies. Am Zool 28: 829–844

Goldstein DL, Nagy KA (1985) Resource utilization by desert quail: time and energy, food and water. Ecology 66: 378–387

Golightly RT, Ohmart RD (1984) Water economy of two desert canids, coyote and kit fox. J Mammal 65: 51–58

Golley FB (1960) Energy dynamics of a food chain of an old-field community. Ecol Monogr 30: 187–206

Gorecki A (1968) Metabolic rate and energy budget in the bank vole. Acta Theriol 13: 341–365

Gorecki A (1969) Metabolic rate and energy budget of the striped field mouse. Acta Theriol 14: 181–190

Gorecki A (1971) Metabolism and energy budget in the harvest mouse. Acta Theriol 16: 213–220

Goyal SP, Ghosh PK (1983) Body weight exponents of metabolic rate and minimal thermal conductance in burrowing desert rodents. J Arid Environ 6: 43–52

Goyal SP, Ghosh PK (1993) Burrow structure of two gerbil species of Thar desert, India. Acta Theriol 8: 453–456

Goyal SP, Ghosh PK, Prakash I (1981) Significance of body fat in relation to basal metabolic rate in some Indian desert rodents. J Arid Environ 4: 59–62

Graaf GD, Nel JAJ (1965) On the tunnel system of Brant's Karoo rat, *Parotomys brantsii*, in the Kalahari Gemsbok National Park. Koedoe 9: 136–139

Green B, Rowe-Rowe DT (1987) Water and energy metabolism in free-living multi-mammate mice, *Praomys natalensis*, during summer. S Afr J Zool 22: 14–17

Greenwald L (1989) The significance of renal relative medullary thickness. Physiol Zool 62: 1005–1014

Greger R, Land F, Deetjen P (1975) Handling of allantoin by the rat kidney. Clearance and micropuncture data. Pfluegers Arch Eur J Physiol 357: 201–207

Grenot CJ, Buscarlet LA (1988) Validation and use of isotope turnover to measure metabolism in free-ranging vertebrates. J Arid Environ 14: 211–232

Grenot C, Serrano V (1979) Vitesse de renouvellement d'eau chez cinq espèces de rongers déserticoles et sympatriques étudiées à la saison sèche dans leur milieu naturel (Desert de Chihuahua, Mexique). C R Acad Sci Paris D288: 104

Grenot C, Pascal M, Buscarlet L, Francaz JM, Sellami M (1984) Water and energy balance in the water vole (*Arvicola terrestris* Sherman) in the laboratory and in the field (Haut-Doubs, France).Comp Biochem Physiol 78A: 185–196

Grodzinski W (1971) Energy flow through populations of small mammals in the Alaskan taiga forest. Acta Theriol 16: 231–275

Grodzinski W, Weiner J (1984) Energetics of small and large mammals. Acta Zool Fenn 172: 7–10

Grodzinski W, Wunder BA (1975) Ecological energetics of small mammals. In: Golley FB, Petrusewicz K, Ryszkowski L (eds) Small mammals: their productivity and population dynamics. Cambridge University Press, Cambridge, pp 173–204

Grubbs DE (1980) Tritiated water turnover in free-living desert rodents. Comp Biochem Physiol 66A: 89–98

Gubernick DJ, Alberts JR (1983) Maternal licking of young: resource exchange and proximate controls. Physiol Behav 31: 593–601

Gubernick DJ, Alberts JR (1985) Maternal licking by virgin and lactating rats: water transfer from pups. Physiol Behav 34: 501–506

Gubernick DJ, Alberts JR (1987) "Resource" exchange in the biparental California mouse (*Peromyscus californicus*): water transfer from pups to parents. J Comp Psychol 101: 328–334

Hafner JC (1993) Macroevolutionary diversification in heteromyid rodents: heterochrony and adaptation in phylogeny. In: Genoways HH, Brown JH (eds) Biology of the Heteromyidae, Spec Publ No 10. American Society of Mammalogists, Brigham Young University, Provo, Utah, pp 291–318

Hagelstein KA, Folk GE (1978) Effects of photoperiod, cold acclimation and melatonin on the white rat. Comp Biochem Physiol 62C: 225–229

Hagerman AE (1989) Chemistry of tannin-protein complexation. In: Hemingway RW, Karchesy JJ (eds) Chemistry and significance of condensed tannins. Plenum, New York, pp 323–333

Hails CJ, Bryant DM (1979) Reproductive energetics of a free-living bird. J Anim Ecol 48: 471–482

Haim A (1981) Heat production and dissipation in a South-African diurnal murid *Lemniscomys griselda*. S Afr J Zool 16: 67–70

Haim A (1984) Adaptive variations in heat production within gerbils (genus *Gerbillus*) from different habitats. Oecologia (Berl) 61: 49–52

Haim A (1987) Thermoregulation and metabolism of Wagner's gerbil (*Gerbillus dasyurus*): a rock dwelling rodent adapted to arid and mesic environments. J Therm Biol 12: 45–48

Haim A, Borut A (1981) Heat production and dissipation in golden spiny mice *Acomys russatus* from two extreme habitats. J Comp Physiol 142B: 445–450

Haim A, Borut A (1986) Reduced heat production in the bushy tailed gerbil *Seeketamys calurus* as an adaptation to arid environments. Mammalia 50: 27–33

Haim A, Fairall N (1986) Geographical variations in heat production and dissipation within two populations of *Rhabdomys pumilio* (Muridae). Comp Biochem Physiol 84A: 111–112

Haim A, Fairall N(1987) Bioenergetics of a herbivorous rodent *Otomys irroratus* (Brants, 1827). Physiol Zool 60: 305–309

Haim A, Harari J (1992) A comparative study of heat prodution and thermoregulation in two sympatric gerbils (*Gerbillus gerbillus* and *G. pyramidum*). Isr J Zool 38; 363–372

Haim A, Izhaki I (1993) The ecological significance of resting metabolic rate and non-shivering thermogenesis for rodents. J Therm Biol 18: 71–81

Haim A, Le Fourie R (1980) Heat production in nocturnal (*Praomys natalensis*) and diurnal (*Rhabdomys pumilio*) South African murids. S Afr J Zool 15: 91–94

Haim A, Levi G (1990) Role of body temperature in seasonal acclimatization: photoperiod-induced rhythms and heat production in *Meriones crassus*. J Exp Zool 256: 237–241

Haim A, Heth G, Avnon Z, Nevo E (1984) Adaptive physiological variation in non-shivering thermogenesis and its significance in speciation. J Comp Physiol 154B: 145–147

Haim, A, Pelaot I, Sela A (1986) Comparison of ecophysiological parameters between two *Apodemus* species coexisting in the same habitat. In: Dubinsky Z, Steinberger Y (eds) Environmental quality and ecosystem stability, vol III A/B. Bar-Ilan University Press, Tel-Aviv, pp 33–40

Haim A, Skinner JD, Robinson TJ (1987) Bioenergetics, thermoregulation and urine analysis of squirrels of the genus *Xerus* from an arid environment. S Afr J Zool 22: 45–49

Haines H, McKenna M (1988) Glomerular filtration rate in wild mice, *Mus musculus*, acclimated to water scarcity. Comp Biochem Physiol 89A: 339–341

Haines H, Schmidt-Nielsen B (1977) Kidney function in spiny mice *Acomys cahirinus* acclimated to water restriction. Bull Mt Desert Isl Biol Lab 17: 94–95

Haines H, Ciskowski C, Harms V (1973) Acclimation to chronic water restriction in the wild house mouse *Mus musculus*. Physiol Zool 46: 110–128

Hainsworth FR (1967) Saliva spreading, activity, and body temperature regulation in the rat. Am J Physiol 212: 1288–1292

Hainsworth FR (1968) Evaporative water loss from rats in the heat. Am J Physiol 214: 979–982

Hainsworth FR (1995) Optimal body temperatures with shuttling: desert antelope ground squirrels. Anim Behav 49: 107–116

Hainsworth FR, Stricker EM (1970) Salivary cooling by rats in the heat, Chap 41. In: Hardy JD, Pharo Gagge A, Stolwijk JAJ (eds) Physiological and behavioral temperature regulation. Charles C Thomas, Springfield, pp 611–626

Hainsworth FR, Stricker EM (1972) Evaporative cooling in the rat: further consideration of functional differences between salivary glands. Can J Physiol Pharmacol 50: 172–175

Hainsworth FR, Stricker EM, Epstein AN (1968) Water metabolism of rats in the heat: dehydration and drinking. Am J Physiol 214: 983–989

Handler JS, Orloff J (1973) The mechanism of action of antidiuretic hormone, Chap 24. In: Orloff J, Berliner RW (eds) Handbook of physiology, Sect 8, Renal physiology. American Physiological Society, Washington, DC, pp 791–814

Hansard SL (1964) Total body water in farm animals. Am J Physiol 206: 1369–1372

Hanski I (1985) What does a shrew do in an energy crisis? In: Smith RH, Sibley RM (eds) Behavioural ecology: symposia of British Ecological Society. Blackwell, Oxford, pp 247–252

Happold DCD (1984) Small mammals, chap 17. In: Cloudsley-Thompson JL (ed) Sahara Desert: key environments. Pergamon Press, London, pp 251–257

Harmeyer J, Martens H (1980) Aspects of urea metabolism in ruminants with reference to the goat. J Dairy Sci 63: 1707–1728

Harris DV, Walker JM, Berger RJ (1984) A continuum of slow-wave sleep and shallow torpor in the pocket mouse *Perognathus longimembris*. Physiol Zool 57: 428–434

Harris GD, Huppi HD, Gessaman JA (1985) The thermal conductance of winter and summer pelage of *Lepus californicus*. J Therm Biol 10: 79–81

Hart JS (1971) Rodents. In: Whittow GC (ed) Comparative physiology of thermoregulation. Academic Press, New York, pp 1–149

Hartman LA, Morton ML (1973) Plasma volume changes in ground squirrels during water deprivation. Comp Biochem Physiol 46A: 79–94

Harvey PH (1986) Energetic costs of reproduction. Nature 321: 648–649

Harvey PH, Nee S (1993) New uses for new phylogenies. Eur Rev 1: 11–19

Harvey PH, Purvis A (1991) Comparative methods for explaining adaptations. Nature 352: 619–624

Harvey PH, Pagel MD, Rees JA (1991) Mammalian metabolism and life histories. Am Nat 137: 556–566

Hayden P, Lindberg RG (1970) Hypoxia-induced torpor in pocket mice (genus: *Perognatus*). Comp Biochem Physiol 33: 167–179

Hayes JP (1989) Altitudinal and seasonal effects on aerobic metabolism of deer mice. J Comp Physiol 159: 453–459

Hayes JP, Chappel MA (1986) Effects of cold acclimation on maximum oxygen consumption during cold exposure and treadmill exercise in deer mice. *Peromyscus maniculatus*. Physiol Zool 59: 473–481

Hayssen V (1984) Basal metabolic rate and the intrinsic rate of increase: an empirical and theoretical reexamination. Oecologia (Berl) 64: 419–421

Hayssen V, Lacy RC (1985) Basal metabolic rates in mammals: taxonomic differences in the allometry of BMR and body mass. Comp Biochem Physiol 81A: 741–754

Heldmaier G, Hoffman K (1974) Melatonin stimulates growth of brown adipose tissue. Nature 247: 224–225

Heldmaier G, Steinlechner S (1981) Seasonal control of energy requirements for thermoregulation in the Djungarian hamster (*Phodupus sungorus*), living in natural photoperiod. J Comp Physiol 142: 429–437

Heldmaier G, Steinlechner S, Rafael J, Latteier B (1982) Photoperiod and ambient temperature cues for seasonal thermogenic adaptation in Djungarian hamster, *Phodupus sungorus*. Int J Biometeorol 26: 339–345

Heldmaier G, Boeckler A, Buchberger A, Klaus S, Puchalski W, Steinlechner S (1986) Seasonal variation in thermogenesis. In: Heller HC, Musacchia XJ, Wang LCH (eds) Living in the cold. Elsevier, New York, pp 361–372

Heldmaier G, Steinlechner S, Ruf T, Wiesinger H, Klingenspor M (1989) Photoperiod and thermoregulation in vertebrates: body temperature and thermogenic acclimation. J Biol Rhythms 4: 251–256

Helversen OV, Reyer HU (1984) Nectar intake and energy expenditure in a flower-visiting bat. Oecologia (Berl) 63: 178–184

Hemingway A (1963) Shivering. Physiol Rev 43: 397–422

Henneman WW (1983) Relationship among body mass, metabolic rate and the intrinsic rate of natural increase in mammals. Oecologia (Berl) 56: 104–108

Henneman WW (1984a) Intrinsic rates of natural increase of altricial and precocial eutherian mammals: the potential price of precociality. Oikos 43: 363–367

Henneman WW (1984b) Commentary. Oecologia (Berl) 64: 421–423

Herreid CF, Kessel B (1967) Thermal conductance in birds and mammals. Comp Biochem Physiol 21: 405–414

Heusner AA (1982) Energy metabolism and body size. I. Is the 0.75 mass exponent of Kleiber's equation a statistical artefact? Respir Physiol 48: 1–12

Heusner AA (1991) Size and power in mammals. J Exp Biol 160: 25–54

Hewitt S (1981) Plasticity of real function in the Australian desert rodent *Notomys alexis*. Comp Biochem Physiol 69A: 297–304

Hill RW (1978) Exhalant air temperatures in the Virginia opposum. J Therm Biol 3: 219–221

Hill RW, Christian DP, Veghte JH (1980) Pinna temperature in exercising jackrabbits, *Lepus californicus*. J Mammal 61: 30–38

Hillenius WJ (1992) The evolution of nasal turbinates and mammalian endothermy. Paleobiology 18: 17–29

Hinds DS (1973) Acclimatization of thermoregulation in the desert cottontail, *Sylvilagus audubonii*. J Mammal 54: 708–728

Hinds DS (1977) Acclimatization of thermoregulation in desert-inhabiting jackrabbits (*Lepus alleni* and *Lepus californicus*). Ecology 58: 246–264

Hinds DS, MacMillen RE (1983) Oxygen consumption, evaporative water loss and conductance in thirteen species of heteromyid rodents. J Comp Physiol 153: 152–164

Hinds DS, MacMillen RE (1984) Energy scaling in marsupials and eutherians. Science 225: 335–337

Hinds DS, MacMillen RE (1985) Scaling of energy metabolism and aerobic capacity in hetermoyid rodents. Physiol Zool 58: 282–298

Hinds DS, Rice-Warner CN (1992) Maximum metabolism and aerobic capacity in hetermoyid and other rodents. Physiol Zool 65: 188–214

Hofmann RR (1989) Evolutionary steps of ecophysiological adaptation and diversification of ruminants: a comparative view of their digestive system. Oecologia (Berl) 78: 443–457

Holleman DF, Dieterich RA (1973) Body water content and turnover in several species of rodents, as evaluated by the tritiated water method. J Mammal 54: 456–465

Holleman DF, White RG, Feist DD (1982) Seasonal energy and water metabolism in free-living Alaskan voles. J Mammal 63: 293–296

Horowitz M, Adler JH (1983) Plasma volume regulation during heat stress: albumin synthesis vs capillary permeability. A comparison between desert and non-desert species. Comp Biochem Physiol 75A: 105–110

Horowitz M, Borut A (1970) Effect of acute dehydration on body fluid compartments in three rodent species, *Rattus norvegicus, Acomys cahirinus* and *Meriones crassus*. Comp Biochem Physiol 35: 283–290

Horowitz M, Borut A (1975) Blood volume regulation in dehydrated rodents: plasma proteins turnover and sedimentation coefficients. Comp Biochem Physiol 51A: 827–831

Horowitz M, Borut A (1994) The spiny mouse (*Acomys cahirinus*) – a rodent prototype for studying plasma volume regulation during thermal dehydration. Isr J Zool 40: 117–125

Horowitz M, Mani D (1978) Potential secretory activity of salivary glands and electrolyte concentrations in the saliva of two rodent species during heat acclimatization. Comp Biochem Physiol 59A: 151–154

Horowitz M, Nadel ER (1984) Effect of plasma volume on thermoregulation in the dog. Pfluegers Arch Eur J Physiol 400: 211–213

Horowitz M, Samueloff S (1979) Plasma water shifts during thermal dehydration. J Appl Physiol 47: 738–744

Horowitz M, Samueloff S (1987) Interactions between circulation and plasma fluids during heat stress, chap 17. In: Samueloff S, Yousef MK (eds) Adaptive physiology to stressful environment. CRC Press, Boca Raton, pp 139–149

Horowitz M, Samueloff S (1988) Cardiac output in thermally dehydrated rodents. Am J Physiol 254: R109–R116

Horowitz M, Samueloff S, Adler JH (1978) Acute dehydration: body water distribution in acclimated and non-acclimated *P. obesus*. J Appl Physiol 44: 585–589

Horowitz M, Bar-Ilan DH, Samueloff S (1985) Redistribution of cardiac output in anesthetized thermally dehydrated rats. Comp Biochem Physiol 81A: 103–107

Hudson JW (1962) The role of water in the biology of the antelope ground squirrel *Citellus leucurus*. Univ Calif Publ Zool 64: 1–56

Hulbert AJ, Dawson (1974) Water metabolism in perameloid marsupials from different enviroments. Comp Biochem Physiol 47A: 617–633

Hulbert AJ, MacMillen RE (1988) The influence of ambient temperature, seed compostion and body size on water balance and seed selection in coexisting heteromyid rodents. Oecologia (Berl) 75: 521–526

Hume ID (1989) Optimal digestive strategies in mammalian herbivores. Physiol Zool 62: 1145–1163

Hume ID, Rubsamen K, von Engelhardt W (1980) Nitrogen metabolism and urea kinetics in the rock hyrax (*Procavia habessinica*). J Comp Physiol 138: 307–314

Huntley AC, Costa DP, Rubin RD (1984) The contribution of countercurrent heat exchange to water balance in the northern elephant seal. J Exp Biol 113: 447–454

Huxley JS (1924) Constant differential growth-ratios and their significance. Nature 114: 895–896

Huxley JS (1932) Problems of relative growth. Methuen, London

Hyvarinen H (1984) Winter strategy of voles and shrews in Finland. In: Merritt JF (ed) Winter ecology of small mammals. Spec Publ Carnegic Museum of Natural History, Pittsburgh, No 10, pp 139–148

Ilan M, Yom-Tov Y (1990) Diel activity pattern of a diurnal desert rodent, *Psammomys obesus*. J Mammal 71: 66–69

Ilany G (1983) Blanford's fox, *Vulpes cana* Blanford, 1877, a new species in Israel. Isr J Zool 32: 150

Irving L, Krog H, Monson M (1957) Metabolism of varying hare in winter. Physiol Zool 28: 173–185

Iverson SL, Turner BN (1974) Winter weight dynamics in *Microtus pennsylvanicus*. Ecology 55: 1030–1041

Ivlev VS (1961) Experimental ecology of the feeding of fishes. Yale University Press, New Haven

Jackson DC, Schmidt-Nielsen K (1964) Countercurrent heat exchange in the respiratory passages. Proc Acad Sci USA 51: 1192–1197

Jacobs J (1974) Quantitative measurement of food selection. A modification of the forage ratio and Ivlev's electivity index. Oecologia (Berl)14: 413–417

Janecek LL, Schlitter DA, Rautenbach IL (1991) A genic comparison of spiny mice, genus *Acomys*. J Mammal 72: 542–552

Jansky L (1973) Nonshivering thermogenesis and its thermoregulatory significance. Biol Rev 48: 85–132

Jansman AJM, Frohlich AA, Marquardt RR (1994) Production of proline-rich proteins by the parotid glands of rats is enhanced by feeding diets containing tannins from faba beans (*Vicia faba* L. J Nutr 124: 249–258

Jean Y, Bergeron J-M (1986) Can voles (*Microtus pennsylvanicus*) be poisoned by secondary metabolites of commonly eaten foods? Can J Zool 64: 158–162

Jenkins SH (1988) Comments on relationships between native seed preferences of shrub-steppe granivores and seed nutritional characteristics. Oecologia (Berl) 75: 521–526

Jones PJH, Winthrop AL, Schoeller DA, Sawyer PR, Smith J, Fuller RM, Heim T (1987) Validation of doubly-labelled water for assessing energy expenditure in infants. Pediatr Res 21: 242–246

Jorgensen CB (1983) Ecological physiology: background and perspectives. Comp Biochem Physiol 75A: 5–7

Kaczmarski F (1966) Bioenergetics of pregnancy and lactation in the bank vole. Acta Theriol 11: 409–417

Kam M, Degen AA (1988) Water, electrolyte and nitrogen balances of fat sand rats (*Psammomys obesus*) when consuming the saltbush *Atriplex halimus*. J Zool (Lond) 215: 453–462

Kam M, Degen AA (1991) Diet selection and energy and water budgets of the common spiny mouse *Acomys cahirinus*. J Zool (Lond) 225: 285–292

Kam M, Degen AA (1992) Effect of air temperature on energy and water balance of *Psammomys obesus*: J Mammal 73: 207–214

Kam M, Degen AA (1993a) Energetics of lactation and growth in the fat sand rat, *Psammomys obesus*: new perspectives of resource partitioning and the effect of litter size. J Theor Biol 162: 353–369

Kam M, Degen AA (1993b) Effect of dietary preformed water on energy and water budgets of two sympatric desert rodents, *Acomys russatus* and *Acomys cahirinus*. J Zool (Lond) 231: 51–59

Kam M, Degen AA (1994) Body mass and growth rate of fat sand rat (*Psammomys obesus*) pups: effect of litter size and water content of *Atriplex halimus* consumed by pregnant and lactating females. Funct Ecol 8: 351–357

Kam M, Degen AA, Nagy KA (1987) Seasonal energy, water, and food consumption of Negev chukars and sand partridges. Ecology 68: 1029–1037

Kamau JMZ, Maina JN, Maloiy GMO (1984) The design and role of the nasal passages in the temperature regulation in the dik-dik antelope (*Rhynchotragus kirkii*) with observations on the carotid rete. Respir Physiol 56: 183–194

Karasov WH (1981) Daily energy expenditure and the cost of activity in a free-living mammal. Oecologia (Berl) 51: 253–259

Karasov WH (1983) Water flux and water requirements in free-living antelope ground squirrels *Ammospermophilus leucurus*. Physiol Zool 56: 94–105

Karasov WH (1985) Nutrient constraints in the feeding ecology of an omnivore in a seasonal environment. Oecologia (Berl) 66: 280–290

Kasirer-Israeli H, Choshniak I, Shkolnik A (1994) Dehydration and rehydration in donkeys: the role of the hind gut as a water reservoir. J Basic Clin Physiol Pharmacol 5: 89–100

Kay FR, Whitford WG (1978) The burrow environment of the banner-tailed kangaroo rat, *Dipodomys spectabilis*, in south-central New Mexico. Am Midl Nat 99: 270–279

Kenagy GJ (1972) Saltbush leaves: excision of hypersaline tissue by a kangaroo rat. Science 178: 1094–1096

Kenagy GJ (1973a) Daily and seasonal patterns of activity and energetics in a heteromyid rodent community. Ecology 54: 1201–1219

Kenagy GJ (1973b) Adaptations for leaf eating in the Great Basin kangaroo rat , Dipodomys merriami. Oecologia (Berl) 12: 383–412

Kenagy GJ, Bartholomew GA (1985) Seasonal reproductive patterns in five coexisting California desert species. Ecol Monogr 55: 371–397

Kenagy GJ, Smith CB (1973) Radioisotopic measurement of depth and determination of temperatures in burrows of heteromyid rodents. In: Nelson DJ (ed) Proc 3rd Natl Symp Radioecology, Natl Rech Inf Serv, Springfield, Virginia pp 265–273

Kenagy GJ, Sharbaugh SM, Nagy KA (1989a) Annual cycle of energy and time expenditure in a golden-mantled ground squirrel population. Oecologia (Berl) 78: 269–282

Kenagy GJ, Stevenson RD, Masman D (1989b) Energy requirements for lactation and postnatal growth in captive golden-mantled ground squirrels. Physiol Zool 62: 470–487

Kenagy GJ, Masman D, Sharbaugh SM, Nagy KA (1990) Energy expenditure during lactation in relation to litter size in free-living golden-mantled ground squirrels. J Anim Ecol 59: 73–88

Kennedy PM, Milligan LP (1980) The degradation and utilization of endogenous urea in the gastrointestinal tract of ruminants: a review. Can J Anim Sci 60: 205–221

Kennerley TE (1964) Microenvironmental conditions of the pocket gopher burrow. Tex J Sci 16: 395–441

Kerley GIH (1989) Diet of small mammals from the Karoo, South Africa. S Afr J Wildl Res 19: 67–72

Kerley GIH (1991) Seed removal by rodents, birds and ants in the semi-arid Karoo, South Africa. J Arid Environ 20: 63–69

Kerley GIH (1992a) Small mammal seed consumption in the Karoo, South Africa: further evidence for divergence in desert biotic processes. Oecologia (Berl) 89: 471–475

Kerley GIH (1992b) Trophic status of small mammals in the semi-arid Karoo, South Africa. J Zool (Lond) 226: 563–572

Kerley GIH, Erasmus T (1991) What do mice select for in seeds? Oecologia (Berl) 86: 261–267

Kerley GIH, Whitford WG (1994) Desert-dwelling small mammals as granivores: intercontinental variations. Aust J Zool 42: 543–555

Kerley GIH, Knight MH, ErasmusT (1990) Small mammal microhabitat use and diet in the southern Kalahari, South Africa. S Afr J Wildl 20: 123–126

Khokhlova IS, Krasnov BR, Shenbrot GI, Degen AA (1994) Patterns affecting seasonal body mass change in several species of rodents in the Ramon erosion cirque, Negev Highlands, Israel. Zool Zhur 73: 106–114 (in Russian with English summary)

Khokhlova IS, Degen AA, Kam M (1995) Body size, gender, seed husking and energy requirements in two species of desert gerbilline rodents, Meriones crassus and Gerbillus henleyi. Funct Ecol 9: 720–724

Kirmiz JP (1962) Adaptation to a desert environment. Butterworth, London

Kleiber M (1932) Body size and metabolism. Hilgardia 6: 315–353

Kleiber M (1975) The fire of life: an introduction to animal energetics. Robert E Krieger, Huntington

Klein PD, James WPT, Wong WW, Irving CS, Murgatroyd PR, Cabrera M, Dallosso HM, Klein ER, Nichols BL (1984) Calorimetric validation of the doubly-labelled water method for determination of energy expenditure in man. Hum Nutr Clin Nutr 38: 95–106

Klir JJ, Heath JE (1992) An infrared thermographic study of surface temperature in relation to external thermal stress in three species of foxes: the red fox (Vulpes vulpes), arctic fox (Alopex lagopus), and kit fox (Vulpes macrotis). Physiol Zool 65: 1011–1021

Knight MH, Skinner JD (1981) Thermoregulatory, reproductive and behavioural adaptations of the big eared desert mouse, Malacothrix typica to its arid environment. J Arid Environ 4: 137–145

Konig B, Markl H (1987) Maternal care in house mice. I. The weaning strategy as a means for parental manipulation of offspring quality. Behav Ecol Sociobiol 20: 1–9

Kooyman GL (1963) Milk analysis of the Kangaroo rat, Dipodomys merriami. Science 142: 1467–1468

Korn H (1989) The annual cycle in body weight of small mammals from the Transvaal, South Africa, as an adaptation to a subtropical seasonal environment. J Zool (Lond) 218: 223–231

Koteja P (1991) On the relation between basal and field metabolic rates in birds and mammals Funct Ecol 5: 56–64

Kotler BP, Brown JS (1990) Rates of seed harvest by two species of gerbilline rodents. J Mammal 71: 591–596

Kotler BP, Brown JS, Subach A (1993) Mechanisms of species coexistence of optimal foragers: temporal partitioning by two species of sand dune gerbils. Oikos 67: 548–556

Kotler BP, Brown JS, Mitchell WA (1994) The role of predation in shaping the behaviour, morphology and community organization of desert rodents. Aust J Zool 42: 449–466

Krebs CJ (1989) Ecological methodology. Harper and Row, New York

Kriz W (1981) Structural organization of the renal medulla: comparative and functional aspects. Am J Physiol 241: R3–R16

Kucheruk VV(1983) Burrows of mammals-their structure, topology and use. Fauna Ecol Rodents 15: 5–54 (in Russian with English summary)

Kunz TH, Oftedal OT, Robson SK, Kretzmann MB, Kirk C (1995) Changes in milk composition during lactation in three species of insectivorous bats. J Comp Physiol B 164: 543–551

Kurta A, Kunz TH, Nagy KA (1990) Energetics and water flux of free-ranging big brown bats (*Eptesicus fuscus*) during pregnancy and lactation. J Mammal 71: 59–65

Langford A (1983) Pattern of nocturnal activity of male *Dipodomys ordii* (Heteromyidae). Southwest Nat 28: 341–346

Langman VA (1985) Nasal heat exchange in a northern ungulate, the reindeer (*Rangifer tarandus*). Respir Physiol 59: 279–287

Langman VA, Maloiy GMO, Schmidt-Nielsen K, Schroter RC (1979) Nasal heat exchange in the giraffe and other large mammals. Respir Physiol 37: 325–333

Langman VA, Baudinette RV, Taylor CR (1981) Maximum aerobic capacity of wild and domestic canids compared. Fed Proc 40: 432

Laragh JH, Sealey JE (1973) The renin–angiotensin-aldosterone hormonal system and regulation of sodium, potassium, and blood pressure homeostasis, chap 26. In: Orloff J, Berliner RW (eds) Handbook of physiology, Sect 8, Renal physiology. American Physiological Society, Washington, DC, pp 831–908

Lederman SA, Rosso P (1981) Effects of fasting during pregnancy on maternal and fetal weight and body composition in well–nourished and undernourished rats. J Nutr 111: 1823–1832

Lee JS, Lifson N (1960) Measurement of total energy and material balance in rats by means of doubly-labelled water. Am J Physiol 199: 238–242

Lee WB, Houston DC (1993a) The role of coprophagy in digestion in voles (*Microtus agrestis* and *Clethrionomys glareolus*). Funct Ecol 7: 427–432

Lee WB, Houston DC (1993b) The effect of diet quality on gut anatomy in British voles (Microtinae). J Comp Physiol B 163: 337–339

Lefebvre EA (1964) The use of D_2O^{18} for measuring energy metabolism in *Columba livia* at rest and in flight. Auk 81: 403–416

Leon M, Woodside B (1983) Energetic limits on reproduction: maternal food intake. Physiol Behav 30: 945–957

Leopold AS (1961) The desert. Time Incorporated, New York

Lifson N, Lee JS (1961) Estimation of material balance of totally fasted rats by doubly-labelled water. Am J Physiol 200: 85–88

Lifson N, McClintock RM (1966) Theory of use of the turnover rates of body water for measuring energy and material balance. J Theor Biol 12: 46–74

Lifson N, Gordon GB, Visscher MB, Nier AO (1949) The fate of utilized molecular oxygen and the source of oxygen of respiratory carbon dioxide: studied with the aid of heavy oxygen. J Biol Chem 180: 803–811

Lifson N, Gordon GB, McClintock RM (1955) Measurement of total carbon dixide production by means of D_2O^{18}. J Appl Physiol 7: 704–710

Lifson N, Little WS, Levitt DG, Henderson RM (1975) D_2O^{18} method for CO_2 output on small mammals and economic feasibility in man. J Appl Physiol 39: 657–664

Lindroth RL (1988) Adaptations of mammalian herbivores to plant chemical defenses. In: Spencer KC (ed) Chemical mediation of coevolution. Academic Press, San Diego, pp 415–445

Lindroth RL, Batzli GO (1984) Plant phenolics as chemical defenses: effects of natural products on surival and growth of prairie voles. J Chem Ecol 10: 229–244

Lindstedt SL (1980a) Energetics and water economy of the smallest desert mammal. Physiol Zool 53: 82–97

Lindstedt SL (1980b) Regulated hypothermia in the desert shrew. J Comp Physiol 137: 173–176

Lindstedt SL, Boyce MS (1985) Seasonality, fasting endurance, and body size in mammals. Am Nat 125: 873–878

Little WS, Lifson N (1975) Validation study of D_2O^{18} method for determination of CO_2 output of the eastern chipmunk (*Tamias striatus*). Comp Biochem Physiol 50A: 55–56

Livesey G, Elia M (1988) Estimation of energy expenditure, net carbohydrate utilization, and net fat oxidation and synthesis by indirect calorimetry: evaluation of errors with special reference to the detailed composition of fuels. Am J Clin Nutr 47: 607–628

Lochmiller RL, Whelan JB, Kirkpatrick RL (1982) Energetic cost of lactation in *Microtus pinetorum*. J Mammal 63: 475–481

Lourens S, Nel JAJ (1990) Winter activity of bat-eared foxes *Octocyon megalotis* on the Cape West Coast. S Afr J Zool 25: 124–132

Lynch GR, White S, Grundel R, Berger M (1978) Effects of photoperiod, melatonin administration and thyroid block on spontaneous daily torpor and temperature regulation in the white–footed mouse, *Peromyscus leucopus*. J Comp Physiol 125: 157–163

Macfarlane WV (1964) Terrestrial animals in dry heat: ungulates. In: Dill DB, Adolph EFA, Wilberg CC (eds) Handbook of physiology, Sec 4. Adaptation to the environment. American Physiological Society, Washington, DC, pp 509–539

Macfarlane WV, Howard B (1972) Comparative water and energy economy of wild and domestic mammals. Symp Zool Soc (Lond) 31: 261–296

Macfarlane WV, Morris RJ, Howard B (1956) Water economy of tropical Merino sheep. Nature (Lond) 178: 304–305

Macfarlane WV, Robinson KW, Howard B, Budtz-Olsen OE (1959) Extracellular fluid distribution in tropical merino sheep. Aust J Agric Res 10: 269–286

Macfarlane WV, Morris RJ, Howard B, MacDonald J, Budtz-Olsen OE (1961) Water and electrolyte changes in tropical Merino sheep exposed to dehydration during summer. Aust J Sci 25: 112

MacMillen RE (1964) Population ecology, water relations, and social behavior of a southern California semidesert rodent fauna. Univ Calif Publ Zool 71: 1–66

MacMillen RE (1983) Water regulation in *Peromyscus*. J Mammal 64: 38–47

MacMillen RE, Christopher EA (1975) The water relations of two populations of noncaptive desert rodents. In: Hadley NF (ed) Environmental physiology of desert organisms. Dowden, Hutchinson and Ross, Shroudsberg, pp 117–137

MacMillen RE, Hinds DS (1983) Water regulatory efficiency in heteromyid rodents: a model and its application. Ecology 64: 152–164

MacMillen RE, Lee AK (1967) Australian desert mice: independence of exogenous water. Science 158: 383–385

MacMillen RE, Lee AK (1969) Water metabolism of Australian hopping mice. Comp Biochem Physiol 28: 493–514

MacMillen RE, Lee AK (1970) Energy metabolism and pulmocutaneous water loss of Australian hopping mice. Comp Biochem Physiol 35A: 355–369

MacMillen RE, Nelson JE (1969) Bioenergetics and body size in dasyurid marsupials. Am J Physiol 217: 1246–1251

Makkar HPS (1993) Antinutritional factors in food for livestock. In: Animal production in developing countries. Occasional publ no 16, Brit Soc Anim Prod, Penicuik, pp 69–85

Maloiy GMO (1970) Water economy of the Somali donkey. Am J Physiol 219: 1522–1527

Maloiy GMO (1972) Renal salt and water excretion in the camel (*Camelus dromedarius*). Symp Zool Soc (Lond) 31: 243–259

Maloiy GMO, Boarer CDH (1971) Response of the Somali donkey to dehydration: hematological changes. Am J Physiol 221: 37–41

Maloiy GMO, Hopcraft D (1971) Thermoregulation and water relations of two East African antelopes: the hartebeest and impala. Comp Biochem Physiol 38A: 525–534

Maloiy GMO, Taylor CR (1971) Water requirements of African goats and haired-sheep. J Agric Sci (Camb) 77: 203–208

Maloiy GMO, Kamau JMZ, Shkolnik A, Meir M, Arieli R (1982) Thermoregulation and metabolism in a small desert carnivore: the Fennec fox (*Fennecus zerda*) (Mammalia). J Zool (Lond) 198: 279–291

Maren TH (1967) Carbonic anhydrase: chemistry, physiology and inhibition. Physiol Rev 47: 597–768

Mares MA (1993) Heteromyids and their ecological counterparts: a pandesertic view of rodent ecology and evolution. In: Genoways HH, Brown JH (eds) Biology of the Heteromyidae. Spec Publ No.10. American Society of Mammalogists, Brigham Young University, Provo, Uttah, pp 652–715

Mares MA, Rosenzweig ML (1978) Granivory in North and South American deserts: rodents, birds, and ants. Ecology 59: 235–241

Martin R (1984) Scaling effects and adaptive strategies in mammalian lactation. Symp Zool Soc (Lond) 51: 87–117

Masman D (1986) The annual cycle of the kestrel, *Falco tinnunculus*, a study in behavioural energetics. Ph D. Thesis. Univ Groningen, Groningen

Masuda A, Oishi T (1988) Effects of photoperiod and temperature on body weight, food intake, food storage, and pelage color in the Djungarian hamster, *Phodopus sungorus*. J Exp Zool 248: 133–139

Mattingly DK, McClure PA (1982) Energetics of reproduction in large-littered cotton rats (*Sigmodon hispidis*). Ecology 63: 183–195

Mattingly DK, McClure PA (1985) Energy allocation during lactation in cotton rats (*Sigmodon hispidis*) on a restricted diet. Ecology 66: 928–937

McClintock R, Lifson N (1957) Applicability of the D_2O^{18} method to the measurement of the total carbon dioxide output of obese mice. J Biol Chem 226: 153–156

McClintock R, Lifson N (1958a) CO_2 output of mice measured by D_2O^{18} under conditions of isotope re-entry. Am J Physiol 192: 721–725

McClintock R, Lifson N (1958b) Determination of the total CO_2 output of rats by the D_2O^{18} method. Am J Physiol 192: 76–78

McFarland WN, Wimsatt WA (1969) Renal function and its relation to the ecology of the vampire bat, *Desmodus rotundus*. Comp Biochem Physiol 28: 985–1006

McGinnies WG, Goldman BJ, Paylore P (eds) (1970) Deserts of the world. University of Arizona Press, Tucson

McNab BK (1966) The metabolism of fossorial rodents: a study of convergence. Ecology 47: 712–733

McNab BK (1971) On the ecological significance of Bergmann's rule. Ecology 52: 845–854

McNab BK (1980a) On estimating thermal conductance in endotherms. Physiol Zool 53: 145–156

McNab BK (1980b) Food habits, energetics, and the population biology of mammals. Am Nat 116: 106–124

McNab BK (1983) Energetics, body size and the limits to endothermy. J Zool (Lond) 199: 1–29

McNab BK (1984) Commentary. Oecologia (Berl) 64: 423–424

McNab BK (1986a) The influence of food habits on the energetics of eutherian mammals. Ecol Monogr 56: 1–19

McNab BK (1986b) Food habits, energetics, and the reproduction of marsupials. J Zool (Lond) 208: 595–614

McNab BK (1987) The reproduction of marsupial and eutheiran mammals in relation to energy expenditure. Symp Zool Soc (Lond) 57: 29–39

McNab BK (1988) Complications inherent in scaling the basal rate of metabolism in mammals. Q Rev Biol 63: 25–54

McNab BK (1991) The energy expenditure of shrews. In: Findley JS, Yates TL (eds) The biology of the soricidae. The Museum of Southwestern Biology, University of New Mexico, Albuquerque, pp 35–45

McNab BK (1992a) The comparative energetics of rigid endothermy: the Arvicolidae. J Zool (Lond) 227: 585–606

McNab BK (1992b) A statistical analysis of mammalian rates of metabolism. Funct Ecol 6: 672–679

McNab BK, Morrison PR (1963) Body temperature and metabolism in subspecies of *Peromyscus* from arid and mesic environments. Ecol Monogr 33: 63–82

Mehanso H, Butler LG, Carlson DM (1987) Dietary tannins and salivary proline-rich proteins: interactions, induction and defense mechanisms. Annu Rev Nutr 7: 423–440

Meigs P (1953) World distribution of arid and semi-arid homoclines. In: Review of research of arid zone hydrology. Arid Zone Programme, 1, UNESCO, Paris, pp 203–209

Meir M, Shkolnik A (1984) Water economy of slughi, canaan and pointer dogs. J Arid Environ 7: 93–99

Mendelssohn H (1982) Wolves in Israel. In: Harrington FH, Pacquet PC (eds) Wolves in the world. Noyes, New Jersey, pp 173–197

Mendelssohn H, Yom-Tov Y, Ilany G, Meninger D (1987) On the occurrence of Blanford's fox, *Vulpes cana* Blanford, 1877, in Israel and Sinai. Mammalia 51: 459–462

Merritt JF (1995) Seasonal thermogenesis and changes in body mass of masked shrews, *Sorex cinereus*. J. Mammal 76: 1020–1035

Meyer MW, Karasov WH (1991) Chemical aspects of herbivory in arid and semiarid habitats. In: Palo RT, Robbins CT (eds) Plant defenses against mammalian herbivory. CRC Press, Boca Raton, pp 167–187

Meyerson-McCormick R, Cranford JA, Akers RM (1990) Milk yield and composition in the punare *Thrichomys apereoides*. Comp Biochem Physiol 96A: 211–214

Mezhzherin VA (1964) Dehnel's phenomenon and its possible explanation. Acta Theriol 8: 95–114

Mezhzherin VA, Melnikova GL (1966) Adaptive significance of some seasonal changes in some morphological indices in shrews. Acta Theriol 11: 503–521

Migula P (1969) Bioenergetics of pregnancy and lactation in European common vole. Acta Theriol 14: 167–179

Millar JS (1977) Adaptive features of mammalian reproduction. Evolution 31: 370–386

Millar JS (1978) Energetics of reproduction in *Peromyscus leucopus*: the cost of lactation. Ecology 59: 1055–1061

Millar JS, Hickling GJ (1990) Fasting endurance and the evolution of mammalian body size. Funct Ecol 4: 5–12

Millar JS, Hickling GJ (1991) Body size and the evolution of mammalian life histories. Funct Ecol 5: 588–593

Mogharabi F, Haines H (1973) Dehydration and body fluid regulation in the thirteen-lined ground squirrel and laboratory rat. Am J Physiol 224: 1218–1222

Morgan KH, Price MV (1992) Foraging in heteromyid rodents: the energy of scratch-digging. Ecology 73: 2260–2272

Morris KD, Bradshaw SD (1981) Water and sodium turnover in coastal and inland populations of the ash-grey mouse, *Pseudomys albocinereus* (Gould), in Western Australia. Aust J Zool 29: 519–533

Morrison S (1953) Total expenditure of energy by pregnant rats. J Physiol 122: 479

Morton SR (1979) Diversity of desert-dwelling mammals: a comparison of Australia and North America. J Mammal 60: 253–264

Morton SR (1980) Field and laboratory studies of water metabolism in *Sminthopsis crassicaudata* (Marsupialia, Dasyuridae). Aust J Zool 28: 213–227

Morton SR (1982) Dasyurid marsupials of the Australia arid zone: an ecological review, chap 12. In: Archer M (ed) Carnivorous marsupials. Royal Zoological Society, New South Wales, Sydney, pp 117–130

Morton SR (1985) Granivory in arid regions: comparison of Australia and North and South America. Ecology 66: 1859–1866

Morton SR, Baynes A (1985) Small mammal assemblages in arid Australia: a reappraisal. Aust Mammal 8: 159–169

Morton SR, Lee AK (1978) Thermoregulation and metabolism in *Planigale maculata* (Marsupialia: Dasyuridae). J Therm Biol 3: 117–120

Morton SR, MacMillen RE, (1982) Seeds as sources of preformed water for desert-dwelling granivores. J Arid Environ 5: 61–67

Morton SR, Hinds DS, MacMillen RE (1980) Cheek pouch capacity in heteromyid rodents. Oecologia (Berl) 46: 143–146

Mount LE (1979) Adaptation to thermal environment. Edward Arnold, London

Mullen RK (1970) Respiratory metabolism and body water turnover rates of *Perognathus formosus* in its natural environment. Comp Biochem Physiol 32A: 379–390

Mullen RK (1971a) Energy metabolism of *Peromyscus crinitus* in its natural environment. J Mammal 52: 633–635

Mullen RK (1971b) Energy metabolism and body water turnover rates of two species of free-living kangaroo rats, *Dipodomys merriami* and *Dipodomys microps*. Comp Biochem Physiol 39A: 379–390

Mullen RK, Chew RM (1973) Estimating the energy metabolism of free-living *Perognathus formosus*: a comparison of direct and indirect methods. Ecology 54: 633–637

Muller EF, Kamau JMZ, Maloiy GMO (1979) O_2-uptake, thermoregulation and heart rate in the springhare (*Pedestes capensis*). J Comp Physiol 133: 187–191

Munkacsi I, Palkovits M (1977) Measurements on the kidney and vasa recta of various mammals in relation to urine concentrating capacity. Acta Anat 98: 456–468

Munger JC, Karasov WH (1989) Sublethal parasites and host energy budgets: tapeworm infection in white-footed mice. Ecology 70: 904–921

Murdiati TB, McSweeny CS, Lowry JB (1992) Metabolism in sheep of gallic acid, tannic acid and hydrolyzable tannin from *Terminalia oblongata*. Aust J Agric Res 43: 1307–1319

Murray BR, Dickman CR (1994a) Granivory and microhabitat use in Australian desert rodents: are seeds important? Oecologia (Berl) 99: 216–225

Murray BR, Dickman CR (1994b) Food preferences and seed selection in two species of Australian desert rodents. Wildl Res 21: 647–655

Murray BR, Hume ID, Dickman CR (1995) Digestive tract characteristics of the spinifex hopping-mouse. *Notomys alexis*, and the sandy island mouse, *Pseudomys hermannsburgensis*, in relation to diet. Aust Mammal 18: 93–97

Mutze GJ, Green B, Newgrain K (1991) Water flux and energy use in wild house mice (*Mus domesticus*) and the impact of seasonal aridity on breeding and population levels. Oecologia (Berl) 88: 529–538

Myrcha A, Ryszkowski L, Walkowa W (1969) Bioenergetics of pregnancy and lactation in the white mouse. Acta Theriol 12: 161–166

Nagy KA (1980) CO_2 production in animals: analysis of potential errors in the doubly labeled water method. Am J Physiol 238: R466–R473

Nagy KA (1983) The doubly labeled water $^3HH^{18}O$ method: a guide to its use. UCLA Publ No 12–1417

Nagy KA (1987) Field metabolic rate and food requirement scaling in mammals and birds. Ecol Monogr 57: 111–128

Nagy KA (1994a) Seasonal water, energy and food use by free-living, arid-habitat mammals. Aust J Zool 42: 55–63

Nagy KA (1994b) Field bioenergetics of mammals: what determines field metabolic rates? Aust J Zool 42: 43–53

Nagy KA, Costa DP (1980) Water flux in animals: analysis of potential errors in the tritiated water method. Am J Physiol 238: R454–R465

Nagy KA, Gruchacz MJ (1994) Seasonal water and energy metabolism of the desert-dwelling kangaroo rat (*Dipodomys merriami*). Physiol Zool 67: 1461–1478

Nagy KA, Knight MH (1994) Energy, water, and food use by springbok antelope (*Antidorcas marsupialis*) in the Kalahari desert. J Mammal 75: 860–872

Nagy KA, Martin RW (1985) Field metabolic rate, water flux, and food consumption and time budget of kaolas, *Phascolarctos cinereus* (Marsupialia: Phascolarctidae) in Victoria. Aust J Zool 42: 655–665

Nagy KA, Milton K (1979) Aspects of dietary quality, nutrient assimilation, and water balance in wild howler monkeys (*Alouatta palliata*). Oecologia (Berl) 39: 249–258

Nagy KA, Montgomery GG (1980) Field metabolic rate, water flux, and food consumption in three-toed sloths (*Bradypus variegatus*). J Mammal 61: 465–472

Nagy KA, Peterson CC (1988) Scaling of water flux rate in animals. Calif Univ publ Zool 120: 1–172

Nagy KA, Suckling GC (1985) Field energetics and water balance of sugar gliders, *Petaurus breviceps* (Marsupialia: Petauridae). Aust J Zool 33: 683–691

Nagy KA, Shoemaker VH, Costa WR (1976) Water, electrolyte, and nitrogen budgets of jackrabbit (*Lepus californicus*) in the Mohave Desert. Physiol Zool 49: 351–363

Nagy KA, Seymour RS, Lee AK, Braithwaite R (1978) Energy and water budgets in free-living *Antechinus stuartti* (Marsupialia: Dasyuridae). J Mammal 59: 60–68

Nagy KA, Lee AK, Martin MR, Fleming MR (1988) Field metabolic rate and food requirement of a small small dasyurid marsupial, *Sminthopsis crassicaudata*. Aust J Zool 36: 293–299

Nagy KA, Bradley AJ, Morris KD (1990) Field metabolic rates, water fluxes, and feeding rates of quokkas, *Setonix brachyurus*, and tammars, *Macropus eugenii*, in western Australia. Aust J Zool 37: 353–360

Nagy KA, Bradshaw SD, Clay BT (1991) Field metabolic rate, water flux, and food requirements of short-nosed bandicoots, *Isoodon obesulus* (Marsupialia: Permelidae). Aust J Zool 39: 299–305

Naumov NP, Lobachev VS (1975) Ecology of desert rodents of the USSR. In: Prakash I, Ghosh PK (eds) Rodents in desert environments. Dr W Junk, The Hague, pp 465–598

Neal BR, Alibhai SK (1991) Reproductive response of *Tetera leucogaster* (Rodentia) to supplemental food and 6-methoxybenzolinone in Zimbabwe. J Zool (Lond) 223: 469–473

Negus NC, Berger PJ (1977) Experimental triggering of reproduction in a natural population of *Microtus montanus*. Science 196: 1230–1231

Nel JAJ (1967) Burrow systems of *Desmodillus auricularis* in the Kalahari Gemsbok National Park. Koedoe 10: 118–121

Nel JAJ (1978) Habitat heterogeneity and changes in small mammal community structure and resource utilization in the southern Kalahari. Bull Carnegie Mus Nat Hist 6: 118–132

Nel JAJ (1990) Foraging and feeding by bat-earned foxes *Otocyon megalotis* in the southwestern Kalahari. Koedoe 33: 9–16

Nel JAJ Mackie AJ (1990) Food and foraging behaviour of bat-eared foxes in the south-eastern Orange Free State. S Afr J Wildl Res 20: 162–166

Nelson JF, Chew RM (1977) Factors affecting seed reserves in the soil of a Mojave desert ecosytem, Rock Valley, Nye County, Nevada, Am Midl Nat 97: 300–320

Nevo E (1989) Natural selection of body size differentiation in spiny mice, Acomys. Z Saeugetierkd 54: 81–89

Nevo E, Shkolnik A (1974) Adaptive metabolic variation of chromosome forms in mole rats, *Spalax*. Experientia 30: 724–726

Nevo E, Tchernov E, Beiles A (1988) Morphometrics of speciating mole rats: adaptive differentiation in ecological speciation. Z Zool Syst Evolutions forsch 26: 286–314

Nicol SC (1978) Rates of water turnover in marsupials and eutherians: a comparative review with new data on the Tasmanian devil. Aust J Zool 26: 465–473

Noll-Banholzer U (1979a) Body temperature, oxygen consumption, evaporative water loss and heart rate in the fennec. Comp Biochem Physiol 62A: 585–592

Noll–Banholzer U (1979b) Water balance and kidney structure in the fennec. Comp Biochem Physiol 62A: 593–597

Noy-Meir I (1973) Desert ecosytems: environment and producers. Annu Rev Ecol Syst 5: 195–214

Obst BS, Nagy KA, Ricklefs RE (1987) Energy utilization by Wilson's storm-petrel (*Oceanites oceanicus*). Physiol Zool 60: 200–210

Olenev G (1979) Dynamics of the generation structure of bank vole in the periods of decrease and restoration of abundance. In: Population ecology and variability in animals. Institute of Ecology Publications, Swerdlovsk, pp 48–53 (in Russian)

Osawa R, Walsh TP (1993) Visual reading method for detection of bacterial tannase. Appl Environ Microbiol 59: 1251–1252

Osawa R, Bird PS, Harbow DJ, Ogimoto K, Seymour GJ (1993) Microbiological studies of the intestinal microflora of the koala, *Phascolarctos cinereus*. I. Colonisation of the caecal wall by tannin-protein-complex-degrading enterobacteria. Aust J Zool 41: 599–609

Oswald C, McClure PA (1990) Energetics of concurrent pregnancy and lactation in cotton rats and woodrats. J Mammal 71: 500–509

Oswald C, Fonken P, Atkinson D, Palladino M (1993) Lactational water balance and recycling in white-footed mice, red-backed voles, and gerbils. J Mammal 74: 963–970

Owen-Smith N, Cooper SM (1989) Nutritional ecology of a browsing ruminant, the kudu (*Tragelaphus strepsiceros*), through the seasonal cycle. J Zool (Lond) 219: 29–43

Pagel MD, Harvey PH (1988) Recent developments in the analysis of comparative data. Rev Biol 63: 413–440

Paladino FV, King JR (1979) Energetic cost of terrestrial locomotion: biped and quadraped runners compared. Rev Can Biol 38: 321–323

Pekins PJ, Gessaman JA, Lindzey FG (1992) Winter energy requirements of blue grouse. Can J Zool 70: 22–24

Pekins PJ, Lindzey FG, Gessaman JA (1994) Field metabolic rate of blue grouse. Can J Zool 72: 227–231

Peled A (1989) Comparative aspects in the physiology of two gazelle species: the mountain gazelle (*Gazella gazella*) and the desert gazelle (*Gazella dorcas isabella*) M Sc Thesis, Tel Aviv Univ (Hebrew with English Abstr)

Perrin MR (1980) Ecological strategies of two coexisting rodents. S Afr J Sci 76: 487–491

Perrin MR, Swanepoel P (1987) Breeding biology of the bushveld gerbil *Tatera leucogaster* in relation to diet, rainfall and life history theory. S Afr J Zool 22: 218–227

Perrin MR, Boyer H, Boyer DC (1992) Diets of the hairy-footed gerbils *Gerbillus paeba* and *G. tytonis* from the dunes of the Namib Desert. Isr J Zool 38: 373–383

Peters JP, VanSlyke DD (1946) Quantitative clinical chemistry, vol I. Williams and Wilkins, Baltimore

Peters RH (1983) The ecological implications of body size. Cambridge University Press, Cambridge

Peterson CC, Nagy KA, Diamond J (1990) Sustained metabolic scope. Proc Natl Acad Sci 87: 2324–2328

Peterson RM, Batzli GO, Banks EM (1976) Activity and energetics of the brown lemming in its natural habitat. Arct Alp Res 8: 131–138

Petterborg LJ (1978) Effect of photoperiod on body weight in the vole, *Microtus montanus*. Can J Zool 56: 431–435

Pettit TN, Nagy KA, Ellis HI, Whittow GC (1988) Incubation energetics of the Laysan albatross. Oecologia (Berl) 74: 546–550

Pfeiffer EW (1970) Ecological and anatomical factors affecting the gradient of urea and non-urea-solutes in mammalian kidneys. In: Schmidt-Nielsen B, Kerr DWS (eds) Urea and the kidney. Excerpta Medica, Amsterdam, pp 358–365

Phillips PK, Heath JE (1995) Dependency of surface temperature regulation on body size in terrestrial mammals. J Therm Biol 20: 281–289

Pillay N, Willan K, Taylor PJ (1994) Comparative renal morphology of some southern African otomyine rodents. Acta Theriol 39: 37–48

Plakke RK, Pfeiffer EW (1970) Urea, electrolyte and total solute excretion following water deprivation in the opposum (*Didelphis marsupialis virginiana*). Comp Biochem Physiol 34: 325–332

Poppitt SD, Speakman JR, Racey PA (1993) The energetics of reproduction in the common shrew (*Sorex araneus*): a comparison of indirect calorimetry and the doubly labeled water method. Physiol Zool 66: 964–982

Poppitt SD, Speakman JR, Racey PA (1994) Energetics of reproduction in the lesser hedgehog tenrec, *Echinops telfairi* (Martin). Physiol Zool 67: 976–994

Porter WP (1969) Thermal radiation in metabolic chambers. Science 166: 115–117

Porter WP (1989) New animal models and experiments for calculating growth potential at different elevations. Physiol Zool 62: 286–313

Porter WP, Busch RL (1978) Fractional factorial analysis of growth and weaning success in *Peromyscus maniculatus*. Science 202: 907–910

Porter WP, Gates DM (1969) Thermodynamic equilibria of animals with environment. Ecol Monogr 39: 227–244

Porter WP, Munger JC, Stewart WE, Budaraju S, Jaeger J (1994) Endotherm energetics: from a scalable individual-based model to ecological applications. Aust J Zool 42: 125–162

Powers DR, Conley TM (1994) Field metabolic rate and food consumption of two sympatric hummingbird species in southeastern Arizona. Condor 96: 141–150

Powers DR, Nagy KA (1988) Field metabolic rate and food consumption by free-living Anna's hummingbirds (*Calypte anna*). Physiol Zool 53: 70–81

Prakash I (1975) The population ecology of the rodents of the Rajasthan desert, India. In: Prakash I, Ghosh PK (eds) Rodents in desert environments. Dr W Junk, The Hague, pp 75–116

Prentice AM, Black AE, Coward WA, Davies HL, Goldberg GR, Murgatroyd PR, Ashford J, Sawyer M, Whitehead RG (1986) High levels of energy expenditure in obese women. Br Med J 292: 983–987

Prentice AM, Coward WA, Davies HL, Murgatroyd PR, Black AE, Goldberg GR, Ashford J, Sawyer M, Whitehead RG (1985) Unexpected low levels of energy expenditure in healthy women. Lancet 325: 1419–1422

Prentice TC, Siri W, Berlin NI, Hyde GM, Parsons RJ, Joiner EE, Lawrence JH (1952) Studies of total body water determination with tritium. J Clin Nutr 35: 591–594

Proctor DF, Andersen I, Lundqvist GR (1977) Human nasal mucosal function at controlled temperatures. Respir Physiol 30: 109–124

Pucek Z (1970) Seasonal and age change in shrews as an adaptive process. Symp Zool Soc (Lond) 123: 599–691

Purohit GR, Ghosh PK, Taneja GC (1972) Water metabolism in desert sheep. Aust J Agric Res 23: 685–691

Purohit KG (1974a) Observations on size and relative medullary thickness in kidneys of some Australian mammals and their ecophysiological appraisal. Z Angew Zool 61: 495–506

Purohit KG (1974b) Observations on the role of kidney in survival of the Tammar wallaby (*Macropus eugenii*) on sea water. Z. Angew Zool 61: 223–238

Purohit KG (1974c) Observations on histomorphology of kidneys and urine osmolarities in some Australian desert rodents. Zool Anz 193: 221–227

Purohit KG (1975) Histomorphology of kidney in the Indian desert gerbil, *Meriones hurrianae*, and its ecophysiological appraisal. Z Angew Zool 62: 9–22

Rabi T, Cassuto Y (1976) Metabolic adaptations in brown adipose tissue of the hamster in extreme ambient temperatures. Am J Physiol 231: 153–160

Racey PA (1982) Ecology of bat reproduction. In: Kunz TH (ed) Ecology of bats. University Park Press, Baltimore, pp 57–104

Randall JA (1993) Behavioural adaptations of desert rodents (Heteromyidae). Anim Behav 45: 263–287

Randall JA (1994) Convergences and divergences in communication and social organization of desert rodents. Aust J Zool 42: 405–433

Randle H, Haines H (1976) Effect of water deprivation on antidiuresis in *Dipodomys spectabilis* and *Rattus norvegicus*. Comp Biochem Physiol 54A: 21–26

Randolph JC (1980) Daily energy metabolism of two rodents (*Peromyscus leucopus* and *Tamias striatus*) in their natural environment. Physiol Zool 53: 70–81

Randolph JC, Cameron GN, McClure PA (1995) Nutritional requirements for reproduction in the cotton rat, *Sigmodon hispidus*. J Mammal 76: 1113–1126

Randolph PA, Randolph JC, Mattingly K, Foster MM (1977) Energy costs of reproduction in the cotton rat, *Sigmodon hispidus*. Ecology 58: 31–45

Reaka ML, Armitage KB (1976) The water economy of harvest mice from xeric and mesic environments. Physiol Zool 49: 307–327

Reese JB, Haines H (1978) Effects of dehydration on metabolic rate and fluid distribution in the jackrabbit, *Lepus californicus*. Physiol Zool 51: 155–165

Reichman OJ, Price MV (1993) Ecological aspects of Heteromyid foraging. In: Genoways HH, Brown JH (eds) Biology of the Heteromyidae, Spec Pub No 10. American Society of Mammalogists, Brigham Young University, Provo, Utah, pp 539–574

Reichman OJ, Van de Graff (1975) Influence of green vegetation on desert rodent reproduction. J Mammal 53: 503–506

Reichman OJ, Fattaey A, Fattaey K (1986) Management of sterile and mouldy seeds by a desert rodent. Anim Behav 34: 221–225

Reynolds HG (1960) Life history notes on Merriam's kangaroo rat in Southern Arizona. J Mammal 41: 48–58

Richards GC (1979) Variation in water turnover by wild rabbits, *Oryctolagus cuniculus*, in an arid environment, due to season, age group and reproductive condition. Aust Wildl Res 6: 289–296

Richmond CR, Langham WH, Trujillo TT (1962) Comparative metabolism of tritiated water by mammals. J Cell Comp Physiol 59: 45–53

Robbins CT (1983) Wildlife feeding and nutrition, Academic Press, London

Robbins CT, Harley TA, Hagerman AE, Hjeljord O, Baker DL, Schwartz CC, Mautz WW (1987a) Role of tannins in defending plants against ruminants: reduction in protein availability. Ecology 68: 98–107

Robbins CT, Mole S, Hagerman AE, Hanley TA (1987b) Role of tannins in defending plants against ruminants: reduction in dry matter digestion. Ecology 68: 1606–1615

Robbins CT, Hagerman AE, Austin PJ, McArthur C Hanley TA (1991) Variation in mammalian physiological responses to a condensed tannin and its ecological implications. J Mammal 72: 480–486

Roberts SB, Coward WA, Norhia V, Schlingenseipen KH, Lucas A (1986) Comparison of the doubly-labelled water ($^2H_2{}^{18}O$) method with indirect calorimetry and a nutrient balance study for simultaneous determination and energy expenditure, water intake and metabolizable energy intake in preterm infants. Am J Clin Nutr 44: 315–322

Robinson JR (1957) Functions of water in the body. Proc Nutr Soc 16: 108–112

Rodland KD, Hainsworth FR (1974) Evaporative water loss and tissue dehydration of hamsters in the heat. Comp Biochem Physiol 49A: 331–345

Rogovin KA (1985) A comparative analysis of behavior in supergeneric groups of jerboas (Rodentia, Dipodidae). Zool Zhur 64: 1702–1711 (in Russian)

Rogowitz GL (1990) Seasonal energetics of the white-tailed jackrabbit (*Lepus townsendii*). J Mammal 71: 277–285

Rosenmann M, Morrison PR (1963) The physiological response to heat and dehydration in the guanaco: *Lama guanicoe*. Physiol Zool 36: 45–51

Rosenmann M, Morrison P (1974) Maximum oxygen consumption and heat loss facilitation in small homeotherms by He—O_2. Am J Physiol 226: 490–495

Rosenzweig ML (1968) The strategy of body size in mammalian carnivores. Am Midl Nat 80: 299–315

Rosenzweig ML, Sterner PW (1970) Population ecology of desert rodent communities: body size and seed-husking as bases for heteromyid coexistence. Ecology 51: 217–224

Rouffignac C de, Bankir L, Rionel N (1981) Renal function and concentrating ability in a desert rodent: the gundi (*Ctenodactylus vali*). Pfluegers Arch Eur J Physiol 390: 138–144

Roxburgh L, Perrin MR (1994) Temperature regulation and activity pattern of the round-eared elephant shrew *Macroscelides proboscideus*. J Therm Biol 19: 13–20

Rubsamen K, Hume ID, von Engelhardt V (1982) Physiology of the rock hyrax. Comp Biochem Physiol 72A: 271–277

Ruf T, Heldmaier G (1992) The impact of daily torpor on energy requirements in the Djungarian hamster, *Phodus sungorus*. Physiol Zool 65: 994–1010

Ruf T, Stieglitz A, Steinlechner S, Blank JL, Heldmaier G (1993) Cold exposure and food restriction facilitate physiological responses to short photoperiod in Djungarian hamsters (*Phodopus sungorus*). J Exp Zool 267: 104–112

Rumbaugh GE, Carlson GP, Harrold DH (1982) Urinary production in the healthy horse and in horses deprived of feed and water. Am J Vet Res 43: 735–737

Ru-yung S, Shao-liang J (1984) Relation between average daily metabolic rate and resting metabolic rate of the Mongolian gerbil (*Meriones unguiculatus*). Oecologia (Berl) 65: 122–124

Sadleir RMFS (1969) The ecology of reproduction in wild and domestic animals. Methuen, London

Sale JB (1965) The feeding behaviour of rock hyraxes. E Afr Wildl J 3: 185–188

Sale JB (1966) The habitat of the rock hyrax. J E Afr Nat Hist Soc 25: 215–214

Saunders DK, Parrish JW (1987) Assimilated energy of seeds consumed by scaled quail in Kansas. J Wildl Manage 51: 787–790

Savage RE (1931) The relation between the feeding of the herring off the east coast of England and the plankton of the surrounding waters. Fishery investigations, Ministry of Agriculture, Food and Fisheries, London Ser 2, 12: 1–88

Schmid WD (1976) Temperature gradients in the nasal passage of some small mammals. Comp Biochem Physiol 54A: 305–308

Schmidly DJ, Wilkens KT, Derr JN (1993) Biogeography. In: Genoways HH, Brown JH (eds) Biology of the Heteromyidae, Spec Publ No 10. American Society of Mammalogists, Brigham Young University, Provo, Utah pp 319–356

Schmidt-Nielsen B (1979) Urinary concentrating processes in vertebrates. Yale J Biol Med 52: 545–561

Schmidt-Nielsen B (1988) Excretory mechanisms in the animal kingdom: examples of the principle "the whole is greater than the sum of its parts." Physiol Zool 61: 312–321

Schmidt-Nielsen B, O'Dell R (1961) Structure and concentrating mechanism in the mammalian kidney. Am J Physiol 200: 1119–1124

Schmidt-Nielsen B, Schmidt-Nielsen K (1951) A complete acount of the water metabolism in kangaroo rats and an experimental verification. J Cell Comp Physiol 38: 165–182

Schmidt-Nielsen B, Schmidt-Nielsen K, Houpt TR, Jarnum SA (1956) Water balace of the camel. Am J Physiol 185: 185–194

Schmidt-Nielsen K (1964) Desert animals: physiological problems of heat and water. Clarendon Press, Oxford

Schmidt-Nielsen K (1975a) Animal physiology: adaptations and environment. Cambridge University Press, London

Schmidt-Nielsen K (1975b) Desert rodents: physiological problems of desert life. In: Prakash I, Ghosh PK (eds) Rodents in desert environments. Dr W Junk, The Hague, pp 379–388

Schmidt-Nielsen K (1984) Scaling: why is animal size so important? Cambridge University Press, Cambridge

Schmidt-Nielsen K, Haines HB (1964) Water balance in a carnivorous desert rodent, the grasshopper mouse. Physiol Zool 37: 259–263

Schmidt-Nielsen K, Schmidt-Nielsen B (1952) Water metabolism of desert mammals. Physiol Rev 32: 135–166

Schmidt-Nielsen K, Schmidt-Nielsen B, Schneiderman H (1948) Salt excretion in desert mammals. Am J Physiol 154: 163–166

Schmidt-Nielsen K, Hainsworth FR, Murrish DE (1970) Counter-current heat exchange in the respiratory passages: effect on water and heat balance. Respir Physiol 9: 263–276

Schmidt-Nielsen K, Schroter RC, Shkolnik A (1981) Desaturation of exhaled air in camels. Proc Royal Soc Lond B211: 305–319

Schoeller DA, Van Santen E (1982) Measurements of energy expenditure in humans by doubly-labelled water method. J Appl Physiol 53: 955–959

Schoeller DA, Webb P (1984) Five-day comparsion of doubly-labelled water method with respiratory gas exchange. J Clin Nutr 40: 153–158

Schoeller DA, Van Santen E, Peterson DW, Dietz WH, Jaspan J, Klein PD (1980) Total body water measurements in humans with ^{18}O and ^{2}H labeled water. Am J Clin Nutr 33: 2686–2693

Schoeller DA, Kushner RF, Jones PJH (1986a) Validation of doubly-labelled water for measuring energy expenditure during parenteral nutrition. Am J Clin Nutr 44: 291–298

Schoeller DA, Ravussin E, Schutz Y, Acheson KJ, Baertschi P, Jequier E (1986b) Energy expenditure by double-labelled water: validation in humans and proposed calculations. Am J Physiol 250: R823–R830

Schoen A (1968) Studies on the water-balance of the East African goat. E Afr Agric For J 34: 256–262

Seeherman H, Taylor CR, Maloiy GMO, Armstrong RB (1981) Design of the mammalian respiratory system. II. Measuring maximum aerobic capacity. Respir Physiol 44: 11–23

Senay LC (1972) Changes in plasma volume and protein content during exposures of working men to various temperatures before and after acclimatization to heat: separation of the roles of cutaneous and skeletal muscle circulation. J Physiol 210: 617–635

Senay LC, Christensen ML (1964) Changes in blood plasma during progressive dehydration. J Appl Physiol 21: 1136–1140

Shenbrot GI, Sokolov VE, Heptner VG (1992) Mammals of Russia and adjacent regions: jerboas. Nauka, Moscow (in Russian)

Shenbrot G, Krasnov B, Khokhlova I (1994a) On the biology of Gerbillus henleyi (Rodentia: Gerbillidae) in the Negev Highlands, Israel. Mammalia 58: 581–589

Shenbrot GI, Rogovin KA, Heske EJ (1994b) Comparison of niche-packing and community organization in desert rodents in Asia and North America. Aust J Zool 42: 479–499

Shkolnik A (1971a) Diurnal activity in a small desert rodent. J Biometeoral 15: 115–120

Shkolnik A (1971b) Adaptation of animals to desert conditions, chap 18. In: Evenari M, Shanan L, Tadmor N (eds) The Negev: the challenge of a desert. Harvard University Press, Cambridge pp 301–323

Shkolnik A, Borut A (1969) Temperature and water relations in two species of spiny mice (Acomys). J Mammal 50: 245–255

Shkolnik A, Choshniak I (1987) Water depletion and rapid rehydration in the hot and dry terrestrial environment. In: Dejours P, Bolis L, Taylor CR, Weibel ER (eds) Comparative physiology: life in water and on land. Fidia Reseach Series, vol 9. Springer, Berlin Heidelberg New York, pp 141–151

Shkolnik A, Schmidt-Nielsen K (1976) Temperature regulation in hedgehogs from temperate and desert environments. Physiol Zool 49: 56–64

Shkolnik A, Borut A, Choshniak I (1972) Water economy of the Bedouin goat. Symp Zool Soc (Lond) 31: 229–242

Shkolnik A, Maltz E, Choshniak I (1980) The role of the ruminant's digestive tract as a water reservoir. In: Ruckenbusch Y, Thivend P (eds) Digestive physiology and metabolism in ruminants. MTP Press, Lancaster, pp 731–742

Shmida A (1985) Biogeography of the desert flora, chap 2. In: Evenari M, Noy-Meir I, Goodall DW (eds) Ecosystems of the world. 12A Hot deserts and the shrublands. Elsevier, Amsterdam, pp 23–77

Shoemaker VH, Nagy KA, Costa WR (1976) Energy utilization and temperature regulation by jackrabbits (*Lepus californicus*) in the Mohave desert. Physiol Zool 49: 364–375

Shuman TW, Robel RJ, Dayton AD, Zimmerman JL (1988) Apparent metabolizable energy content of foods used by mourning doves. J Wildl Manage 52: 481–483

Sibley RM (1991) The life-history approach to physiological ecology. Funct Ecol 5: 184–191

Siebert BD, Macfarlane WV (1975) Dehydration in desert cattle and camels. Physiol Zool 48: 36–48

Silanikove N (1986) Interrelationships betwen feed quality, digestibility, feed consumption, and energy requirements in desert (Bedouin) and temperate goats. J Dairy Sci 69: 2157–2162

Silanikove N (1992) Effects of water scarcity and hot environment on appetite and digestion in ruminants: a review. Livest Prod Sci 30: 175–194

Silanikove N, Brosh A (1989) Lignocellulose degradation and subsequent metabolism of lignin fermentation products by the desert black Bedouin goat fed on wheat straw as a single-component diet. Br J Nutr 62: 509–520

Silanikove N, Tagari H, Shkolnik A (1993) Comparison of rate of passage, fermentation rate and efficiency of digestion of high fiber diet in desert Bedouin goats compared to Swiss Saanan goats. Small Ruminant Res 12: 45–60

Smith AP, Nagy KA, Fleming MR, Green B (1982) Energy requirements and water turnover in free-living Leadbeater's possums, *Gymnobelidus leadbeateri* (Marsupialia: Petauridae). Aust J Zool 30: 737–749

Smith RE, Horwitz BA (1969) Brown fat and thermogenesis. Physiol Rev 49: 330–425

Smith RJ (1980) Rethinking allometry. J Theor Biol 87: 97–111

Smith RJ (1984) Allometric scaling in comparative biology: problems of concept and method. Am J Physiol 246: R152–R160

Soholt LF (1973) Consumption of primary production by a population of kangaroo rats (*Dipodomys merriami*) in the Mojave desert. Ecol Monogr 43: 357–376

Speakman JR (1987) The equilibrium concentration of oxygen-18 in body water: implications for the accuracy of the doubly-labelled water technique and a potential new method of measuring RQ in free-living animals. J Theor Biol 127: 79–95

Speakman JR (1988) The doubly-labelled water technique for measurement of energy expenditure in free-living animals. Sci Prog Oxf 72: 227–237

Speakman JR (1990) Principles, problems and a paradox with the measurements of energy expenditure of free-living subjects using doubly-labelled water. Stat Med 9: 1365–1380

Speakman JR (1992) Evolution of animal body size: a cautionary note on assessments of the role of energetics. Funct Ecol 6: 495–486

Speakman JR (1993a) Evolution of body size: predictions from energetics. Funct Ecol 7: 134

Speakman JR (1993b) How should we calculate CO_2 production in doubly labelled water studies of animals? Funct Ecol 7: 746–750

Speakman JR, Racey PA (1987) The energetics of pregnancy and lactation in brown long-eared bats (*Plecotus auritus*). In: Fenton MB, Racey PA, Raynor JMV (eds) Recent advances in the study of bats. Cambridge University Press, Cambridge, pp 367–393

Speakman JR, Racey PA (1988) Validation of the doubly labelled water technique in small insectivorous bats by comparison with indirect calorimetry. Physiol Zool 61: 514–526

Speakman JR, Racey PA, Burnett AM (1991) Metabolic and behavioral consequences of the procedures of the doubly labelled water technique on white (MFI) mice. J Exp Biol 157: 123–132

Speakman JR, McDevitt RM, Cole KR (1993a) Measurement of basal metabolic rates: don't lose sight of reality in the quest for comparability. Physiol Zool 66: 1045–1049

Speakman JR, Nair KS, Goran MI (1993b) Revised equations for calculating CO_2 production from doubly labeled water in humans. Am J Physiol (Endocrinol Metab 27) 264: E912–E917

Sperber I (1944) Studies on the mammalian kidney. Zool Bidr Upps 22: 249–431

Stebbins LL (1978) Some aspects of overwintering in *Peromyscus maniculatus*. Can J Zool 56: 386–390

Stephenson PJ, Racey PA (1993a) Reproductive energetics of the Tenrecidae (Mammalia: Insectivora). I. The large-eared tenrec, *Geogale aurita*. Physiol Zool 66: 643–663

Stephenson PJ, Racey PA (1993b) Reproductive energetics of the Tenrecidae (Mammalia: Insectivora). II. The shrew-tenrecs, *Microgale* spp. Physiol Zool 66: 664–685

Stephenson PJ, Speakman JR, Racey PA (1994) Field metabolic rate in two species of shrew-tenrec, *Microgale dobsoni* and *M. talazaci*. Comp Biochem Physiol 107A: 283–287

Steudel K, Porter WP, Sher D (1994) The biophysics of Bergmann's rule: a comparison of the effects of pelage and body size variation on metabolic rate. Can J Zool 72: 70–77

Stevens CE (1989) Evolution of vertebrate herbivores. In: Skadhauge E, Norgaad (eds) Pro Int Symp on comparative aspects of digestion in ruminants and hindgut fermenters. Acta Vet Scand Suppl 86: 9–19

Studier EH, Baca TP (1968) Atmospheric conditions in artificial rodent burrows. Southwest Nat 13: 401–410

Sud YC, Chao WC, Walker GK (1993) Dependence of rainfall on vegetation: theoretical considerations, simulation experiments, observations, and inferences from simulated atmospheric soundings. J Arid Environ 25: 5–18

Taneja GC (1973) Water economy in sheep. In: Mayland HF (ed) Proceedings: water-animal relations. University of Idaho Press, Kimberly, pp 95–132

Taylor CR (1970) Dehydration and heat: effect on temperature regulation of East African ungulates. Am J Physiol 219: 1136–1139

Taylor CR (1982) Scaling limits of metabolism to body size: implications of animal design. In: Johanson K, Bolis L (eds) A companion to animal physiology. Cambridge University Press, Cambridge, pp 161–179

Taylor CR, Lyman CP (1967) A comparative study of the environmental physiology of an East-African antelope, the eland, and the Hereford steer. Physiol Zool 40: 280–295

Taylor CR, Schmidt-Nielsen K, Raab JL (1970) Scaling of energetic cost of running to body size in mammals. Am J Physiol 219: 1104–1107

Tennent DM (1945) A study of the water losses through the skin in the rat. Am J Physiol 145: 436–440

Thomas DW, Samson C, Bergeron J-M (1988) Metabolic costs associated with the ingestion of plant phenolics by *Microtus pennsylvanicus*. J Mammal 69: 512–515

Thompson SD (1985) Bipedal hopping and seed-dispersion selection by heteromyid rodents: the role of locomotion energetics. Ecology 66: 220–229

Thompson SD, Nicoll ME (1986) Basal metabolic rate and energetics of reproduction in therian mammals. Nature 321: 690–693

Thompson SD, MacMillen RE, Burke EM, Taylor CR (1980) The energetic cost of bipedal hopping in small mammals. Nature 287: 223–224

Thornthwaite CW (1948) An approach towards a rational classification of climate. Geogr Rev 38: 55–94

Tisher CC (1971) Relationship between renal structure and concentrating ability in the rhesus monkey. Am J Physiol 220: 1100–1106

Tolkamp BJ, Brouwer BO (1993) Statistical review of digestion in goats compared with other ruminants. Small Ruminant Res 11: 107–123

Torres-Mura JC, Lemus ML, Contreras LC (1989) Herbivorous specialization of the South American desert rodent *Tympanoctomys barrerae*. J Mammal 70: 646–648

Tracy CR (1988) A view of animal autecology. Physiol Ecol Sect Newsl Ecol Soc Am 1: 1–3

Trojan M (1977) Water balance and renal adaptations in four Palaearctic hamsters. Naturwissenschaften 64: 591–592

Tucker VA (1965) Oxygen consumption, thermal conductance, and torpor in the California pocket mouse *Perognathus californicus*. J Cell Comp Physiol 65: 393–403

Tucker VA (1966) Diurnal torpor and its relation to food consumption and weight changes in the California pocket mouse *Perognathus californicus*. Ecology 47: 245–252

UNESCO (1977) Map of the world distribution of arid regions. MAB technical note 7, UNESCO, Paris

Vallet J, Rouanet J-M Besancon P (1994) Dietary grape seed tannins: effects on nutritional balance and on some enzymatic activities along the crypt-villus axis of rat small intestine. Ann Nutr Metab 38: 75–84 van Heerden J, Dauth J (1987) Aspects of adaptation to an arid environment in free-living squirrels *Xerus inauris*. J Arid Environ 13: 83–89

Verzar F, Keith J, Parchet V (1953) Temperatur und Feuchtigkeit der Luft in den Atemwegen. Pfluegers Arch Eur 7 Physiol 257: 400–416

Viro P (1987) Size variation in cohorts of the bank vole, *Clethrionomys glareolus* (Schreber 1970) in northern Poland. Z Saeugetierk 52: 76–88

Vispo CR, Bakken GS (1993) The influence of thermal conditions on the surface activity of thirteen-lined ground squirrels. Ecology 74: 377–389

Voltura MB, Wunder BA (1994) Physiological responses of the Mexican woodrat (*Neotoma mexicana*) to condensed tannins. Am Midl Nat 132: 405–409

Walsberg GE (1988) The significance of fur structure for solar heat gain in the rock squirrel, *Spermophilus variegatus*. J Exp Biol 138: 243–257

Walsberg GE, Schmidt CA (1989) Seasonal adjustment of solar heat gain in a desert mammal by altering coat properties independently of surface coloration. J Exp Biol 142: 387–400

Wang LCH, Hudson JW (1970) Some physiological aspects of temperature regulation in the normothermic and torpid hispid pocket mouse, *Perognathus hispidus*. Comp Biochem Physiol 32: 275–293

Wang LCH, Jones DL, MacArthur RA, Fuller WA (1973) Adaptation to cold: energy metabolism in an atypical lagomorph, the arctic hare (*Lepus arcticus*). Can J Zool 51: 841–846

Weathers WW (1981) Physiological thermoregulation in heat-stressed birds: consequences of body size. Physiol Zool 54: 345–361

Weathers WW, Nagy KA (1984) Daily energy expenditure and water flux in black-rumped waxbills (*Estrilda troglodytes*). Comp Biochem Physiol 77A: 453–458

Weathers WW, Stiles FG (1989) Energetics and water balance in free-living hummingbirds. Condor 91: 324–331

Weathers WW, Buttemer WA, Hayworth AM, Nagy KA (1984) An evaluation of time-budget estimates of daily energy expenditure in birds. Auk 101: 459–472

Weaver D, Walker L, Alcorn D, Skinner S (1994) The contributions of renin and vasopressin to the adaptation of the Australian spinifex hopping mouse (*Notomys alexis*) to free water deprivation. Comp Biochem Physiol 108A: 107–116

Webster AJF (1979) Energy metabolism and requirements. In: Church (ed) Digestive physiology and nutrition of ruminants. O & B Books, Corvallis, pp 210–229

Weiner J (1987) Limits to energy budget and tactics in energy investments during reproduction in the Djungarian hamster (*Phodopus sungorus sungorus* Pallas 1770). Symp Zool Soc (Lond) 57: 167–187

Weiner J (1989) Metabolic constraints to mammalian energy budgets. Acta Theriol 34: 3–35

Weiner J, Gorecki A (1981) Standard metabolic rate and thermoregulation in five species of Mongolian small mammals. J Comp Physiol B 145: 127–132

Weisser F, Lacy FB, Weber H, Jamison RL (1970) Renal function in the chinchilla. Am J Physiol 219: 1706–1713

Weissenberg S (1977) Comparative physiology in two populations of common spiny mouse (*Acomys cahirinus*) from desert and Mediterranean regions M Sc Thesis, Tel Aviv Univ, Tel Aviv (Hebrew with English abstract)

Welch WR (1984) Temperature and humidity of expired air: interspecific comparisons and significance for loss of respiratory heat and water from endotherms. Physiol Zool 57: 366–375

Westerterp KR, Bryant DM (1984) Energetics of free existence in swallows and martins (Hirundinidae) during breeding: a comparative study using doubly-labelled water. Oecologia (Berl) 62: 376–381

Westerterp KR, de Poer JO, Saris WHM, Schoffelen P, ten Hoor F (1985) Measurement of energy expenditure using doubly-labelled water. Int J Sports Med 5: 74–75

Wickler SJ (1980) Maximal thermogenic capacity and body temperatures of white-footed mice (*Peromyscus*) in summer and winter. Physiol Zool 53: 338–346

Wickler SJ (1981) Nonshivering thermogenesis in skeletal muscle of seasonally acclimatized mice, *Peromyscus*. Am J Physiol (Regulatory Integrative Comp Physiol) 241: R185–R189

Williams JB (1985) Validation of the doubly-labelled water technique for measuring energy metabolism in starlings and sparrows. Comp Biochem Physiol 80A: 349–353

Williams JB (1993) Energetics of incubation in free-living orange-breasted sunbirds in South Africa. Condor 95: 115–126

Williams JB, Nagy KA (1984) Daily energy expenditure of savannah sparrows: comparison of time energy budget and doubly-labelled water estimates. Auk 101: 221–229

Wilson DE, Reeder DM (1993) Mammal species of the world, 2nd edn. Smithsonian Institution Press, Washington

Wilson RT (1989) Ecophysiology of the camelidae and desert ruminants. Springer, Berlin Heidelberg New York

Withers PC (1982) Effect of diet and assimilation efficiency on water balance for two desert rodents. J Arid Environ 5: 375–384

Withers PC (1983) Seasonal reproduction by small mammals of the Namib desert. Mammalia 47: 195–204

Withers PC, Casey TM, Casey KK (1979a) Allometry of respiratory and haematological parameters of arctic mammals. Comp Biochem Physiol 64A: 343–350

Withers PC, Lee AK, Martin RW (1979b) Metabolism, respiration and evaporative water loss in the Australian hopping-mouse Notomys alexis (Rodentia: Muridae). Aust J Zool 27: 195–204

Withers PC, Louw GN, Henschel J (1980) Energetics and water relations of Namib desert rodents. S Afr J Zool 15: 131–137

Wolff JO, Bateman GC (1978) Effects of food availability and ambient temperature on torpor cycles of Perognathus flavus (Heteromyidae). J Mammal 59: 707–716

Wood RA, Nagy KA, MacDonald NS, Wakakuwa ST, Beckman RJ, Kaaz H (1975) Determination of oxygen-18 in water contained in biological samples by charged particle activation. Anal Chem 47: 646–650

Wright JW, Donlon K (1979) Inhibition of starvation-induced body dehydration by saline consumption in rats and gerbils. Behav Neural Biol 25: 535–554

Wright JW, Harding JW (1980) Body dehydration in xeric adapted rodents: does the renin-angiotensin system play a role? Comp Biochem Physiol 66A: 181–188

Wunder BA (1985) Energetics and thermoregulation. In: Tamarin RH (ed) Biology of the new world Microtus. Spec Publ of the American Society of Mammalogists, No 8. Brigham Young University, Provo, Utah, pp 812–844

Wunder BA, Dobkin DS, Gettinger RD (1977) Shifts of thermogenesis in the prairie vole (Microtus ohrogaster). Oecologia (Berl) 29: 11–26

Yaakobi D, Shkolnik A (1974) Structure and concentrating capacity in kidneys of three species of hedgehog. Am J Physiol 226: 948–952

Yahav S, Choshniak I (1989) Energy metabolism and nitrogen balance in the fat jird (Meriones crassus) and the Levant vole (Microtus guentheri). J Arid Environ 16: 315–332

Yahav S, Haim A, Shkolnik A (1982) Thermoregulation and activity in Apodemus mystacinus (Mammalia: Muridae) on Mount Carmel. Isr J Zool 31: 157–158

Yokota SD, Benyajati S, Dantzler WH (1985) Comparative aspects of glomerular filtration in vertebrates. Renal Physiol 8: 193–221

Yom-Tov Y (1991) Character displacement in the psammophile Gerbillidae of Israel. Oikos 60: 173–179

Yousef MK, Dill DB, Mayes MG (1970) Shifts in fluids during dehydration in the burro, Equus asinus. J Appl Physiol 29: 345–349

Zurofsky Y, Shkolnik A, Ovadia M (1984) Conservation of blood plasma fluids in hamadryas baboon after thermal dehydration. J Appl Physiol 57: 763–771

Subject Index

Printing: Saladruck, Berlin
Binding: Buchbinderei Lüderitz & Bauer, Berlin